Ceramic Injection Molding

MATERIALS TECHNOLOGY SERIES

Series editor: Renée G. Ford

This Materials Technology Series is dedicated to state-of-the-art areas of materials synthesis and processing as related to the applications of the technology. By thorough presentation of the science underlying the technology, it is anticipated that these books will be of practical value both for materials scientists and engineers in industry and for engineering students, to acquaint them with developments at the forefront of materials technology that have potential commercial significance.

Ceramic Injection Molding

BEEBHAS C. MUTSUDDY

Senior Specialist at Dow Corning Corporation, Michigan, USA

and

RENÉE G. FORD

Editor-in-Chief of the journal *Materials Technology. A Strategic Industry Report on the Synthesis, Processing and Applications of Advanced Materials,* based in New York, USA

CHAPMAN & HALL

London · Glasgow · Weinheim · New York · Tokyo · Melbourne · Madras

Published by Chapman & Hall,
2–6 Boundary Row, London SE1 8HN, UK

Chapman & Hall, 2–6 Boundary Row, London SE1 8HN, UK

Blackie Academic & Professional, Wester Cleddens Road, Bishopbriggs, Glasgow G64 2NZ, UK

Chapman & Hall GmbH, Pappelallee 3, 69469 Weinheim, Germany

Chapman & Hall USA, One Penn Plaza, 41st Floor, New York NY 10119, USA

Chapman & Hall Japan, ITP-Japan, Kyowa Building 3F, 2-2-1 Hirakawacho, Chiyoda-ku, Tokyo 102, Japan

Chapman & Hall Australia, Thomas Nelson Australia, 102 Dodds Street, South Melbourne, Victoria 3205, Australia

Chapman & Hall India, R. Seshadri, 32 Second Main Road, CIT East, Madras 600 035, India

First edition 1995

Typeset in 10/12 point Times Roman by Digitype, USA
Printed in Great Britain at the University Press, Cambridge

ISBN 0 412 53810 5

A catalogue record for this book is available from the British Library

Library of Congress Catalog Card Number: 94-68996

∞ Printed on acid-free text paper, manufactured in accordance with ANSI/NISO Z39.48-1992 (Permanence of Paper)

Contents

Foreword

Powder injection molding is a relatively new manufacturing process that combines the shape complexity and productivity of plastic injection molding with the performance attributes of ceramics, metals, and carbides. It is capable of producing a wide range of component sizes and shapes, and is well suited to the production of complex-shaped ceramics. In this process, a high concentration of powder is mixed with a binder to form a moderate-viscosity feedstock. This feedstock is molded in equipment that is very similar to that used for polymer injection molding. The molding step involves concurrent heating and pressurization of the feedstock and requires careful attention. Computer modeling and microprocessor controls help to minimize defects in molding. After molding the binder is extracted from the component, which is heated to the sintering temperature to attain a high final product density. The process is applicable to a wide range of established and emerging materials, and can achieve competitive final properties.

An important recent trend has been the change of status of powder injection molding from being considered a "black-art" to becoming a respected topic for materials processing research. Further, the technology has been adapted as a fundamental route to manufacturing complex shapes from many common engineering materials. Accordingly, much growth is anticipated as powder injection molding becomes an accepted manufacturing technique. The growth will include the fabrication of automotive components, artificial bones, structural parts, refractory hardware, and bearings with high wear resistance. There is likely to be further growth in applications as the process attributes are better appreciated by the engineering community.

Often there is a long incubation period before a new material processing route attains widespread acceptance. This reflects initial problems with credibility and stabilization. Engineers must be trained as to appropriate applications, and they require standards and design data that can only be provided over time. Further, process and design engineers must be trained in the fundamentals to encourage maximum benefit in practice. This book on ceramic injection molding by Drs Mutsuddy and Ford highlights several aspects of the technology and provides fundamental teachings relevant to its future growth. Accordingly, there is opportunity to apply the fundamentals to improve the technology as currently practiced in industry. The clear presentation of both scientific and technological fundamentals provides an important contribution along these lines. The book includes a perspective on the history of the process, a description of fundamental tests and tools needed to study powders and binders, and it defines the important steps needed to practise the powder injection molding process. Specific attention is given to ceramics and the required powder characteristics, while the authors also cover the basics of particle packing, binder design, mixture formulation, and the rheological behavior of powder–binder mixtures. Throughout this book the authors present models for the process steps, and details on molding, tool design, process control, debinding, and sintering, thus providing important information needed by the ceramic community. One novel aspect of the book is its use of example systems as illustrations of process details.

The book is a valuable contribution to our understanding of powder injection molding and will aid in advancing both the practice and the science of ceramics. Drs Mutsuddy and Ford are to be congratulated for their perseverance in putting together this landmark presentation.

RANDALL M. GERMAN
Brush Chair in Materials
Pennsylvania State University
July 1992

Preface

The past 20 years have seen a significant increase in research and development efforts in advanced ceramics, with strong industrial interest and optimistic forecasts for commercial growth. However, while the sales of electronic ceramics have grown more than 30 percent annually in the worldwide market since 1975, engineering or structural ceramics have not lived up to commercial expectations despite significant technological advances. A major barrier has been their high cost of processing and manufacture.

Forming is a key process step in determining the cost competitiveness of engineering ceramics as high performance products. Typically, ceramics shrink to about two-thirds of their green (unfired) volume, during sintering. Therefore, it is extremely difficult to fabricate ceramics to a net shape with required dimensional tolerances. Furthermore, because ceramics are brittle and hard, machining is difficult and can introduce flaws that lead to failure in use. During the past 10 to 15 years, the goal has been to develop net shape-forming processes that produce a final product which requires little or no machining.

The major forming methods include slip casting, extrusion, uniaxial pressing, isostatic pressing, tape casting, and injection molding. Recently, injection molding has received increased attention as a way to form complex ceramic parts in both traditional and advanced ceramics. Although injection molding has been used in ceramic manufacturing since about 1930, until quite recently the technology has been relatively unknown and even proprietary.

This book comprehensively covers the process steps involved in ceramic injection molding. Specific binder systems and molding and dewaxing processes are discussed to exemplify approaches to solving the kinds of problems that are encountered in actual practice.

The book's seven chapters cover the following topics: an overview of the ceramic injection molding process, ceramic powder processing and characterization, binder selection and formulation, mix preparation, molding and mold design, binder removal or dewaxing, and sintering as related to product evaluation.

The first chapter presents an overview of the state-of-the-art in ceramic injection molding. The subsequent chapters first discuss the reasons *why* particular characteristics or basic functions are essential, then *how* these functions should be developed or controlled, followed by an explanation of *what* one can reasonably expect to achieve. The experimental data presented are representative of the results of tests designed to demonstrate acceptable solutions to problems encountered in the ceramic injection molding process. Thus this book takes a *how to* approach. Wherever possible the solutions to problems are generalized for maximum usefulness.

The overall objective is to provide a convenient and authoritative source of information and guidance about the materials, equipment, and testing procedures for processing ceramic powder plus various additives into suitable formulations for successfully fabricating ceramic products by injection molding. Although this book is aimed at production executives, engineers, and technologists in ceramic manufacturing, it has value for a wider audience. It should prove particularly helpful to engineers, however specialized, who are working in ceramic manufacturing for the first time. It can also be used by companies as a reference for their customers, and for those interested in obtaining a general knowledge of the subject.

The applied aspects of ceramic injection molding are emphasized. However, it is hoped that the underlying theory is presented in sufficient detail to serve as a text or reference book for a course in ceramic fabrication, particularly where the instructor has had some practical operating experience with which to supplement the data.

As a consequence of the worldwide interest in ceramic injection molding, in November 1988 Battelle Columbus Laboratories initiated an industrial group research program to determine the important factors in the ceramic injection molding process and in molding advanced ceramics. Some of the experimental data resulting from this program are included in this book. However, the major part of the technical information presented is compiled from recent research in the late 1980s and early 1990s. Since the Battelle study, significant advances have been made in the technology, with improvements in the molding process, binder formulation, and binder removal. All the most recent developments are included in this book.

The authors wish to thank Battelle and the 48 companies in the United

States, Japan, and Europe that supported Battelle's multiclient program. Among those who contributed to the original study are Dr J.H. Adair, Dr G.K. Bansal, Mr R.K. Beal, Dr R.B. Bennett , Dr D.M. Bigg, Mr P.A. Depoutiloff, Mr W.H. Duckworth, Ms R.A. Holman, Mr R.B. Iden, Mr P.V. Johnson, Mr D.C. Makarius, Ms C.J. Markhoff, Mr L.G. McCoy, Dr G.L. Messing, Dr D.E. Niesz, Dr B.K. Parekh, Mr J.H. Peterson, Mr A.J. Roese, Dr J.R. Schorr, Dr D.K. Shetty, Mr C.R. Spilker, Mr W.B. Zimmerman. We would also like to thank all those researchers and companies who so generously supplied data and permission to use the illustrations that appear in this book, with special thanks to Diamonite Products, Howmet Turbine Components Corp., and the Carborundum Co.

We especially wish to thank Professor Randall M. German, who after reviewing the book graciously agreed to write the Foreword.

We hope this book will provide the information and guidance needed to readily adapt ceramic injection molding for your particular applications.

BEEBHAS C. MUTSUDDY
RENÉE G. FORD
1994

1

Overview of injection molding

Ceramic, which is made of earth, shaped with water, dried in air, and made durable by fire, represents the ideal combination of the four elements that the ancients believed compose the world. The most abundant and accessible material on the earth's crust is clay. Early humans were in constant touch with it.

Potters certainly have existed at least as long as civilization. Much of our knowledge of prehistory derives from clay and stone objects (see Figure 1-1) that were fashioned and fired before there were written records. Ceramic objects first appeared between 15,000 B.C. and 10,000 B.C. with the dawn of the Neolithic age. Nomads were aware of ceramic but did not make and use ceramic objects because of their fragility. The appearance of ceramic and its development mark an important stage in the progress of humankind. Figure 1-2 traces the history of ceramic materials and tools (Reh 1989).

No individual or race can claim credit for originating the shaping of ceramic. Crude forms of pottery apparently were developed for daily use in many places throughout early civilization. Relics still being discovered in ancient tombs in the Orient and the Near East, as well as in Egypt, suggest that each group originated its early work independently.

The introduction of the potter's wheel, probably the first machine, marked the beginning of the mechanization of ceramic forming. The earliest known use of a potter's wheel occurred in the Mesopotamian town of Worka around 5000 B.C. Wheels also were used during the Indus Valley civilization from 3250 B.C. to 3000 B.C., and the Mayans of Central America used them for making ceramic toys. At the same time the Chinese were making exquisite porcelain objects. In Europe, the Romans of the first and second centuries A.D. developed pottery to a very sophisticated level and

FIGURE 1-1 Much of our knowledge of prehistory is derived from clay and stone objects.

spread the skill of making high-quality ceramics throughout their vast empire, which included England, France, Spain, and parts of Germany.

Only in the twentieth century has shaping ceramics for industrial purposes become an important business. As early as 1929 patents were granted for ceramic molding (Miller 1929; Triggs 1933). As the manufacture of automobiles and thus internal combustion engines started to grow, there was a need for large numbers of spark plugs, which had fairly complex shapes. New shape-forming techniques had to be developed to accommodate the demand. By 1937, the AC Spark Plug Division of General Motors

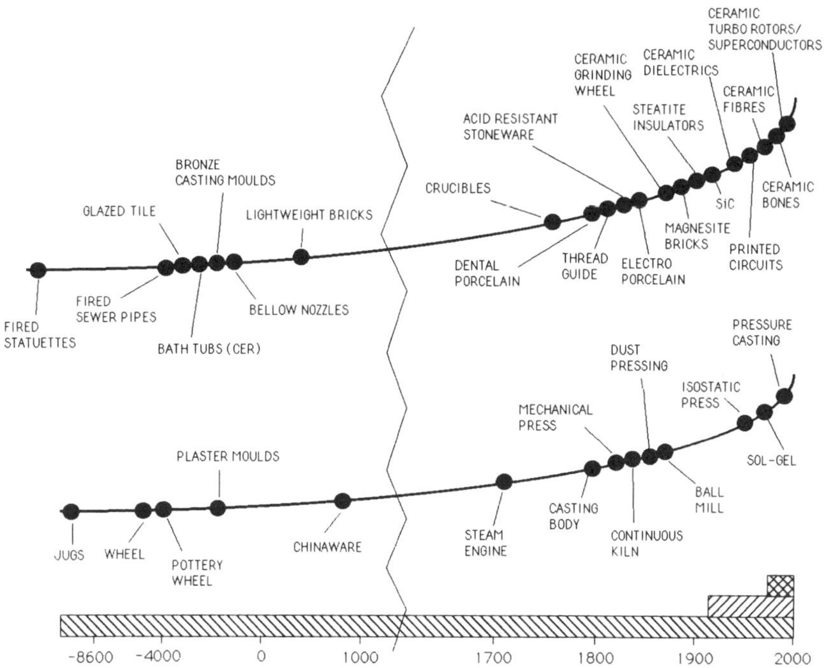

FIGURE 1-2 Landmarks in the history of the technology of ceramics.

was producing spark plugs in large quantities using injection molding (Schwartzwalder 1938). A cross section of the machine used is shown in Figure 1-3.

Through the 1940s and 1950s development efforts particularly concerned with binder selection, flow characteristics (rheology), and dimensional accuracy continued in Germany, England, Japan, and the United States, largely for spark plug insulators, refractory products, and some porcelain objects (Hennicke and Neuenfeld 1968). The injection molding of various shapes made of the ceramic powder alumina (Al_2O_3) was described in the literature, but the technology was highly proprietary (Bahn and Blechschmidt 1959).

Renewed interest in ceramic applications was generated by the advent of jet propulsion systems for aircraft. The fabrication of complicated jet engine components required radical changes in foundry practice. Casting the new alloys demanded high temperatures as well as the kind of complex ceramic cores shown in Figure 1-4. The core manufacturing industry provided another opportunity for the increased development of the injection molding process. By the early 1960s, the spark plug manufacturers began to

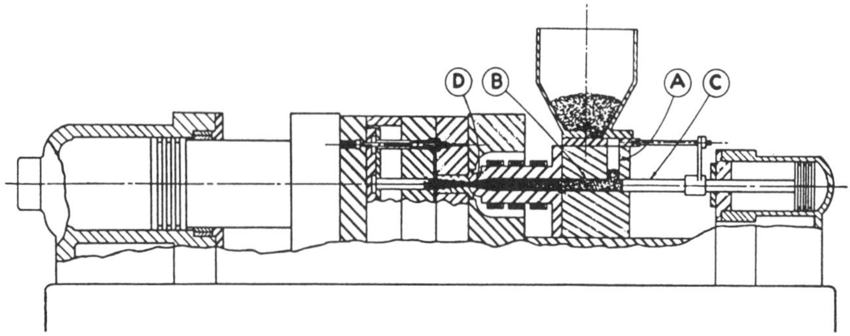

FIGURE 1-3 Cross section of the injection molding machine used for early versions of the spark plug.

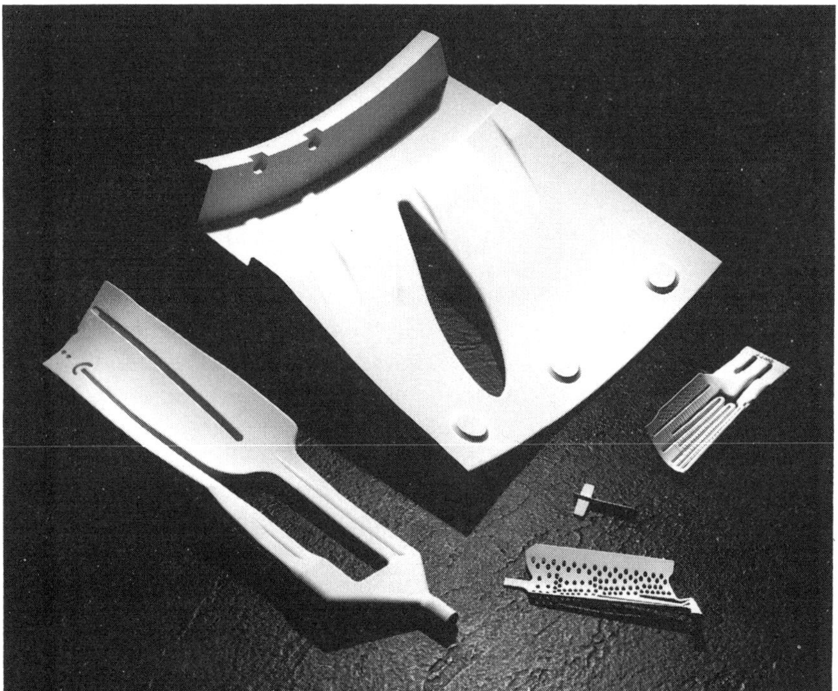

FIGURE 1-4 Ceramic cores used for the investment casting of alloys. Courtesy of Howmet Turbine Components Corp.

shift to an alternative forming method known as dry bag pressing, while injection molding became established as a routine forming technique for manufacturing cores (Randolph 1961).

During the 1950s and 1960s fairly extensive but extremely proprietary developmental work in injection molding was actively pursued both in the United States and in Europe. At this time the synthetic fiber and textile industries were growing rapidly. Higher speeds plus the mechanization of fiber production caused excessive wear in the metallic thread guides that were used exclusively in this industry. Switching to ceramic thread guides became an obvious choice, and as it has turned out, a highly successful one. Injection molding is ideally suited for the manufacture of thread guides, which are small and come in a variety of complex shapes, some of which are shown in Figure 1-5. For similar reasons ceramic nozzles have been gradually introduced into the welding industry. As can be seen in Figure 1-6, these nozzles have a complex geometry with internal threads. They too are being successfully fabricated by injection molding. Other injection molded

FIGURE 1-5 Injection molded thread guides. Courtesy of Diamonite Products.

FIGURE 1-6 Nozzles and assorted complex shapes. Courtesy of Howmet Turbine Components Corp.

wear parts, shown in Figure 1-6, include bushings and wire drawing dies. A number of electronic parts also are being produced by injection molding. Despite these examples of successful applications, the market for injection molded products has remained both relatively small and fragmented.

In 1970, the Advanced Research Projects Agency of the U.S. Department of Defense initiated a program to demonstrate the use of ceramics in heat engines. The production of dense silicon nitride and dense silicon carbide components for use in the hot flow path of vehicular gas-turbine engines and some diesel engines is being extensively investigated. Furthermore, these complex-shaped components must meet strict dimensional tolerances, be highly reliable, and be produced cost-competitively in quantity. These requirements have renewed interest in injection molding. The heat-engine-related components most suited for injection molding include:

- Radial rotors (Figure 1-7)
- Vanes (Figure 1-8)
- Turbocharging rotors (Figure 1-9)
- Combustion prechamber inserts (Figure 1-10)

The companies currently involved with ceramic injection molding either are small or represent only a small fraction of the total business of a large company. Consequently, the acceptance of this process as a routine production method has been extremely slow (Edirisinghe and Evans 1986). Undoubtedly, the age of ceramics will depend on the ability to produce complex shapes in large quantities, reproducibly and economically. The injection molding process therefore will play a key role.

The state of the art of injection molding is described in this chapter. Information gathered from a literature and patent search among other sources is used as the basis for expanding the discussion in subsequent chapters.

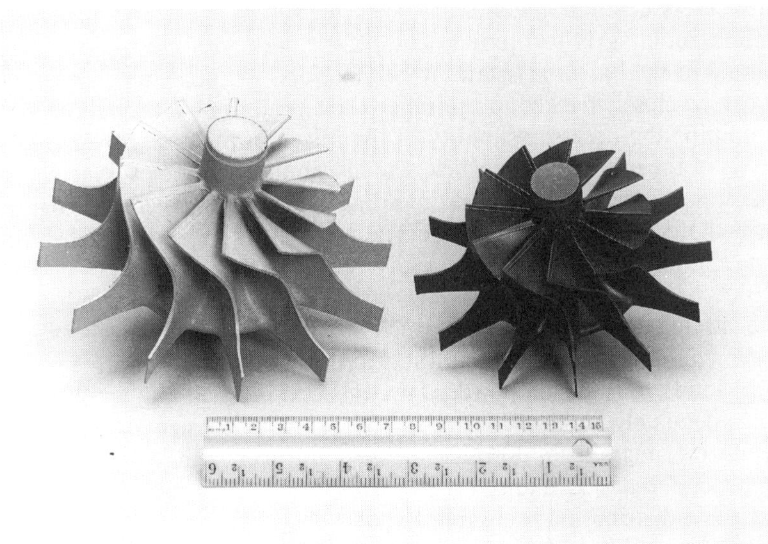

FIGURE 1-7 Silicon carbide radial rotors. Courtesy of Carborundum/BP Corp.

FIGURE 1-8 Silicon carbide vanes. Courtesy of Carborundum/BP Corp.

TAILORING THE POWDER

Almost every ceramic material has been injected molded. The objective of injection molding is to produce complex shapes with close dimensional accuracy in the "as fired" state. It is, therefore, essential to use a powder that can be closely packed to minimize the amount of binder needed while maintaining the necessary fluidity of the injection molding mix. From an analysis of the influence of particle size distribution on the packing characteristics and filler in an injection molding mix, it was concluded that one intermediate milling time can yield a powder with good flow behavior during molding (Strivens 1963).

Work on the injection molding of silicon for reaction bonding has demonstrated the importance of particle size distribution (Mangels 1978). It was shown that for a constant composition containing 73.5 volume percent (vol %) solids, only silicon powder finely ground in a ball mill for 140 hours can be adequately molded (Mangels 1978). The particle size distribution for this powder ranges from 30 micrometers (μm) to less than 1 μm.

In general, it has been found that—irrespective of the mean size—as the particle size distribution becomes narrower, the powder becomes less moldable (Mangels 1979). Alternatively, extremely high loadings of the ceramic powder filler and thus high density of the "green" (presintered) part, indi-

FIGURE 1-9 Turbocharger rotors. Courtesy of Battelle Columbus.

cating maximum packing, have been achieved using a very broad size distribution (Corbett and Shaffer 1988). It also has been shown that bimodal distribution (the presence of two different size fractions in a powder) provides improved packing volume as well as good flow behavior in highly filled suspensions (Mutsuddy 1983).

Although the above discussion emphasizes the importance of the packing characteristics of the ceramic powders in the final mix formulations, few guidelines have been developed. Furthermore, the trend in the ceramics industry is to use highly reactive powders (mean particle size less than 1 μm) to achieve better end properties and to reduce the sintering temperature and time. Such powders tend to agglomerate, which can have a deleterious effect on particle packing. In recent years, however, silicon nitride, silicon carbide, and alumina (aluminum oxide), all with particle sizes less than 1 μm and specific surface areas between 5 m^2/g and 20 m^2/g, have been successfully injection molded (Tanaka et al. 1985; Kamiya et al. 1985; Wada and Oyama 1986; Bandyopadhyay and French 1986; Heinrich et al. 1988).

FIGURE 1-10 Combustion prechamber inserts. Courtesy of Battelle Columbus.

The ability to injection mold whisker-reinforced ceramic composites has met with less success until recently because during injection molding the whiskers become oriented, which is not desirable for achieving optimum mechanical properties in the composite, and excessive damage to the whiskers during the process makes it difficult to maintain the aspect ratio required for achieving optimal properties in the composite part. However, recently success has been reported for injection molding dense complex parts of alumina reinforced with 15 vol% silicon carbide whiskers at high pressures (from 2,068 to 27,580 kPa, depending on the geometry of the part). The sintered parts are stronger and tougher than similar injection molded unreinforced alumina parts, indicating that although the whiskers become aligned in the flow direction to some degree, the hardness and fracture toughness of the composite parts apparently are not affected (Griffin and Danforth 1992). Injection molding ceramic composites reinforced with ceramic particles in a range of sizes has not been found to markedly improve their toughness and reliability compared with injection molded monolithic parts.

DESIGNING A BINDER SYSTEM

The amount of binder required markedly affects the properties of the part being molded during subsequent processing. For optimal results the binder content should be kept to a minimum. In addition to reducing costs and burnout time, a low binder content aids in controlling dimensional variation during binder burnout, in reducing shrinkage during sintering, and in attaining a high sintered density.

In a study of a variety of thermoplastic binder systems (ceramic powder–polymer suspensions), it was concluded that such systems should possess three flow characteristics: a flow index smaller than one (power law),[1] fluidity greater than 10 Pa·sec at a shear rate of about 100 sec^{-1}, and a relatively low dependence of the viscosity on temperture at the same shear rate. Furthermore, shrinkages in the binder system due to phase changes or thermal contraction should be kept to a minimum (Edirisinghe and Evans 1989).

Almost all the natural and synthetic organic compounds that are commercially available have been tried as binders and are covered by patents. Low molecular weight polyethylene, (5,000–20,000), a particular grade (unspecified) of polystyrene, and, more recently, paraffin and microcrystalline waxes with small additions of other organic modifiers have been used successfully. The object is to achieve mixes that do not separate but flow freely during molding, can be readily removed from the mold without sticking, and burn out from the molded shape without causing it to crack or rupture.

A binder system that will meet the above criteria has been sought since 1932. In general, wax-based formulations (including one containing a liquid epoxy as a plasticizer) have received wide acceptance (Wiech 1978, 1981a; Johnsson et al. 1984; French et al. 1984, 1987; Peltsman and Peltsman 1984). Useful thermoplastic formulations include: atactic polypropylene with or without polyethylene or ethylene–vinyl acetate but containing a phthalate plasticizer and a lubricant such as stearic acid or paraffin wax (Saito et al. 1976; Pett et al. 1981); polystyrene ethylene–acrylic acid

[1]The power law relates the shear stress, τ and the shear rate, $\dot{\gamma}$, for pseudoplastic materials in the following way:

$$\tau = K\,\dot{\gamma}^{n}$$

where the exponent, n, defined as the power law exponent, is a number greater or smaller than one, and K is the consistency coefficient.

copolymers plus polysulphone for molding silicon carbide (Ohnsorg 1979, 1980; Smoak and Storm 1979; Sugano 1983); various elastomeric formulations containing rubber-related block polymers mixed with a paraffin wax plasticizer, a processing aid such as stearic acid, and an anti-oxidant (Pett et al. 1981); and ethylene–vinyl acetate copolymer with various combinations of fatty acids plus a coupling agent containing an amine functional group (Mutsuddy 1983, 1988).

Other advantageous binder materials include polybutyl methacrylate in combination with certain additives, which improves the adhesion of the binder to the powder surface, reduces the binder removal time, and minimizes the formation of internal defects during removal (May 1974; Crossley et al. 1975; Tanaka et al. 1985; Thomas and Evans 1988); and a polyacetal binder can be used that volatilizes evenly and completely during binder removal without creating any internal defects (Farrow and Conciatori 1986).

Thermosetting resins also have been used as binders, particularly with silicon carbide powders (Fitchmun 1976; Whalen et al. 1978; Willermet et al. 1978; Wirth et al. 1978; Whalen and Johnson 1981). Their major advantage is the availability of residual carbon after pyrolysis and the relative ease of binder removal by thermal processes. However, unlike thermoplastics, thermosets cannot be recycled.

Metalorganic polymeric precursors also can be used as binders (Verbeek 1974). After injection molding, thermal treatment decomposes this type of binder to a very reactive ceramic residue that surrounds the ceramic particles in the ceramic powder–binder mix. The result is a more highly densified part with an improved microstructure. A further advantage is that the debinding time is reduced from several days to a few hours. Thus far, carbosilane precursors have been used as binders for injection molding silicon carbide. Although the chemistry and pyrolytic properties of carbosilane have been investigated, information on the rheology and moldability of its precursors as binders for injection molding is largely proprietary and thus not readily available (Yajima et al. 1975, 1977; Markle et al. 1979; Andersson and Warren 1984; Bunsell et al. 1988).

The major advantage of water-soluble binders is the reduction of debinding time. Water-soluble polymers such as the methyl, hydroxyethyl, and hydroxypropyl methyl celluloses are used for injection molding powders (Rivers 1978, Sarkar and Greminger 1983). The cellulose ethers combined with polyacrylamides produce macro-defect-free cement (Birchall et al. 1980). The most recently developed water-soluble binder formulation is a mixture of ceramic powder with a gel-forming material such as agar, plus water. The water, which constitutes most of the binder, is removed first by controlled humidity drying. The small amount of the organic binder remaining then is removed by thermal processing during sintering. The dewaxing time thus is reduced to several hours from several days (Fanelli and Silvers 1988).

Capillary rheometry is a widely used way to investigate the flow behavior of a ceramic powder–binder formulation. It provides a practical and direct measure of the moldability of the mix. In this method, the apparent viscosity of the mixture is measured as a function of shear rate or temperature or time. To date, little attention has been given to predicting the flow behavior from theoretical models (Mutsuddy et al. 1984).

Despite the extensive and growing number of binder formulations, ceramic injection molding remains more an art than a science. There is still insufficient knowledge about the chemical compatibility of different binder components, about interfacial chemistry, and about the stability of selected ceramic powder–binder compositions at elevated temperatures and under various shear conditions.

PREPARING THE MIX

Mixing a large fraction of ceramic powder with a much smaller fraction of organic binder can create a moldability problem. The flow characteristics of the mix, its subsequent moldability, and the quality of the final parts all depend on the homogeneity of the mix; so the preparation of the mix is a key factor. The mixing process is a complex phenomenon and its importance in developing ceramic injection molding has not been fully appreciated.

Two methods of mixing the binder with the ceramic powder have been developed (Howatt 1948; Wainer 1952; Taylor 1966; Waugh 1970). In one, the ceramic powder plus the organic binder are kneaded in a hot mixer to a homogeneous consistency at an elevated temperature. A twin-roll mill, which is well suited for processing high-viscosity materials, can be used as a mixer. It consists of two counterrotating differential-speed rollers with an adjustable nip, and imposes intense shear stresses on the material as it passes through the nip. Because there is little transverse mixing, an operator is needed to displace the strip of material transversely as the material passes through the nip. A twin-roll mill is used to produce moldable cement paste with a very high content of solids (Birchall et al. 1981). Also two types of extruder can be used for mixing: a single-screw or a twin-screw extruder. The latter, however, achieves a greater degree of dispersion (Edirisinghe and Evans 1989).

In the second mixing method, the organic binder is dissolved in a suitable solvent and the resulting solution is mixed with the ceramic powder in a ball mill or a blade-type mixer. The solvent then is removed by heating the mix for a sufficient time above the boiling point of the solvent. These two mixing methods are the most widely used.

In general, indices that can be used to quantify the extent of mixing are based on the standard deviation or the variance of the composition of spot samples taken at intervals from the mixture. However, this mathematical method cannot be used for controlling the process but only for checking its results afterward. Unfortunately, evaluation of the quality of the mixing process is still largely based on subjective observations.

THE MOLDING PROCESS

The machines presently used for ceramic injection molding are a direct transfer from plastic injection molding, usually without being modified. An injection molding machine also was suggested for making spark plug insulators (Ehlers 1948). An injection molding process flow sheet is shown in Figure 1-11, and a typical molding machine in Figure 1-12.

Injection molding machines, which can be broadly classified as either plunger or screw types, are characterized according to their shot capacity, plasticizing capacity, rate of injection, injection pressure, and mold locking force (clamp pressure). They range in size from those with a shot capacity of a few grams to machines that can deliver several kilograms. Some contain added features such as the capability to preplasticize the ceramic powder mix. Both automatic and semiautomatic as well as fixed or rotary machines are available.

Shot capacity is the maximum weight of material that can be injected per shot, usually expressed in terms of grams of polystyrene or cellulose acetate. This rating depends both on the volume of the cylinder swept during one stroke of the plunger and on the volumetric capacity of the feed mechanism. Thus its determination requires correcting for the bulk density of the mix. A preferable way to express shot capacity is in terms of the volume of material that can be injected by the plunger into a mold at a specific pressure.

One factor affecting the output of a screw-type molding machine is its plasticizing capacity. This is usually defined as the weight per hour (expressed in kg/hr) of material that can be heated to the molding temperature. Many plunger machines are fitted with a torpedo to improve the uniformity of the temperature in the mix and with preplasticizing units to increase the plasticizing capacity. The torpedo, which is in the heated cylinder, forces the viscous mix into closer contact with the cylinder walls. Its fins act to conduct heat from the hot cylinder walls to the torpedo. In some machines the torpedo has a separate heating unit.

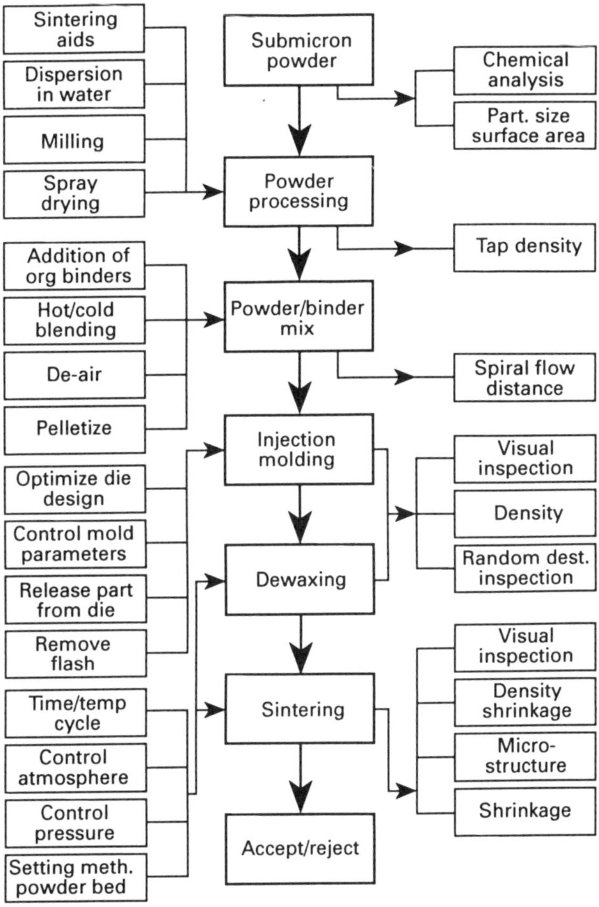

FIGURE 1-11 Injection molding process flow sheet.

The injection rate also is an important factor in determining the output of a machine. It usually is expressed as the volume of material discharged per second through the nozzle during a normal injection cycle, and depends on the pressure, the temperature, the material used, and the smallest aperture (or gate) in the flow channel. Thus with low-viscosity materials and fairly large gates, the speed of the injection plunger may be the limiting factor.

The injection pressure of an injection molding machine (the pressure exerted by the face of the plunger) can vary over a wide range. However,

FIGURE 1-12 Typical injection molding machine.

because of pressure losses in the system, the pressure in the mold cavities is much lower than the plunger pressure. Nevertheless, the actual force exerted on the inside surface of the mold is large even if the projected area of the part (the area of the molded part parallel to the clamps) is large. Therefore, the clamp pressure is an important factor in determining the maximum projected area that can be molded on a given machine. In machines as large as 100 tons or more, the molds normally are closed by large, direct-acting hydraulic rams. In smaller machines (less than 100 tons), a toggle-type action is preferable. Up until about 15 years ago, plunger-type machines were the most widely used. However, there has been a gradual shift to screw-type machines because plunger types have been unavailable in the large sizes (over 100 tons) now in demand.

A number of innovative machines have been designed. One, for example, is a multiple injection molding unit mounted on a tablelike framework with an individual hydraulic cylinder and a material-transfer cylinder for each molding unit. The material-transfer cylinders are all connected to a manifold that in turn is connected to a pressurizing unit. This unit includes a hopper and heater, plus a pressure-transmitting device for pressurizing a plastic mix to a high hydrostatic pressure. The pressurized plastic mix is forced through the manifold into all the material-transfer cylinders, then from each material-transfer cylinder into a respective molding unit. A toggle-type clamping mechanism holds the individual mold in position (Hagerborg 1962).

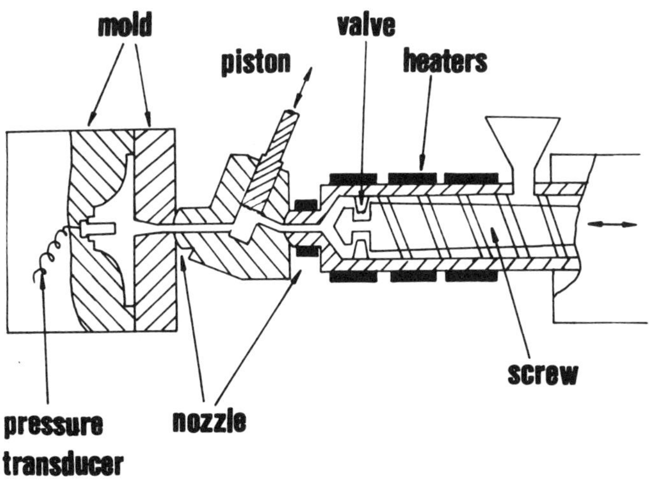

FIGURE 1-13 Newly designed injection molding machine to reduce internal shrinkage. Courtesy of Edirisinghe and Evans.

FIGURE 1-14 Low pressure injection molding machine. Courtesy of Peltsman Corp.

An unusual machine in which a refrigerated ceramic slurry is injected rapidly into a mold at a high pressure before the ceramic slurry sets and becomes rigid was designed for making ceramic parts for electronic applications. Controlling the pH accelerates the setting of the ceramic slurry in the mold (Mellen 1965). Another machine that operates in a vacuum removes internal flaws caused by entrapped air (Huther 1983).

A screw-type machine now is being marketed that is specifically designed to mold highly abrasive ceramics. It uses screws made of an unidentified corrosion- and abrasion-resistant material. The barrels are lined with a nickel-base iron–chromium boride composite (Inoue et al. 1989). To reduce the internal shrinkage in molded parts, a machine has been designed, shown in Figure 1-13, that applies oscillating pressure to the ceramic mixture in the mold (Kunststofftechnik 1979; Edirisinghe and Evans 1987). The pressure on the material is caused to fluctuate over a wide range, resulting in an extension of the heat flow in the sprue. This prolongs the solidification of the material in this section of the mold. The oscillating pressure is applied by a high-pressure valve assembly that can be maintained at high temperatures, mounted between the nozzle and the mold.

Machines for making electronic parts that operate at pressures as low as 0.2 to 0.6 MPa also have been built (Gribovsky 1961; Peltsman and Peltsman 1984). Operating at low pressures and low temperatures is claimed both to extend the life of the machine and to increase operator safety. One design (Figure 1-14) consists of a tank connected with a feeder pipe, both with independent temperature control systems. A planetary mixer, for blending the ceramic powder and binder mixture, is located in the tank, which is connected with a vacuum pump to deair the mix during preparation. The die is clamped to the opening of the feeder pipe with an air cylinder. During molding, air pressure is applied to the surface of the ceramic mix in the tank. The molding cycle is automatically controlled.

DESIGNING THE MOLD

The efficiency of the mold design and the quality of the mold construction are both measures of the success of an injection molding operation. Before designing a mold, however, initial consideration should be given to the design of the part. The determination of whether a part should be manufactured using injection molding should be based on factors such as its size and weight, the thickness of its sections, anticipated shrinkage, tolerances required, the draft (how much it tapers), the presence of threads, radii dimensions, and the number and the size of holes. After a detailed review of all the information available about a part, the mold design should take into consideration factors such as the sprue design, the runners, the number and the location of gates, the parting line (where the two halves of a mold meet), the location of vents, the mode of ejection, and the way that the mold is heated and cooled. Very little has been written on the important subject of mold design (Strivens 1957; Taylor 1966; Gilbert and Pfau 1978; Quackenbush et al. 1982; Stanciu 1984).

Proper selection of a molding machine and an optimum mold design are essential elements of ceramic injection molding. However, its ultimate success strongly depends on achieving a proper balance among the key operating parameters of pressure, temperature, and time cycles. Development of a generalizable formula for optimum processing conditions requires a more fundamental understanding of the rheology of mixes, of fluid dynamics during molding, and of the limitations on the geometry of parts that can be successfully injection molded.

REMOVING THE BINDER

The next critical step in ceramic injection molding is the removal of the organic binder from the molded parts prior to the sintering step. Binder systems generally are composed of several organic ingredients with markedly different melt-viscosity and decomposition ranges. Therefore, the ingredients are removed as slowly as is feasible in order to create capillary passages gradually within the part with increasing temperature and time. This allows the major binder ingredients to escape at higher temperatures without causing the part to crack and fail.

Thermal degradation in air has been the principal method used for binder removal (Mutsuddy 1986). In this process, the heat applied to the part increases the vapor pressure of the minor additives and at the same time softens the major binder component (such as polyethylene). The part must be heated slowly to the maximum temperature to avoid vaporizing the minor additives faster than the vapor can be dissipated through its pores. Otherwise, excessive vapor pressure within the part will cause cracking and blistering, or will distort the softened part if it is not supported. The debinding process can take from 70 to 100 hours.

A number of approaches have been devised to improve upon the conventional thermal process for removing the binder. By controlling the rate of extraction to a preset level of weight loss in a vacuum, 13 weight percent (wt%) of binder can be removed from a 20-mm-thick silicon nitride cylinder in 35 hours. This is close to a 50% reduction in cycle time (Johnsson et al. 1984). The binder also has been successfully removed from molded parts by heat treatment in a pure oxygen atmosphere with a flow rate of 1 liter per gram per hour (Gilissen and Smolders 1984). In another method, the molded parts are packed into a material that can absorb the binder when it is converted to a liquid. At least 45% of the binder can be wicked out in less than 8 hours without causing any imperfections. A fine-grained binder-absorbing material (such as superground alumina) produces the best results (Mann 1978; Watanabe 1982).

Another method for reducing binder removal time is to add an organic compound that sublimes (such as naphtha) to the total binder composition (these compounds have a vapor pressure of 0.1 mm Hg at room temperature and a boiling point no higher than 250°C). The binder then can be sublimed from the molded part within a relatively short time of from 16 hours to 24 hours, with minimal shrinkage. The process is carried out in a partial vacuum, or by moderately heating the parts, or by a combination of both (Bush 1967; Herrmann 1967; Horton 1973). Vacuum distillation and various solvent extraction techniques also have been found to considerably reduce debinding time (Strivens 1957, 1960; Adee 1980; Wiech 1981b).

Controlled heat and gas pressure (for example, argon, helium, nitrogen) also can be used for removing the binder. The gas pressure within the high-pressure chamber is maintained above the binder's vapor pressure, and the binder is condensed in a cooled chamber connected through a valve with the heated high-pressure chamber. The rate of binder removal from the molded part can be regulated by controlling the cooling rate within the cooled chamber or by controlling the rate at which the binder vapor can flow in the atmosphere from the part to the cooled chamber (Wiech 1981b). A novel method extracts the binder from the molded part by using a supercritical fluid,[2] such as carbon dioxide, or a chlorofluorocarbon as the solvent (Nakajima et al. 1988; Nishio et al. 1989). Despite the various approaches investigated to develop a reliable cost-effective method for removing the binder from an injection molded part, no optimum solution has as yet emerged.

A major problem in ceramic injection molding is that of meeting dimensional specifications; tolerances within 0.005 cm are now common, and some applications require tolerances within 0.0013 cm. A linear relationship has been shown between the variation in shrinkage and the binder content (Taylor 1966) and several suggestions have been made for improving dimensional accuracy (Strivens 1963):

---

[2]Supercritical fluids, which are intermediate between gases and liquids, are better solvents than ordinary liquids at relatively low temperatures.

- Careful preparation of the powder to assure a closer packing of particles.
- Maintaining a higher filler loading to enhance the green density of the part.
- Accurate weighing and mixing of the binder ingredients with the ceramic powder.
- Careful control of the debinding and sintering schedule.

Acceptance of a new technology rests primarily on the quality of the product achieved by applying the technique cost-effectively. For the new and promising technology of ceramic injection molding to become economically viable, all the technical variables discussed above must be brought under control.

A basic understanding of such key areas of ceramic injection molding as the rheology of molding mixes, mixing techniques, molding, mold design, and binder removal has been developed over the past 10 years. Based on the information generated it is now possible to produce commercially automotive parts such as turbocharger rotors and prechamber combustion inserts as well as ceramic dental braces, agricultural spray nozzles, and ceramic scissors, among others.

The current market for injection molded ceramic parts in the United States is less than $100 million per year, but it is expected to grow at a rate of 8 to 10% over the next decade as ceramics become increasingly used in heat engines and other applications requiring parts that can withstand high temperatures and corrosive environments. The ultimate widespread adoption of this technology, however, will require solving problems such as removing the organic binder from molded sections with thicknesses greater than 1.27 cm and holding dimensional tolerances within ± 0.001%.

References

Adee, J. M. 1980. Process for Forming Metals Parts with Less than 1 Percent Carbon Content, U.S. Patent 4,225,345.

Andersson, C. H. and R. Warren. 1984. Silicon Carbide Fibers and their Potential for Use in Composite Materials, Part 1, *Composites*, 15, 16–24.

Bahn, R. and H. Blechschmidt. 1959. Forming of Non-plastic Ceramic Materials by Injection Molding, *Silikat Tech.*, 10, 442–445.

Bandyopadhyay, G. and K. W. French. 1986. Fabrication of Near-Net-Shape Silicon Nitride Parts for Engine Application, 31st International Gas Turbine Conference and Exhibit, Düsseldorf, Republic of Germany, June 8–12.

Birchall, J. D., A. J. Howard, and K. Kendall. 1981. Cementitious Product, Eur. Patent 0021682.

Bunsell, A. R. et al. 1988. *Composite Materials Series* (*Fiber Reinforced Composite Materials*), 2 (Ch. 9), 446–478.

Bush, E. A. 1967. Method of Molding Comminuted Nonplastic Inorganic Materials, U.S. Patent 3,346,680.

Corbett, W. J. and P. T. B. Shaffer. 1988. Injection Molded Ceramics, CeramTec '88 Conference, Dearborn MI, ASM International, Metals Park, OH U.S.A.

Crossley, I. A. et al. 1975. Injection of a Ceramic into a Mold, U.S. Patent 3,882,210.

Edirisinghe, M. J. and J. R. G. Evans. 1986. Fabrication of Engineering Ceramics By Injection Molding, *Int. J. High Technology Ceramics*, 2(1/4), 1–31, 249–278.

Edirisinghe, M. J. and J. R. G. Evans 1987. An Oscillating Pressure Unit for Ceramic Injection Molding, *Materials and Design*, 8(5), 284–288; 9(2), 85–93.

Edirisinghe, M. J. and J. R. G. Evans. 1989. Systematic Development of the Ceramic Injection Molding Process, *Materials Science and Engineering*, A109, 17–26.

Ehlers, R. W. 1948. Method of Molding Ceramic Articles, U.S. Patent 2,446,872.

Fanelli, A. J. and R. D. Silvers. 1988. Process for Injection Molding Ceramic Composition Employing an Agaroid Gel-Forming Material to Add Green Strength to a Preform, U.S. Patent 4,734,237.

Farrow, G. and A. B. Conciatori. 1986. Injection Moldable Ceramic Composition Containing a Polyacetal Binder and Process of Molding, U.S. Patent 4,624,812.

Fitchmun, D. R. 1976. Method of Making a Silicon Carbide Article of Less Than Full Density, U.S. Patent 3,947,550.

French, K. W., J. T. Neil, and L. L. Turnbaugh. 1984. Composition for Injection Molding, U.S. Patent 4,456,713.

French, K. W. et al. 1987. Method for Fabricating Large Cross-Section Injection Molded Ceramic Shapes, U.S. Patents 4,708,838 and 4,704,242.

Gilbert, R. T. and I. B. Pfau. 1978. Method of Molding Elongated Thin Wall Articles, U.S. Patent 4,083,903.

Gilissen, R. and A. Smolders. 1987. Binder Removal from Injection Molded Ceramic Bodies, in *High Tech Ceramics*, ed. P. Vincenzini, Elsevier Science Publishers, Amsterdam, Netherlands.

Gribovsky, P. O. 1961. *Hot Molding of Ceramic Parts*, G. E. I., Moscow–Leningrad, 400.

Griffin, C. W. and S. C. Danforth. 1992. Injection Molding of Ceramic Matrix Composites, Phase I SBIR Final Report, 1989.

Hagerborg, J. G. 1962. Injection Molding Machine, U.S. Patent, 3,049,757.

Heinrich, J., E. Backer, and M. Bohmer 1988. Hot Isostatic Pressing of Silicon Nitride Powder Compacts and Reaction Bonded Silicon Nitride, *J. Am. Ceramic Soc.*, 71(1), c-28–c-31.

Hennicke, H. W. and K. Neuenfeld. 1968. Injection Molding as a Shaping Method in Ceramics, *Ber. Deut. Keram. Ges.*, 45, 469–473.

Herrmann, E. R. 1967. Molding Comminuted Nonplastic Inorganic Material, U.S. Patent 3,330,892.

Horton, R. A. 1973. Compositions and Methods for Making Molded Refractory, U.S. Patent 3,769,044.

Howatt, Glen H. 1948. Method of Injection Molding Ceramic Bodies Using Thermoplastic Binder, U.S. Patent 2,434,271.

Huther, Werner. 1983. Apparatus for Injection Molding of Precision Parts, U.S. Patent, 4,412,804.

Inoue, M., Y. Kihara, and Y. Arakida. 1989. Injection Molding Machine for High-Performance Ceramics, *Interceram*, 2, 53–57.

Johnsson, A. et al. 1984. Rate-Controlled Thermal Extraction of Organic Binders from Injection Molded Bodies, in *Advances in Ceramics, Forming of Ceramics*, ed. Mangels and Messing, 9, 241–245, Am. Ceramic Soc.

Kamiya, S. et al. 1985. Silicon Nitride Swirl Chamber for High Power Turbocharged Diesel Engines, *Soc. Automotive Eng.*, 850523.

Kunststofftechnik, Demag. 1979. Method of Apparatus for Molding Plastic Ceramic Materials, Federal Republic of Germany, U.K. Patent 1553924.

Mangels, J. A. 1978. Development of Injection Molded Reaction Bonded Silicon Nitride, in *Ceramics for High Performance Applications—II*, ed. J. J. Burke, E. H. Lenoe, and R. N. Katz, Brook Hill Publishing Company, Chestnut Hill, MA, 113–131.

Mangels, J. A. 1979. Development of Moldable, High Density Reaction Bonded Silicon Nitride, NASA—Lewis Research Center, Contract No. DEN 3-20 (May 1978, August 1978, and January through April 1979).

Mann, D. L. 1978. Injection Molding of Sinterable Silicon-Base, Non-oxide Ceramics, Technical Report AFML-TR-78- 200, Final Report for Period September 1977 to October 1978.

Markle, R. A. et al. 1979. Preparation of Si–N–C Fibers by Controlled Pyrolysis of Novel Organosilicon Precursors, NASA 33223.

May, E. R. W. 1974. Production of Silicon Nitride Material Components, U.S. Patent 3,819,786.

Mellen, E. J. 1965. Injection Molding of Ceramic Cores, U.S. Patent 3,222,435.

Miller, W. J. 1929. Improved Machine for the Manufacture of Pottery, U.K. Patent 283151.

Mutsuddy, B. C. 1983. Influence of Powder Characteristics on the Rheology of Ceramic Injection Molding Mixtures, *Proc. Brit. Ceram. Soc.*, 33, 117–137.

Mutsuddy, B. C. 1988. Formulation of Injection Molding Binder Systems, *Chem. Eng. Comm.*, 1–17.

Mutsuddy, B. C. 1986. Oxidative Removal of Organic Binders from Injection Molded Ceramics, in *Non-oxide Technical and Engineering Ceramics*, ed. Stuart Hampshire, Elsevier Applied Science, 397–408.

Mutsuddy, B. C. et al. 1984. Computer Modeling of Fluid Flow, in *Advances in Ceramics*, ed. Mangels and Messing, 9, 259–264, Am. Ceramic Soc.

Nakajima, N., S. Yasuhara, and M. Ishihara. 1988. Method of Removing Binder Material from a Shaped Ceramic Preform by Extracting with Supercritical Fluid, U.S. Patent 4,731,208.

Nishio, H. et al. 1989. A Novel Concept for Ceramic Injection Molding, 1st Japan International SAMPE Symposium and Exhibition, New Materials and Processes for the Future, Soc. Advance. Mater. and Proc. Engin. (SAMPE), Chibo, Japan.

Ohsorg, R. W. 1979, 1980. Composition and Process for Injection Molding Ceramic Materials, U.S. Patents 4,144,207 and 4,233,256.

Peltsman, I. and M. Peltsman. 1984. Low Pressure Molding of Ceramic Materials, *Interceram*, 4, 56.

Pett, R. A. et al. 1981. Moldable Mixture of Sacrificial Binder and Sinterable Particulate Solids, U.S. Patent 4,265,794.

Quackenbush, C. L., K. French, and J. T. Neil. 1982. Fabrication of Sinterable Silicon Nitride by Injection Molding, *Ceram. Eng. Sci. Proc.* 3, 20–34.

Randolph, R. L. 1961. Injection Molded Ceramics, *Mat. in Design Eng.*, 54(2), 10–12.

Reh, H. 1989. How Realistic Are Market Forecasts for High Performance Ceramics? *Interceram*, 38(2), 25–30.

Rivers, R. D. 1978. Method of Injection Molding Powder Metal Parts, U.S. Patent 4,113,480.

Saito, K., T. Tanaka, and T. Hibino. 1976. Method of Producing a Ceramic Article by Injection Molding, U.S. Patent 4,400,110.

Sarkar, N. and G. K. Greminger. 1983. Methylcellulose Polymers as Multifunctional Processing Aids in Ceramics, *Am. Ceram. Soc. Bull.*, 62, 1280–1288.

Schwartzwalder, K. 1938. Refractory Body and Method of Making Same, U.S. Patent 2,122,960.

Smoak, R. H. and R. S. Storm. 1979. Interactive Development of Injection Molded Sintered Alpha Silicon Carbide Turbine Material, Amer. Soc. of Mech. Eng. Pub. No.79-GT-77.

Stanciu, V. V. 1984. Tooling for Ceramic Injection Molding, in *Advances in Ceramics*, ed. J. Mangels, 9, 239–240, Am. Ceramic Soc.

Strivens, M. A. 1957. Improvements in or Relating to the Formation of Molded Articles from Sinterable Materials, U.K. Patent 808,583.

Strivens, M. A. 1960. Formation of Ceramic Moldings, U.S. Patent 2,939,199.

Strivens, M. A. 1963. Injection Molding of Ceramic Insulating Materials, *Am. Ceram. Soc. Bull.*, 42(1), 13–19.

Sugano, T. 1983. Materials for Injection Molding Silicon Carbide, Proceedings of the 1st Symposium on Research and Development of Basic Technology for Future Industry, Fine Ceramics Project, Japan Technology Association, Tokyo, 67–84.

Tanaka, T. et al. 1985. Injection Molding of Alumina, *Yogyo-Kyokai-Shi*, 93(a), 96–100.

Taylor, H. D. 1966. Injection Molding Intricate Ceramic Shapes, *Am. Ceram. Soc. Bull.*, 45(9), 768–700.

Thomas, M. S. and J. R. G. Evans. 1988. Non-uniform Shrinkage in Ceramic Injection Molding, *Brit. Ceram. Trans. J.*, 87, 22–26.

Triggs, W. A. 1933. Method of Molding Materials, U.K. Patent 400281.

Verbeek, W. 1974. Production of Shaped Articles of Homogeneous Mixtures of Silicon Carbide and Silicon Nitride, U.S. Patent 3,853,567.

Wada, S. and Y. Oyama. 1986. Thermal Extraction of Binder Components from Injection-Molded Bodies. Presented at the 2nd International Conference on Materials and Components for Engines, Lübeck, Federal Republic of Germany, April 14–17.

Wainer, E. 1952. Method of Molding Nonmetallic Powders, U.S. Patents 2,593,507 and 2,593,943.

Watanabe, K. 1982. Ceramic Binder Removal Method, Japanese Patent 100973.

Waugh, A. 1970. Process for Forming Sintered Leachable Objects for Various Shapes, U.S. Patent 3,549,736.

Whalen, T. J. and C. F. Johnson. 1981. Injection Molding of Ceramics, *Am. Ceram. Soc. Bull.*, 60, 216–220.

Whalen, T. J., J. E. Noakes, and L. L. Terner. 1978. Progress on Injection Molded Reaction-Bonded Silicon Carbide, in *Ceramics for High Performance Applications—II*, ed. J. J. Burke, E. H. Lenoe, and R. N. Katz, Brook Hill Publishing Co., Chestnut Hill, MA, 179–192.

Willermet, P. A., R. A. Pett, and T. J. Whalen. 1978. Development and Processing of Injection Molded Reaction-Sintered SiC, *Am. Ceram. Soc. Bull.*, 57, 744–747.

Wiech, R. E. 1978. Manufacture of Parts from Particulate Materials, U.K. Patent 1516079.

Wiech, R.E. 1981a. Method of Making Inelastically Compressible Ductile Particulate Material Article and Subsequent Working Thereof, Eur. Patent Applications 8110211.2, 81100210.4, and 81100209.6.

Wiech, R. E. 1981b. Method and Means for Removing Binder from a Green Body, U.S. Patent 4,305,756.

Wirth, P. et al. 1978. German Patent 27 50 095.

Yajima, S., J. Hayashi, and M. Omori. 1975. Continuous Silicon Carbide Fiber of High Tensile Strength. *Chem. Lett.*, 931–934.

Yajima, S., K. Okamura, and T. Shishido. 1977. SiC Bodies Sintered with Three-Dimensional Cross-Linked Polycarbosilane, *Am. Ceram. Soc. Bull.*, 56(12), 1060–1063.

2

Powder processing and characterization

Powder processing is of paramount importance for injection molding. First, however, the powder's characteristics must be accurately defined. What is desired is a powder that retains the necessary flow properties during molding, sinters readily, and when processed produces complex shapes with dimensional accuracy. To enable the correlation of its behavior with each processing step, a powder must be characterized for its chemistry, physical features (size, distribution, shape), and packing characteristics, before being processed.

This chapter explores the interrelationships of powder characteristics. Techniques and procedures for determining powder characteristics are suggested and described, and the effects of agglomeration and methods to minimize agglomeration are examined. Finally, guidelines for powder processing are suggested.

POWDER PACKING

The concentration of a ceramic powder that can be incorporated into a continuous binder matrix to provide high green density and close dimensional accuracy in the "as fired" state depends on its packing characteristics. For this reason it is particularly important to understand the role of particle packing during injection molding. Furthermore, the consideration of a particle packing method needs to include its impact on the rheology of the powder–binder mix (Mutsuddy 1983). Dilatant flow (increasing viscosity with increasing shear rate), which results from particle–particle interaction, must be prevented from occurring during the forming process to avoid the possibility of powder–binder separation during molding (Funk and Dinger 1986, 1992).

All ceramic processing techniques need to take particle packing into consideration. Efforts to optimize the packing of particles were made as early as 1928 (Furnas 1928). Since then, models have been developed for both discrete and continuous particles (Westman and Hugill 1930, Andreasen 1930, Patankar and Mondal 1980; Funk and Dinger 1992).

This section covers the present state of understanding in particle packing and its relation to injection molding. The specific factors treated include particle shape, particle arrangement, particle size distribution, particle surface roughness, and the degree of agglomeration.

Particle Shape

Particles can be spherically shaped or equiaxed, or they can be in the form of fibers or platelets. Each shape has different packing characteristics. Round particles have been shown to have the highest packing densities (Table 2-1). Milling reduces any roughness on the surface of particles, resulting in rounder particles and thus higher packing densities.

In a completely dispersed system, platelets and fibers would have higher packing densities, assuming that they can be completely oriented. For ceramics composed of very fine particles, less than 1 μm in diameter, it is usually not possible to achieve complete orientation of the reinforcing phase. Only very thin platelets can become sufficiently oriented to produce appreciable differences in density.

A particle's aspect ratio, which is defined as the quotient of its larger dimension (fiber length or platelet diameter) and its smaller dimension (fiber diameter or platelet thickness), also affects the packing density. Fibers with an aspect ratio (L/D) of 2 can pack to 68% of theoretical density, whereas when the aspect ratio is 10, the maximum achievable theoretical

TABLE 2-1. The effect of particle shape on the maximum packing density in a random distribution.

Particle shape	Maximum packing density volume percent
Spheres	63.7
Isotropic irregular particles	63.0
Cylindrical rods*	52.0
Dagger-shaped fibers	33.0

*Length to diameter (L/D) ratio ≈ 5.

density decreases to 42% (Milewski 1974). Furthermore, although a high aspect ratio reduces the tendency for orientation in fibers, the trend is opposite that for platelets: platelets with higher aspect ratios tend to orient more readily. Platelet-shaped particles, however, are rarely the only constituents of a powder. Individual platelets generally are surrounded by platelet stacks, equiaxed grains, and other platelets with a wide variety of aspect ratios.

Packing spheres with fibers also has been evaluated, based on assumptions in the Furnas model (Milewski 1978). It was found that the packing density is a function of the aspect ratio of the fiber (L/D) as well as the ratio of the sphere diameter to the fiber diameter (R). An example of bulk volume as a function of fiber weight fraction for fibers with an L/D of 10 with various ratios of the sphere to fiber diameters (R values) is shown in Figure 2-1. As can be seen, the minimum bulk volume decreases as R increases. Minimum bulk volumes for all the sphere–fiber mixtures in Figure 2-1 are about 0.2 wt% fibers.

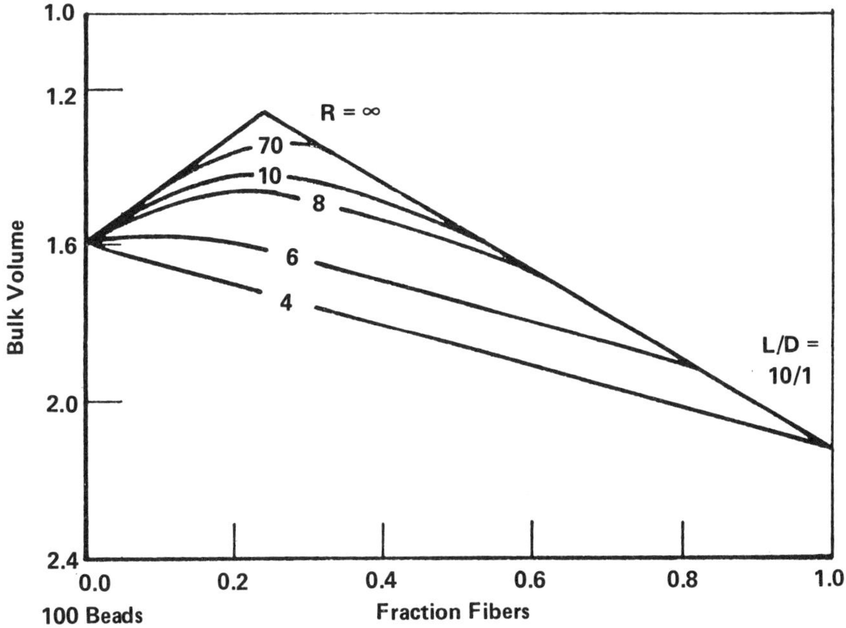

FIGURE 2-1 Bulk volume as a function of fiber weight fraction for fibers with an aspect ratio (L/D) of 10 and with various ratios of the sphere to fiber diameters (R values).

TABLE 2-2. The effect of spatial arrangement on the maximum packing density of identical-size spheres.

Spatial arrangement	Maximum packing density volume percent
Face-centered cubic	74
Double nested	70
Body-centered cubic	68
Orthorhombic	60
Simple cubic	52
Random	64

Particle Arrangement

Particle arrangement is also an important factor in achieving high packing density. Most studies on the effects of particle arrangement have used spherical powders of uniform size. Spheres can nest together in a variety of configurations, each giving a different packing density. This is shown for spheres in Table 2-2 and Figure 2-2. An empirical relationship between the

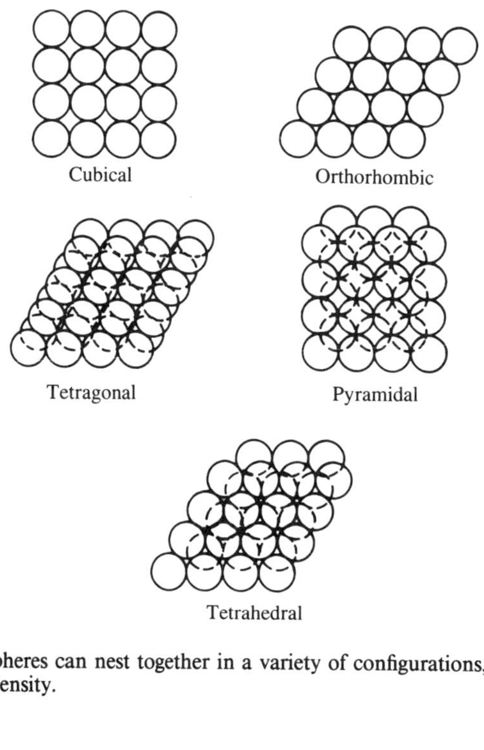

FIGURE 2-2 Spheres can nest together in a variety of configurations, each with a different packing density.

packing density for spherical particles and the number of particle contacts is given by the following equation (Smith et al. 1929):

$$C = 3 + 22\,(\rho_y)^3 \tag{2-1}$$

where C is the number of contacts per particle, and ρ_y is the relative packing density. As indicated in Figure 2-3, there is good agreement between experimentally determined values and values calculated using equation (2-1). However, the slope of the line in Figure 2-3 is dependent on the shape of the particle. For example, dendritic particles have much greater free space between them than spherical particles. Consequently, the maximum pack-

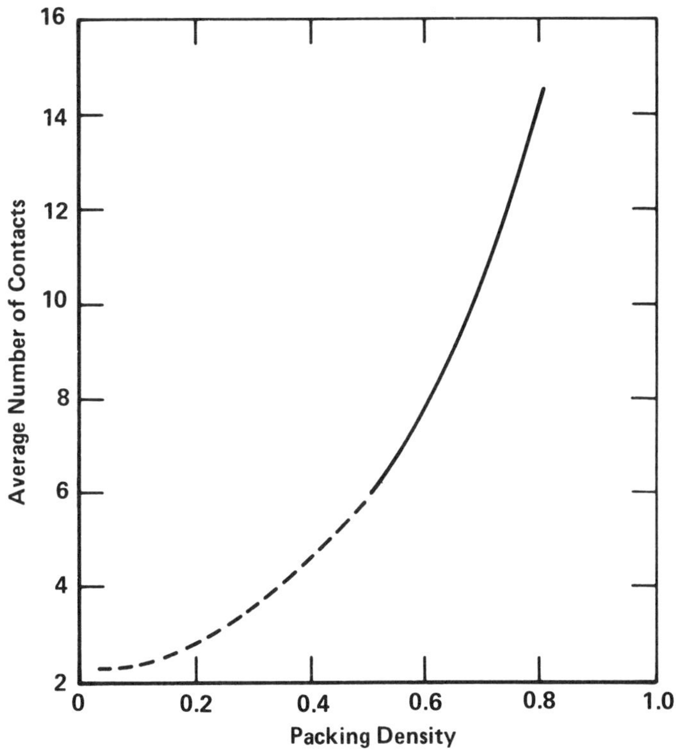

FIGURE 2-3 There is good agreement between experimentally determined values for the empirical relationship between the packing density and the number of particle contacts and the values calculated using equation (2-1).

ing density for dendritic particles is less than that for spherical particles, while a face-centered particle arrangement theoretically should result in the highest density (see Table 2-2).

With commercial powders, the arrangement of the particles, which tend to be widely distributed in size, is generally random. The effect of this randomness on powder packing is difficult to separate from the effect of other parameters such as particle size and distribution.

Particle Size Distribution

Higher packing densities can be achieved if the particle sizes are not uniform. This enables the finer particles to fill the holes between the large particles. The particle size distribution can be broadened in two ways: mixtures of two or more particle sizes can be blended, for example, coarse and fine particles; or a wide continuous distribution of particle sizes can be selected.

When two or more separate particle sizes are mixed, the finer size can be selected to give the best fit in the interstices of the coarser size. A model has been developed for calculating the packing density of a mixture of coarse and fine particles (Furnas 1928). The packing density of the mixture will change as the relative amounts of the coarse and fine fractions in the mixture change.

Figure 2-4 summarizes the relationships that can exist in bimodal particle packing. In region I, fine particles fill the pores in the coarse particle bed. The specific volume of the mixture in region I (V_{I}) is given by:

$$V_{\mathrm{I}} = V_c W_c = V_c (1 - W_f) \tag{2-2}$$

where V_c is the specific volume of the coarse particles, and W_c and W_f are the weight fractions of the coarse and fine particles, respectively.

In region II, coarse particles expand the bed of the fine particles. The specific volume of the mixture in region II (V_{II}) is given by:

$$V_{\mathrm{II}} = V_o W_c + V_f W_f = V_o + (V_f - V_o) W_f \tag{2-3}$$

where V_o is the specific volume of the solid material, and V_f is the specific volume of the fine particles.

The specific volume for binary powder mixtures with large size ratios, as a function of the weight fraction of the fines, is shown in Figure 2-5. The minimum specific volume is at the intersection of regions I and II defined by W_f^*. Simultaneous solution of equations (2-2) and (2-3) gives W_f^* as:

$$W_f^* = \frac{V_c - V_o}{V_f + V_c - V_o} \tag{2-4}$$

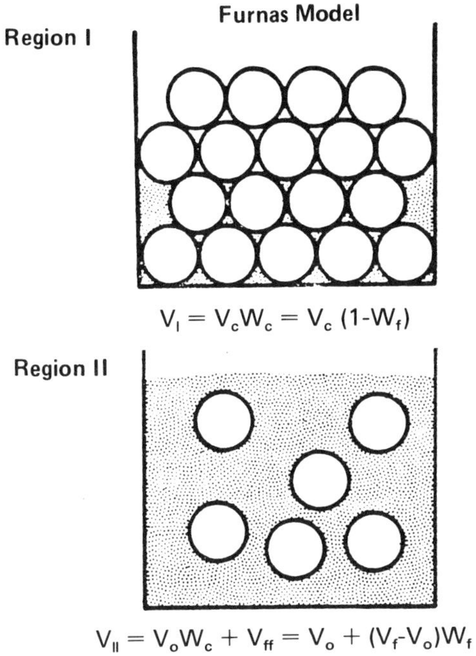

FIGURE 2-4 A summary of the relationships that can exist in bimodal particle packing.

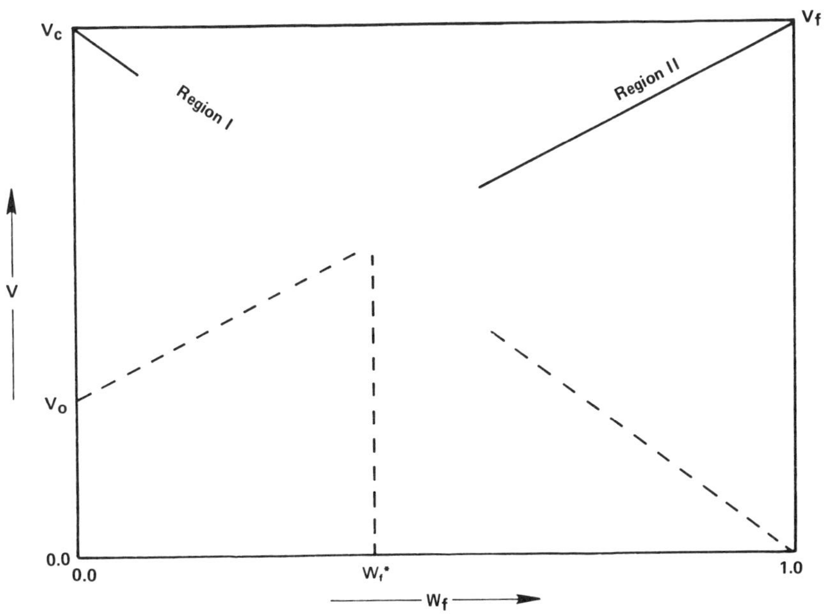

FIGURE 2-5 The specific volume for binary powder mixtures with large size ratios.

At W_f^* the interstices of the coarse particle bed are completely filled with fines.

An optimum composition of a binary mixture for sintering would be one in which the fine weight fraction completely fills the interstices of the coarse bed with solid material when sintered. The weight fraction of fines W_f^m required to optimize densification when the coarse fraction sinters very slowly relative to the fine fraction is given by (Onoda 1977):

$$W_f^m = \frac{W_c V_o}{V_c} \tag{2-5}$$

The size ratios of particles in mixtures, which are required to obtain optimum packing, that is, to achieve the calculated packing density at the composition calculated from the Furnas model, have been empirically studied. A size ratio of at least 7 is needed to arrive near the optimum packing density in binary mixtures of spherical particles, with larger ratios giving still closer agreement with the calculated packing densities (McGeary 1961). For example, homogeneous mixtures of irregularly shaped sand particles reach theoretical packing density at a size ratio of 10. Deviations from theoretical predictions have been shown to result from inhomogeneities in the mixtures (Messing and Onoda 1978).

Increasing the number of particle sizes in the mixture can increase the calculated packing density (Ayer and Soppet 1965). Table 2-3 shows the maximum packing density attainable for random packing of mixtures containing from one to four sizes of spherical particles. The ratios of the sizes are in increments of 7 : 1 (from coarser to finer particles), and the compositions are given in weight percent of each size fraction. Although the maximum packing density attainable for random packing will depend on the individual packing densities of the component sizes, the trend can be clearly seen. With the addition of more size fractions, a larger fraction of pore

TABLE 2-3. The effect of particle size distribution on the maximum packing density of randomly dispersed spheres.

Ratio of particle diameter(s) of the size fractions	Ratio of fractions, weight percent	Maximum packing density, volume percent
	100	64
d, $1/7d$	85/15	86
d, $1/7d$, $1/49d$	75/14/11	95
d, $1/7d$, $1/49d$, $1/343d$	72/14/10/4	98

space can be occupied by particles, and the maximum packing density increases. Little advantage is gained by using more than four sizes.

In most commercial powders, deviations from the idealized powder model in the original Furnas model limit its practical significance. If coarse to fine size ratios are less than about 10, the particles have nonspherical shapes, agglomerates are present, the particles are less than 1 μm in diameter, and the maximum packing densities in binary mixtures are less than the values predicted by theory. The effect of the size ratio on the specific volume is given in Figure 2-6. As the size ratio decreases, the minimum obtainable specific volume increases. There are also shifts in the curves at which W_f^* occurs, that is, where the interstices in the coarse particle bed are completely filled with fine particles. As a consequence, in gap-graded materials, size ratios must be greater than at least 10, or no real advantage is

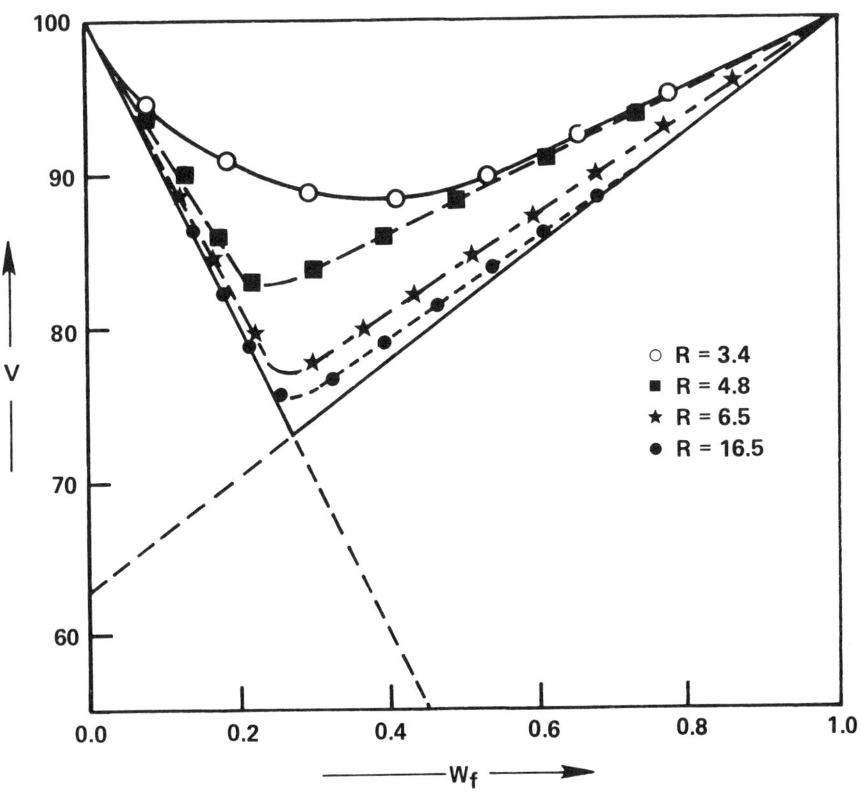

FIGURE 2-6 The effect of particles with different size ratios on the specific packing volume.

gained by gap-grading. In gap-grading, materials are blended by combining two different size fractions, for example, a $<20\,\mu m + >10\,\mu m$ fraction with a $<1\,\mu m + >0.1\,\mu m$ fraction.

Broadening the particle size range in a continuous distribution increases the packing density relative to that for uniformly sized particles. Several empirical studies have indicated trends in packing densities using continuous size distribution. For example, the packing density of spherical glass beads is apparently more sensitive to the range of the size distribution than to the shapes of the particles (Boctar et al. 1965). A wider size range increases the packing density by 6 to 8%. A similar result has been reported for sand grains (Sohn and Moreland 1968).

In ceramic injection molding, an attempt is made to obtain an optimum binder content for molding. The optimum binder content for a ceramic–binder system represents a compromise between maintaining the rheological conditions for molding and minimizing the amount of binder. The rheological conditions for each ceramic–binder system are determined by the concentration of the ceramic as well as the size of the particles and their distribution.

By using mathematical models, it has been demonstrated that if the total volume of powder is split into 25 vol% fine particles and 75 vol% coarse particles, a very high loading of about 70 vol% can be obtained without an increase in viscosity. The proportion of coarse to fine particles that will produce minimum viscosity for a range of total filler loadings is shown in Figure 2-7. A minimum is indicated in the region of approximately 30 vol% fines (Farris 1968).

Particle Surface Roughness

Much of the problem in quantitatively assessing the effect of surface roughness arises from difficulty encountered in quantifying the surface roughness of fine particles. Nevertheless, some general observations can be made on the effect of surface roughness as it relates to packing density and flow behavior.

Basically, surface roughness reduces the distance of closest approach between particles. The effect on packing density is summarized in Figure 2-8a. The "effective" density of a particle is defined as the mass of the particle divided by the volume enclosed by all the particles, including the surface asperities. As the surface roughness increases, the effective density of the individual particles decreases with increasing surface roughness, for a given absolute packing density.

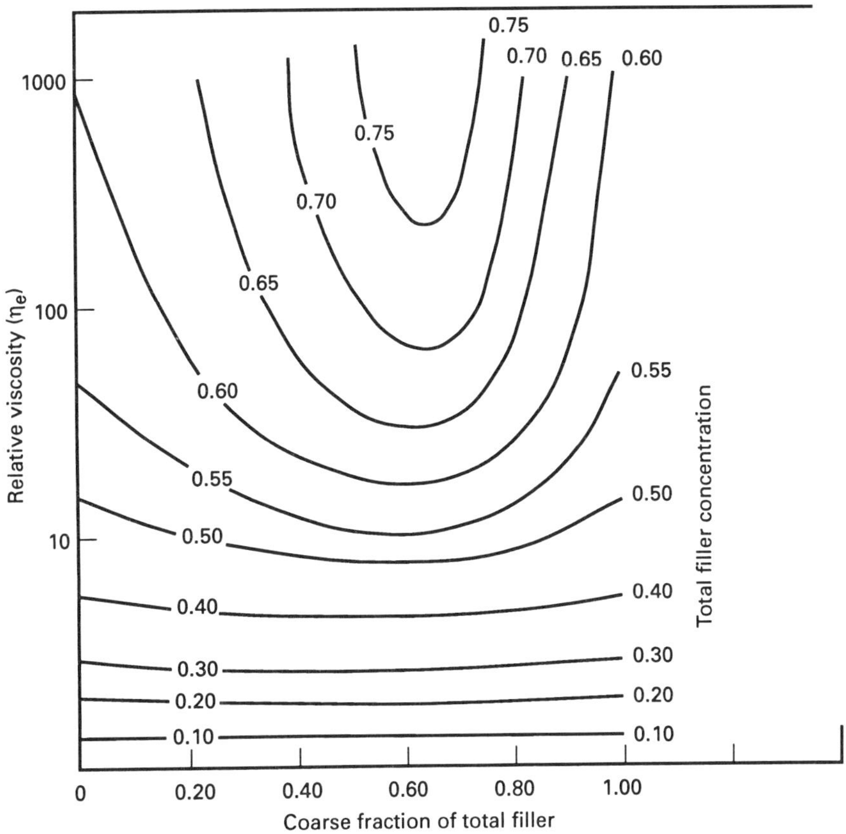

FIGURE 2-7 The proportion of coarse to fine powders in a mixture that will produce minimum viscosity in a range of total filler loadings.

The effect of surface roughness on shear stress–shear rate behavior is depicted in Figure 2-8b. As the surface roughness of the particles increases, the volume of liquid between the particles that is available for shear decreases, and the interaction of asperities between colliding particles causes the work of friction to increase. The result is higher instantaneous viscosities at a given shear rate. Consequently, with increasing surface roughness a pseudoplastic system (where the viscosity decreases as the shear rate increases) can change to a dilatant system (where the viscosity increases as the shear rate increases).

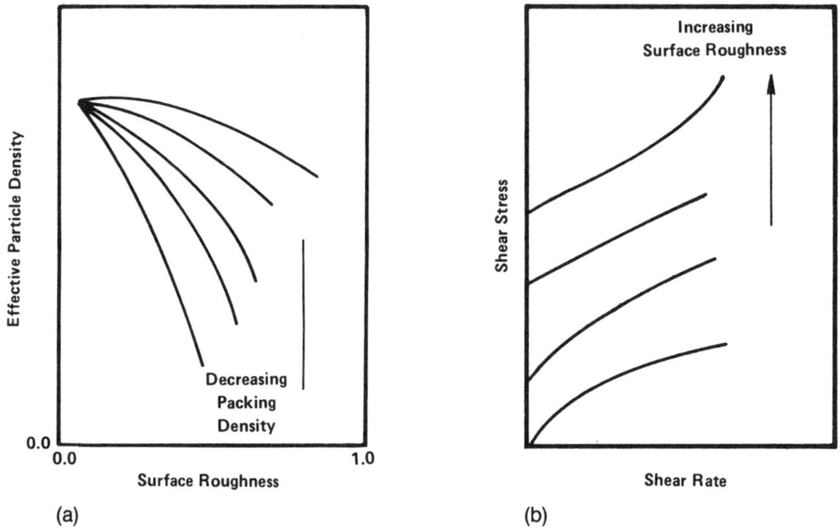

FIGURE 2-8 Effects of particle surface roughness (a) on packing density; (b) on shear rate behavior.

Particle Agglomeration

For a binder that does not wet the particles, the agglomerates remain intact unless subjected to intense shear mixing. This is not always a satisfactory solution because it is energy-intensive, and the particles often coalesce when the shear is removed. A more desirable technique is to chemically treat the surface of the ceramic particles so that the surface tension of the powder becomes compatible with that of the binder. This improves the wettability of the particles by the binder, reduces the shear required to break down the agglomerates, and prevents the particles from coalescing once the shear is discontinued.

Attraction occurs naturally because of the cohesive forces (see Table 2-4) that always act between fine particles, either with or without additives. These natural cohesive (or van der Waals) forces, which are due to fluctuating dipoles, are so weak, however, that to be effective the size of the particles must be small enough that the force of gravity acting on an individual particle is negligible. The gravitational force decreases as the third power of the particle diameter, whereas the natural cohesive force decreases by approximately the first or second power of the diameter. With particles that are 1 μm in diameter, for example, the cohesive force is about six orders of magnitude larger than the force of gravity. Other bonding forces are due to electrostatic forces caused by the tendency of electrons to flow from one

TABLE 2-4. Particle–particle interaction.

Force	Ceramic terminology
Electrostatic	Soft agglomerates*
van der Waals	Soft agglomerates
Polymer bridges	Flocs
Liquid bridges	Soft agglomerates
Capillary liquid	Soft agglomerates
Viscous binder	Hard agglomerates
Solid bridges	Hard agglomerates (aggregates)

*Agglomerates: a general term describing primary particles bonded together by surface forces, liquid bridges, or solid bridges.

surface to another because of differences in the electronic work functions that give rise to a contact potential difference (Rumpf and Schubert 1978).

Table 2-5 indicates the effect of agglomerates on the maximum packing density, assuming a random spatial distribution of unimodal particles. There is a continuous reduction in maximum packing from 64 vol% for nonagglomerated powder to 32 vol% for a powder composed of agglomerated particles consisting of 250 single particles. In addition to these natural bonding mechanisms, probably more important bonding forces are due to liquid bridges, capillary liquids, viscous binders, and solid bridges. These forces are considered more important because they increase the strength of the agglomerates. In this sense, the solid bridges formed by crystallizing or by sintering could be considered the most critical.

TABLE 2-5. The effect of agglomerates on maximum packing in a random distribution.

Primary particles in each agglomerate	Maximum packing density
1	64
3	60
5	53
8	47
12	43
18	39
32	37
57	35
122	34
250	32

The type of agglomerate (hard or soft) and the relative amount of each type in a powder are both important in determining the powder processing required to disperse hard and soft agglomerates to maximize green densities. A convenient method for assessing the relative strengths of agglomerates is to determine the packing density of a powder as a function of the uniaxial pressure (Niesz et al. 1968). Figure 2-9 is a schematic of compaction curves for a well-dispersed powder and for powders containing soft and hard agglomerates. Discontinuities in powder compaction curves indicate where agglomeration fracture occurs.

Particles smaller than 1 μm in diameter are almost always agglomerated. Unless a surfactant is added to reduce the van der Waals forces between fine particles in a powder, agglomerates remain after intensive milling and even can increase. Organic coatings on particles have two effects on particle packing. One effect is that the specific volume is increased because of the coating; however, the relative effect of an organic coating on the specific

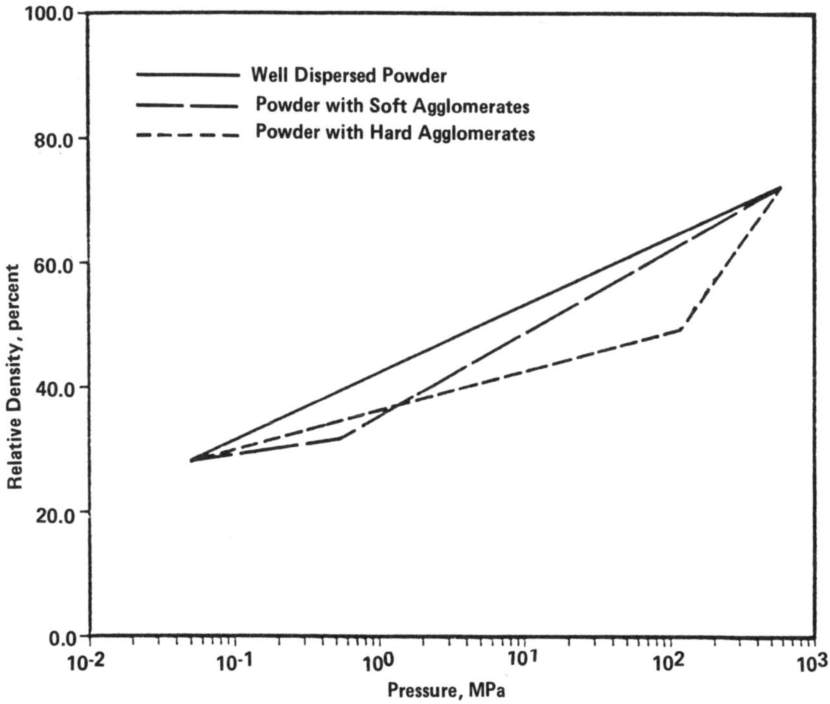

FIGURE 2-9 Compaction curves for a well-dispersed powder and for powders containing soft and hard agglomerates.

volume is less noticeable on coarse particles than on fines because of the effect of the ratio of relative diameter to coating thickness. The other effect is that organic coatings on particles can act as surfactants to disperse agglomerates in powders. This results in lower specific volumes.

Whether the specific volume is increased or reduced by organic coatings depends both on the particular properties of the powder and the organic and on the surface interaction between them. This topic will be discussed in more detail in the next chapter.

POWDER CHARACTERISTICS

Since the earliest injection molding work with alumina (Schwartzwalder 1938), almost all ceramics have been injection molded. However, because every commercially available ceramic powder is supplied in a wide range of forms and grades, it is not possible to present a complete study of each. Thus only those ceramic powders of a specific form and grade that are of immediate interest to most ceramic engineers are discussed in detail here. They include:

- A16-SG (superground) alumina (Al_2O_3)
- Type C calcium oxide and magnesium oxide (CaO/MgO), stabilized zirconia (ZrO_2)
- Bimodal alumina (Al_2O_3)
- Silicon nitride (Si_3N_4)
- Silicon (Si)
- Silicon carbide (SiC)

The suitability of a particular ceramic powder for injection molding initially should be determined from its chemical and physical properties. The following discussion examines in detail some of the powder characteristics described earlier.

Chemical Properties

The A16-SG alumina is a low-soda, high-purity, ultrafine, alpha powder (see Table 2-6). It is a Bayer-processed alumina ground by special milling to the ultimate particle size; consequently, it can be sintered at a relatively low temperature (1550–1650°C) to obtain high-density products. This powder does not contain a grain-growth inhibitor such as MgO.

The zirconia type C powder is composed of 60% cubic and 40% monoclinic zirconia phases. It is produced by vibratory blending (without media) of type A and type B powders. The type A is a high-purity monoclinic crystal form, and type B is a cubic zirconia stabilized with CaO and MgO.

TABLE 2-6. Compositions of commercial ceramic powders, weight percent.

	Types of powders				
Components	A16-SG Al_2O_3	Type C ZrO_2	SN502 Si_3N_4	A-10 SiC	Si
Al_2O_3	99.9				0.75
ZrO_2		95.25			
SiO_2	0.01	0.5			
CaO		3.0			0.001
MgO		1.25			
Fe_2O_3	0.01				0.97
Na_2O	0.08				
Si			Trace		98.27
Amorphous			40.0		
Alpha			57.6	99.5	
Beta			2.4	Trace	
Oxygen			2.7	0.55	
Carbon				30.44	
Chlorine			0.2		

Although these powders are nominally 44 μm (−325 mesh), it should be noted that no grinding is involved in their production. This powder mixture is designed to be composed of 85% of the cubic phase after firing at a recommended schedule.

The SN502 silicon nitride is a submicron-size powder produced by calcining (to remove Cl^- and NH_4^+) an amorphous grade of silicon nitride (SN402), which is prepared by a chemical vapor deposition process. It is typically composed of a mixture of alpha and beta phases; the alpha phase usually is predominant, and oxygen is the major contaminant. Minor amounts of free silicon and chlorine also are found in the powder. The major impurities in the silicon powder are iron and alumina. Most of the iron comes from the milling; the alumina may have been present in quartz, from which silicon is obtained by a reduction process. It has been suggested that iron enhances the rate of nitridation during the reaction bonding process.

The compositional analysis of alpha silicon carbide in Table 2-6 indicates the relative purity of this powder. Although the carbon content is slightly higher (30.44 wt%) than theoretical (29.95 wt%), the excess carbon (0.49 wt%) will not be detrimental. When boron carbide is the sintering aid, it is standard practice to add up to 4.0 wt% carbon. The oxygen content reported is on the high side and is likely to be present as silica on the surface of the silicon carbide powder.

It is known that impurities in the ceramic powders adversely affect the properties of the sintered products. However, what effect the impurities have on the ceramic–binder interaction during mixing and molding, or how they affect the chemical and thermal stability of the binder components, is not yet established.

Physical Properties

As stated earlier, maximum packing densities can be achieved by using binary powder mixtures. The effect of particle size distribution on packing, binder concentration, and rheology of ceramic–binder systems is now examined using a binary alumina powder having a bimodal size distribution with a size ratio of 10. The discussion emphasizes adjusting the physical characteristics of the ceramic powders to improve the packing density and to achieve controlled tolerances in the final product.

The bimodal size distribution was produced by wet mixing 30 wt% of fine powder (1–3 μm) and 70 wt% of coarse powder (9–13 μm). The size ratio is calculated from the average particle size of the coarse fraction divided by the average size of the fine fraction. Physical characteristics of the individual powders and a bimodal mixture are given in Table 2-7. The apparent density of the bimodal powder is intermediate between the densities of the coarse and the fine fractions. When it is tapped to achieve the minimum volume (a standard test to determine the density of an uncompacted powder), the mixture's density is about the same as that observed for the coarse fraction.

The particle size distribution shown in Figure 2-10 and the scanning electron micrograph (SEM) shown in Figure 2-11 do not indicate the presence of any agglomerates that could be responsible for the low tap density of the bimodal powder. Inhomogeneity in the powder mixture could be a cause of the low tap density.

TABLE 2-7. Characteristics of fine, coarse, and bimodal powders.

	Size range, μm		
	1–3 (a)	9–13 (b)	(1–3) (9–13) (30%a–70%b)
Mean particle size (μm)	1.1	11.0	12.6
Surface area (m^2/g)	6.332	0.762	2.355
Apparent density (%) Theoretical	20.1	36.2	26.0
Tap density (%) Theoretical	39.7	52.5	52.2

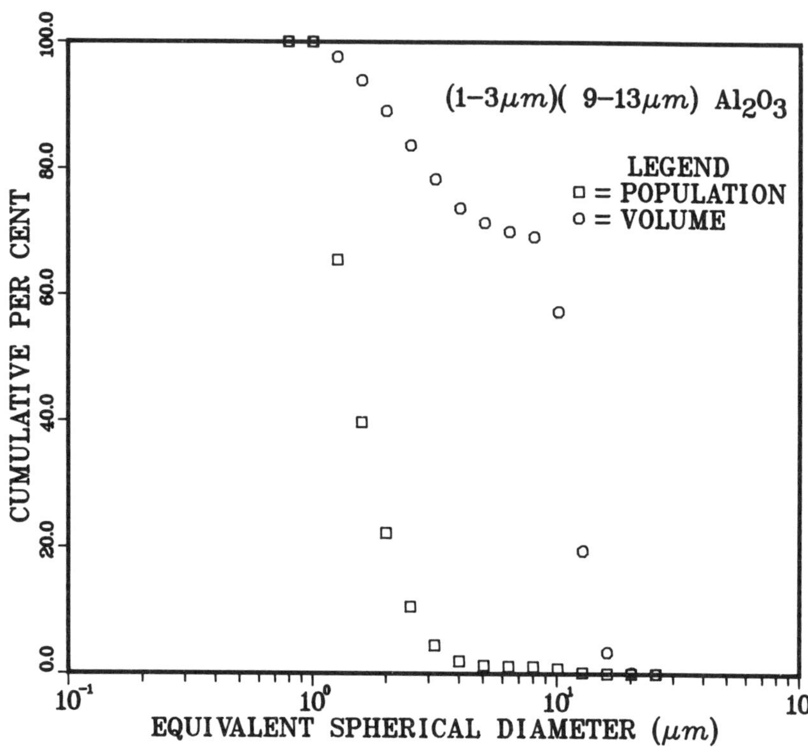

FIGURE 2-10 The particle size distribution for a bimodal powder.

The compaction density for this powder (see Figure 2-12) is high—above 7 MPa. This suggests that under pressure the particles become rearranged to fill the voids more completely. Therefore, it is possible to produce a density near the theoretical density if the frictional forces are reduced and the potential for rearrangement is increased. These conditions can be attained during injection molding, as frictional forces are reduced when the binder coats the particles, and, because the mix is fluid and under pressure, the particles are able to rearrange.

In theory, a bimodal particle size distribution can fill a matrix to a maximum concentration of 86 vol%. However, the relatively high viscosity of the filled systems plus excessive interparticle interaction usually prevent higher filler concentrations from being obtained (Bailey 1966; Whalen et al. 1978).

In most studies with bimodal systems, powders smaller than 1 μm in diameter have not been used. Also most commercial powders used for

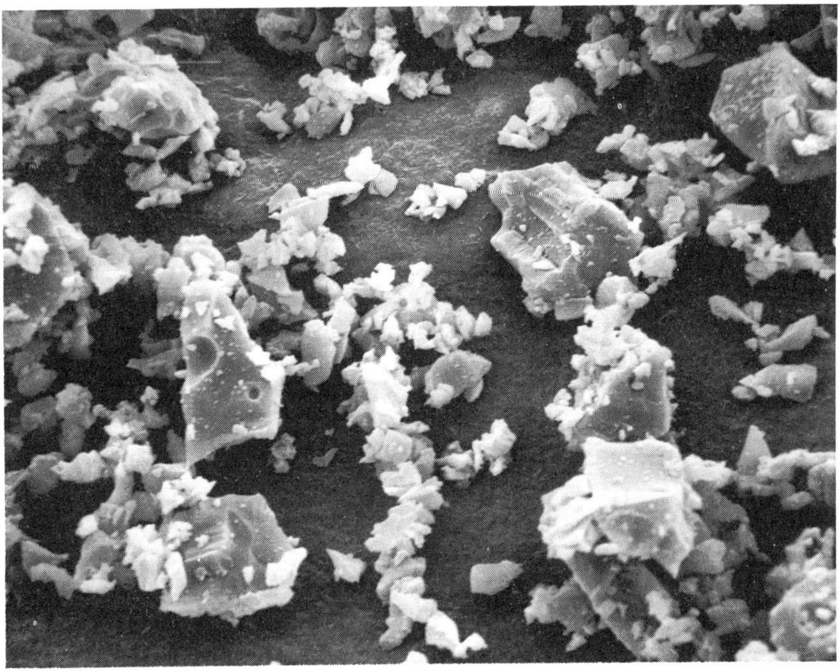

FIGURE 2-11 The SEM of the bimodal powder does not reveal the presence of agglomerates (×2000).

ceramic forming contain a continuous particle size distribution. Thus, it is more pertinent to evaluate the packing behavior of these powders in the light of a continuous distribution rather than a bimodal or trimodal size distribution.

The ceramic powders discussed in this book are broadly classified into two groups. The first group consists of powders with particle sizes under 44 μm, and includes silicon. Powders in the second group have particle sizes under 1 μm, and include A16-SG alumina, type C zirconia, silicon nitride, and silicon carbide.

The Effect of Particle Size

The particle size distributions for A16-SG alumina, type C zirconia, as-received silicon, and milled silicon (dry ball milled for 24 hours) are shown in Figure 2-13. The cumulative percent is plotted both as the volume and as the population against the equivalent diameter. The size distribution range for both A16-SG alumina and type C zirconia is very narrow. Although

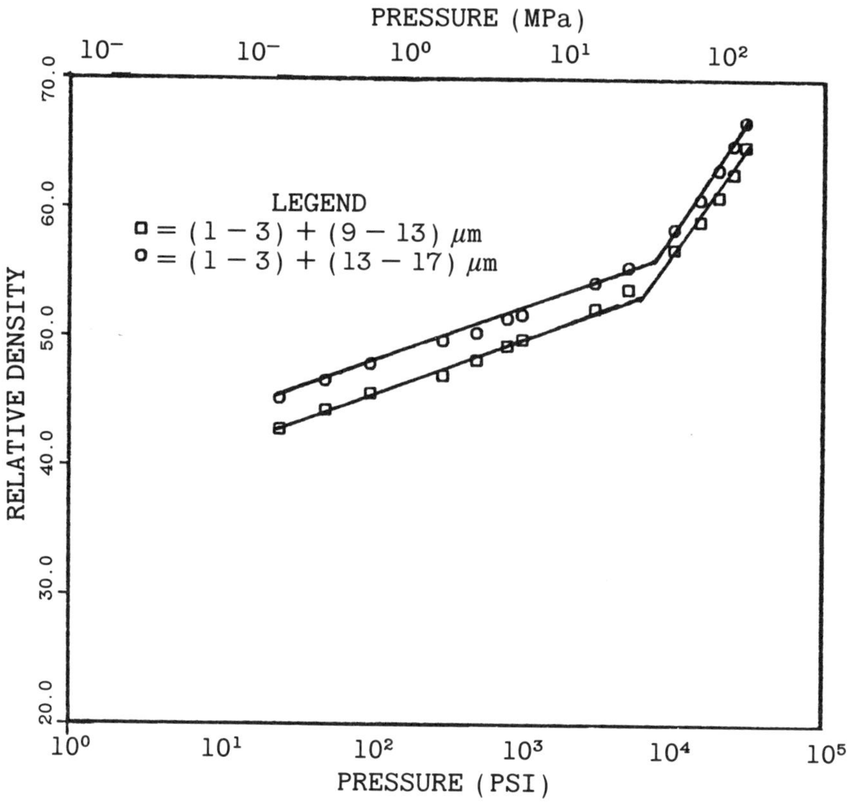

FIGURE 2-12 The high compaction density of this powder suggests that under pressure the particles become rearranged to fill the voids more completely. From Battelle Columbus.

with both powders the particles are crowded within a narrow range, as shown in Table 2-8, type C zirconia has a much higher tap density than A16-SG alumina.

Particle crowding within a narrow range suggests that the packing efficiency depends more on the size of the population than on the range of the size distribution. Nevertheless, it does not adequately explain the difference in the tap densities of A16-SG alumina and type C zirconia. An explanation could be the spherical particle shape of type C zirconia plus a lower amount of agglomeration because of the smaller surface area.

In dealing with powders containing particles 44 μm in diameter or smaller, the effects of the particle size distribution and the range of particle sizes on packing can be different. This is exemplified by an examination of two powders: milled and as-received silicon. Although their effective populations are similar, the as-received silicon has a much wider particle size range than the milled powder. This is reflected in the considerably higher tap density of the as-received silicon compared with the milled silicon (see Table 2-8). The surface area of the milled silicon relative to the as-received silicon is also larger (see Table 2-9).

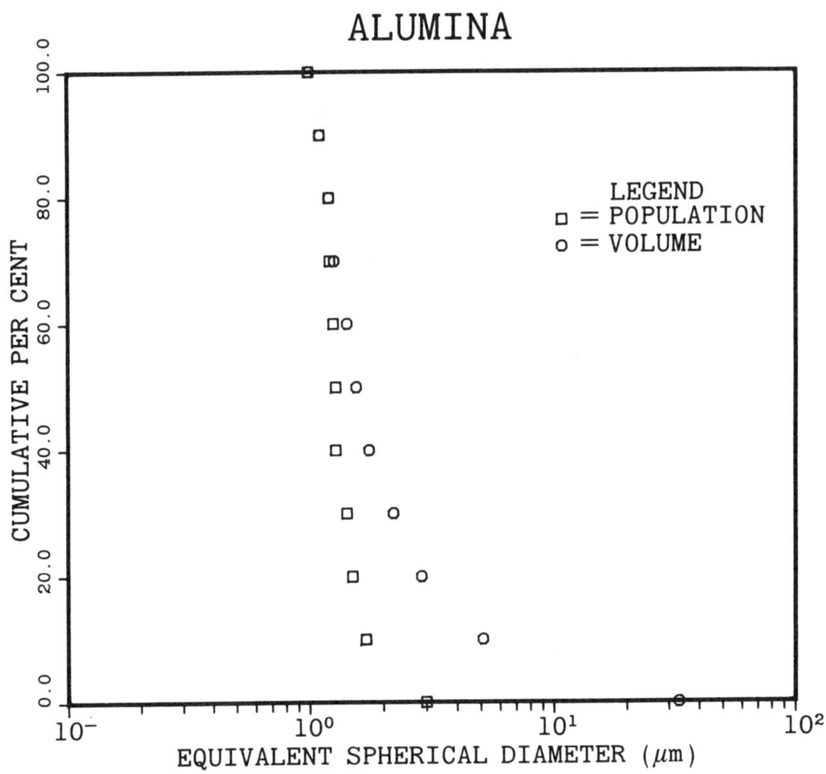

(a)

FIGURE 2-13 The cumulative percent for different powders is plotted as a function of their equivalent diameters with respect both to volume and to population. (a) A16-SG alumina; (b) type C zirconia; (c) as-received silicon; (d) milled silicon. From Battelle Columbus.

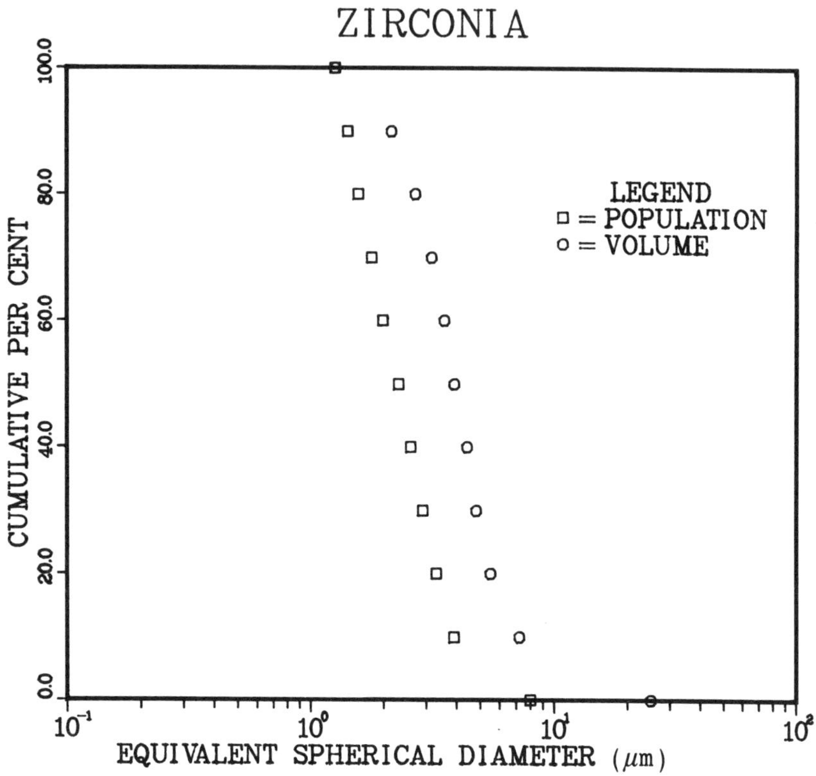

(b)

FIGURE 2-13 *contd*

The effect on packing of a wider range of particle sizes also can be seen from examining the particle size distribution of two silicon powders, one classified at 20 μm and the other at 5 μm (see Figure 2-14). Although there is only a small difference in the size of their populations, the range of particle sizes in the 20-μm cut is noticeably larger than in the 5-μm cut. This could explain the difference in their tap densities (see Table 2-10).

The SN502 silicon nitride powder is excluded from this discussion because a reliable size distribution cannot be obtained from standard methods of particle size analysis with a powder containing a mixture of both needle shapes and equiaxed particles (see Figure 2-15). However, the mean particle size for SN502 silicon nitride can be calculated from specific surface area data.

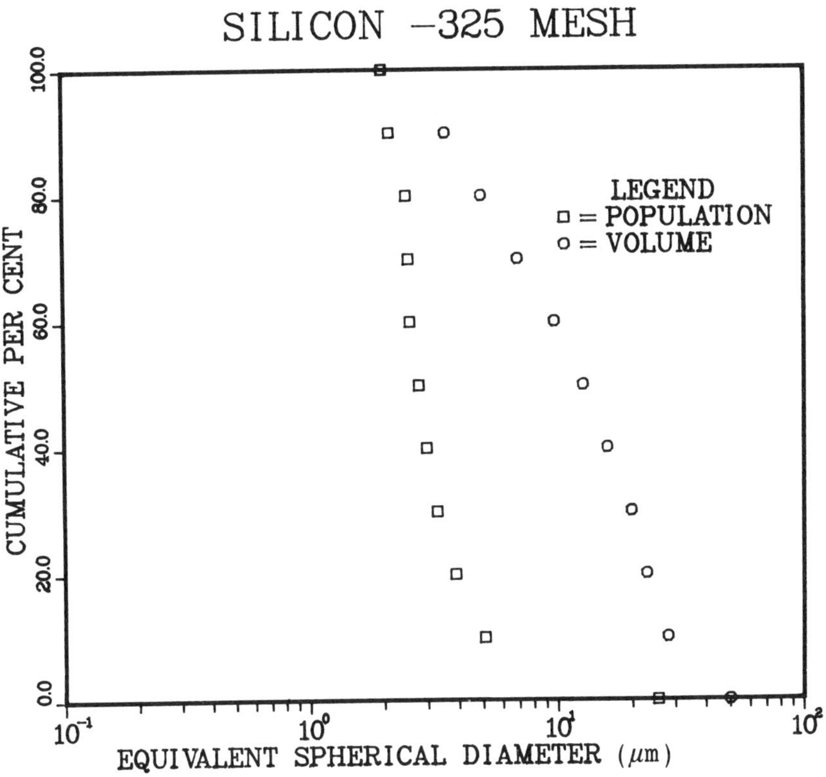

(c)

FIGURE 2-13 *contd*

In summary, the effect on packing of the particle size distribution and the range of particle sizes is more pronounced with 44 μm and finer powders (as seen with silicon), but the particle shape (as seen with zirconia) and the extent of agglomeration (as seen with A16-SG alumina) become dominant packing factors in powders with particle sizes smaller than 1 μm.

Particle Shape

One must consider the shape of the powder particles before determining the powder processing approach that will permit maximum packing density. In the case of type C zirconia, for example, the particles are primarily spherical

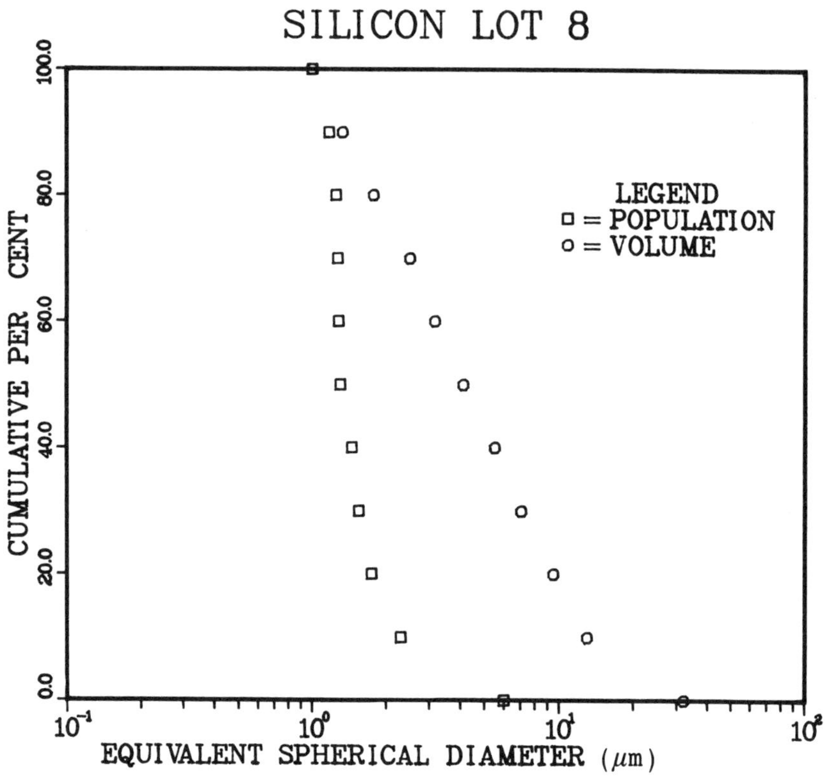

(d)

FIGURE 2-13 *contd*

TABLE 2-8. The packing characteristics of different ceramic powders.

Powder	Measured tap density, percent of theoretical
A16-SG Al_2O_3	28.0
Type C ZrO_2	43.0
As-received Si	54.0
Milled Si	37.0
Si, 5 μm	33.0
Si, 20 μm	43.0
As-received Si_3N_4 SN502	15.0
Milled Si_3N_4 SN502	27.0

TABLE 2-9. The surface areas of various ceramic powders.

Powder	Measured surface area (m^2/g)
A16-SG Al_2O_3	9.078
Type C ZrO_2	1.219
Milled Si_3N_4 SN502	4.389
Si (as-received)	1.24
Si (milled)	3.55

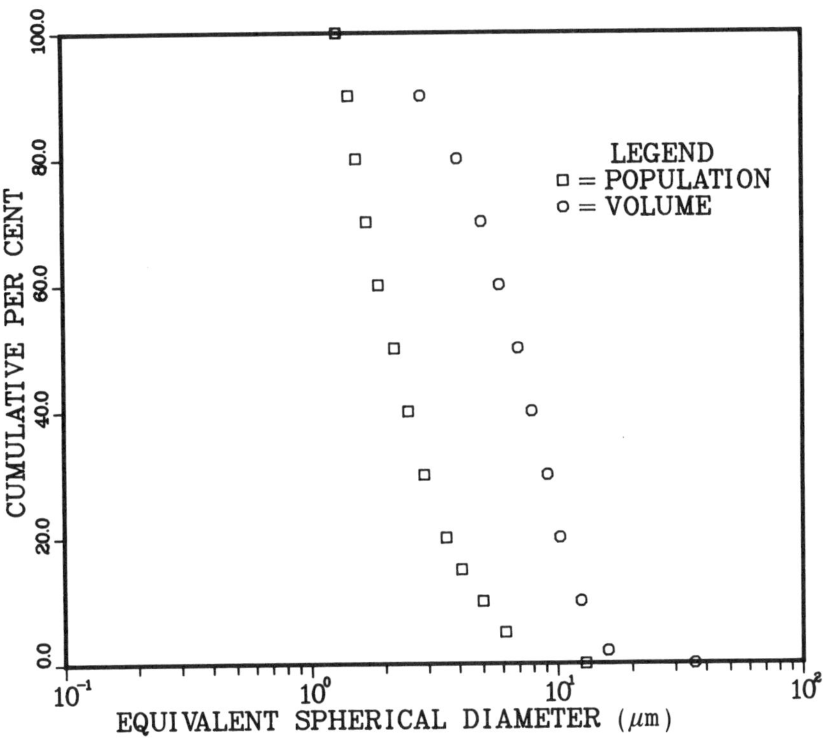

(a)

FIGURE 2-14 The effect on packing of a wider range of particle sizes can be seen from examining the particle size distribution for two silicon powders. (a) 20-μm cut; (b) 5-μm cut.

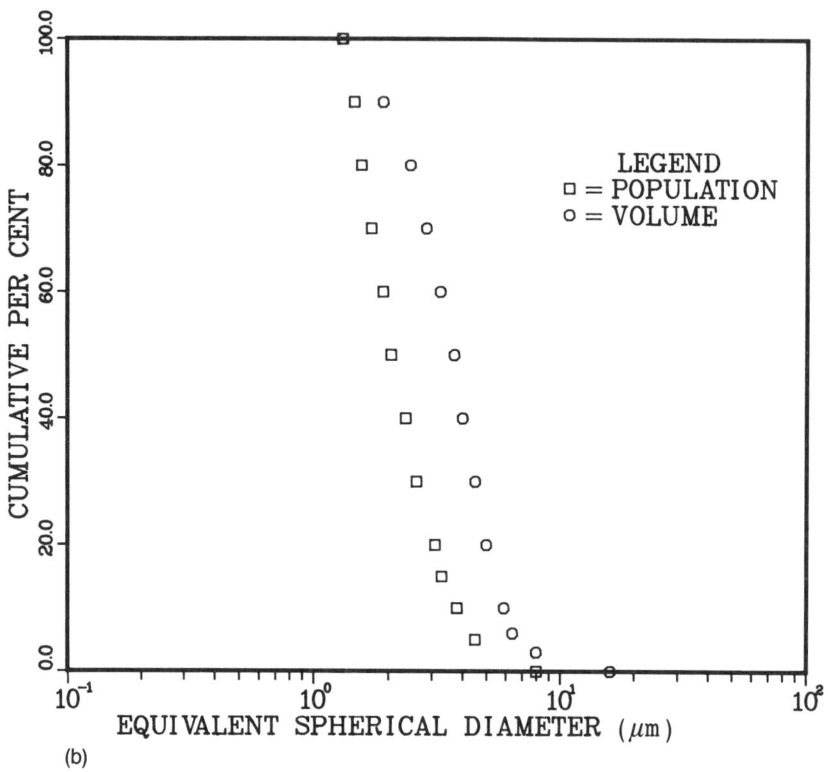

FIGURE 2-14 *contd*

(see Figure 2-16), thus approaching the nearly ideal situation for maximum packing. The use of tap density as a measure of packing shows that the packing density of this powder is only 43%—considerably less than the theoretical value of 63.7%. The low tap density cannot be explained simply as a function of the particle shape; the size of the particles and their distribution also must be taken into consideration. Figure 2-13b shows that

TABLE 2-10. The agglomeration number and tap density of various powders.

Powder	Agglomeration number	Percent of theoretical density
A16-SG Al_2O_3	538.0	28.0
Type C ZrO_2	24.1	43.0
As-received Si	2.7	54.0
Milled Si	7.0	37.0

FIGURE 2-15 SN502 silicon nitride powder contains both needle shapes and equiaxed particles (×6400). From Battelle Columbus, OH.

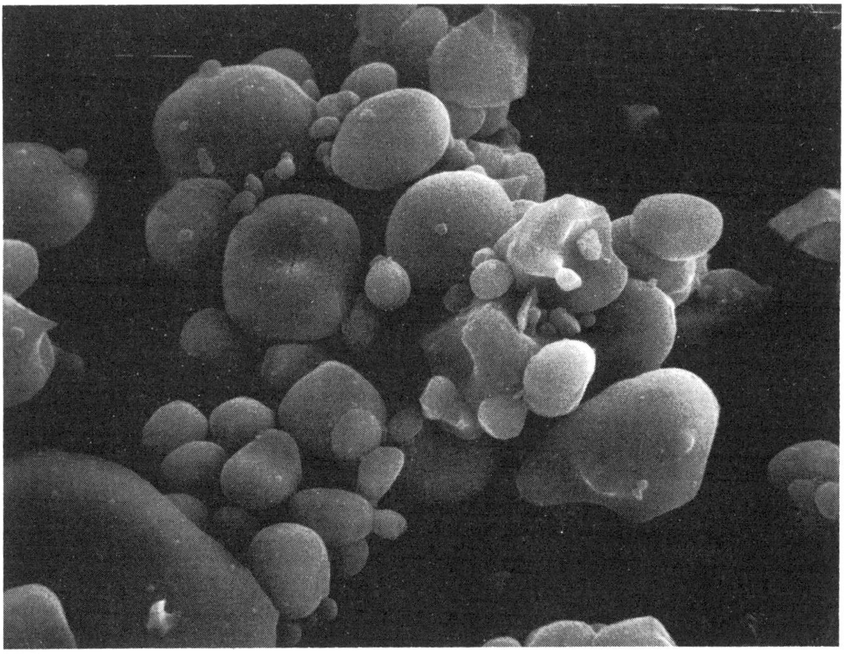

FIGURE 2-16 The SEM of type C zirconia particles shows that they are primarily spherical (×5000). From Battelle Columbus, OH.

FIGURE 2-17 These SEMs of as-received A16-SG alumina powder indicate the presence of agglomerates and the irregular shapes of the particles. (a) ×750 and (b) ×15000.

type C zirconia particles are crowded within a narrow range of particle sizes. If one could extend the particle size range while keeping the particle shape spherical, the packing density could be raised to the theoretical value.

Particles with an irregular shape can be packed to a higher density if the surface unevenness is reduced by processing. A16-SG alumina and silicon both are powders with irregularly shaped particles (see Figures 2-17 and 2-18); as-received silicon has the higher tap density, 54%.

It has been suggested that maximum densities could be achieved with fibers of any aspect ratio (length/diameter). However, the density would depend on the ratio of the diameter of spherically shaped particles to that of elongated or fiber-shaped material. This becomes important in the case of SN502 silicon nitride, which usually is composed of both equiaxed grains and needle-shaped particles. A comparison of the tap densities for the as-received and the milled SN502 silicon nitride powder (see Table 2-8) indicates some increase in density in the milled powder. This could be due to a change in the aspect ratio of the needle-shaped particles after milling,

FIGURE 2-18 As this SEM shows, silicon particles are also irregular in shape.

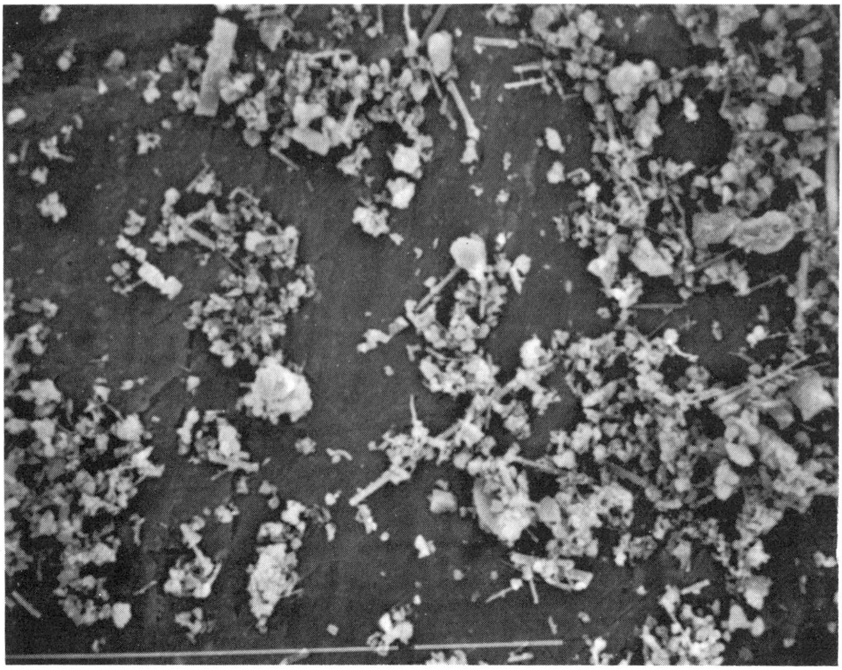

FIGURE 2-19 This SEM shows a change in the aspect ratio of the needle-shaped particles of SN502 silicon nitride after milling (×2000).

which reduces their size from 51.6 μm to 7.2 μm. A reduction in particle size also occurs (see Figure 2-19) from the breakdown of the agglomerates. In general, the shape of a particle affects the packing density if the particle size distribution and range are significantly wide, for example, between 1 and 44 μm.

Dealing with Agglomerates

Two factors need to be considered in dealing with agglomerates in ceramic powders: the extent of agglomeration and the strength of the agglomerates. Several methods are available for estimating the extent of agglomeration (Tremper and Gordon 1978; Tadros 1980; Onoda 1982). A straightforward approach is to compare Coulter counter data (a method of measuring particle size and distribution) with particle sizes calculated from BET[1] specific surface area measurements. Coulter counter data give the total volume distribution of a powder, independent of whether it is composed of primary particles, agglomerates, or both. BET measurements give the diameter of the ultimate or primary particles in the powder. The ratio of the Coulter particle size to the BET particle size gives an estimate of the average number of primary particles in each agglomerate. Assuming that the shape of the particle is spherical, an agglomeration number for each powder can be derived from the following relationship (NATO ASI 1977):

$$AN(50) = \frac{V(50)_{\text{Coulter}}}{V_{\text{BET}}} \tag{2-6}$$

where $AN(50)$ is the agglomeration number, $V(50)_{\text{Coulter}}$ is the Coulter median volume, and V_{BET} is the BET particle volume.

The data from Figure 2-13 and Table 2-9 were used to calculate agglomeration numbers for four powders. The results are summarized in Table 2-10. As indicated by the scanning electron micrographs in Figure 2-17, the A16-SG alumina was extensively agglomerated with an agglomeration number or $AN(50)$ equal to 225. The other powders discussed below had $AN(50)$'s ranging from that of the alumina to a low of 2.7 for as-received silicon. Note that the $AN(50)$ of the milled silicon, 7.0, is substantially higher than that of the as-received silicon. Apparently although milling decreased the ultimate particle size, it also increased the extent of agglomeration. If the $AN(50)$ is less than 1, it can be attributed to several factors: the use of an average volume for a nonnormal particle size distribution, the lack of sphericity in the powder, and the breakup of soft agglomerates in the shear field of the Coulter apparatus or in handling (Matthews and Rhodes 1970). The $AN(50)$ could also be less than 1 if the single particles are porous.

A comparison of the agglomeration numbers and the tap densities given in Table 2-10 with the theoretical predictions in Table 2-5 shows reasonable agreement. Although the agglomeration number cannot be treated as an absolute value, it allows comparisons between the various powders and indicates that processing is needed to minimize the extent of agglomeration.

---

[1]BET is a surface measurement technique developed in the 1930s by Brunauer, Emmett, and Teller. Its basic premise is that a porous material will adsorb a measurable amount of gaseous nitrogen that is proportional to its surface area.

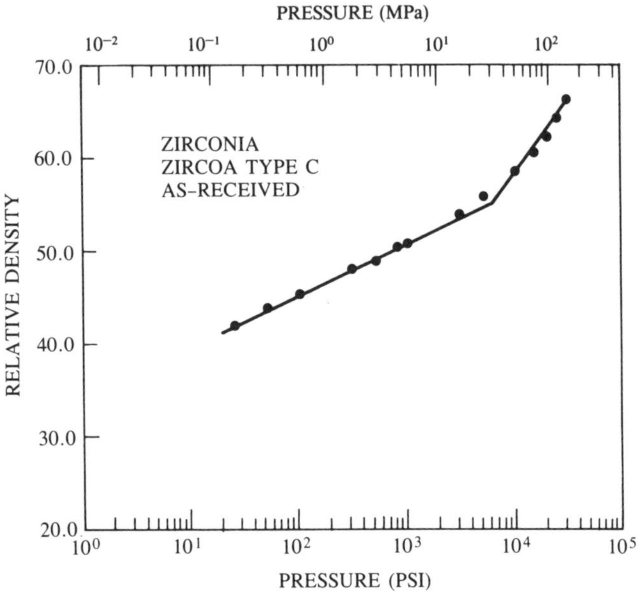

(a)

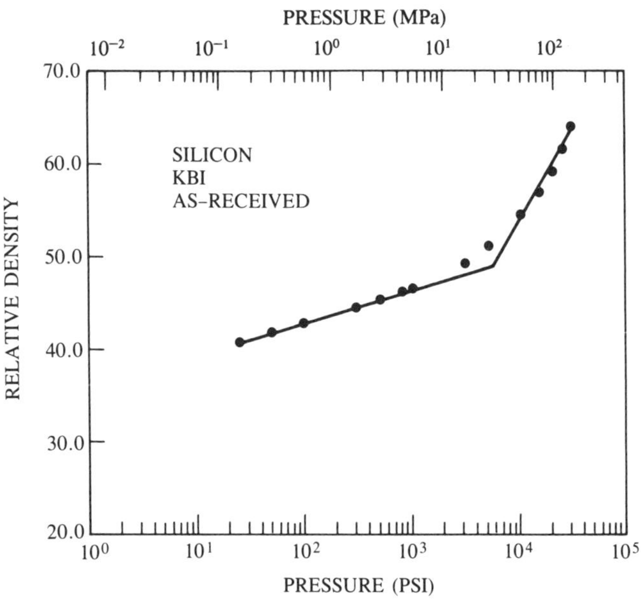

(b)

FIGURE 2-20 Plots of the compaction data for four different powders. The relative density is plotted as a function of the logarithm of the pressure. (a) As-received Zircoa type C zirconia powder; (b) as-received KBI silicon powder; (c) as-received Al6-SG alumina powder; (d) as-received GTE SN502 silicon nitride powder. From Battelle Columbus, OH.

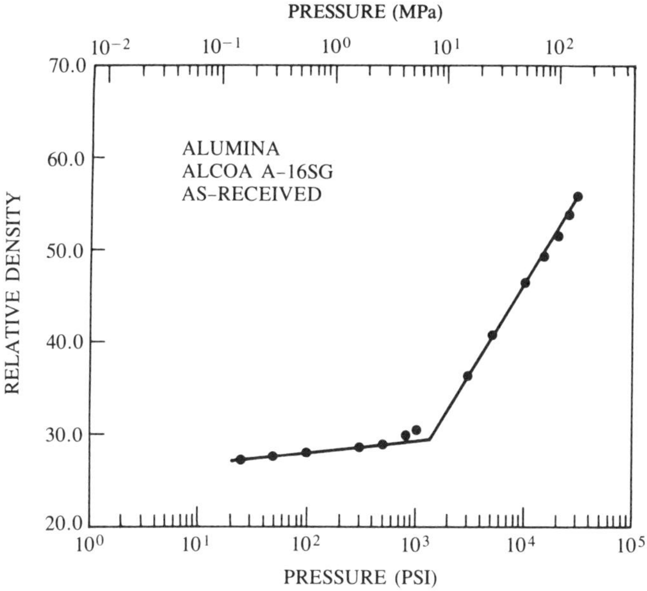

(c)

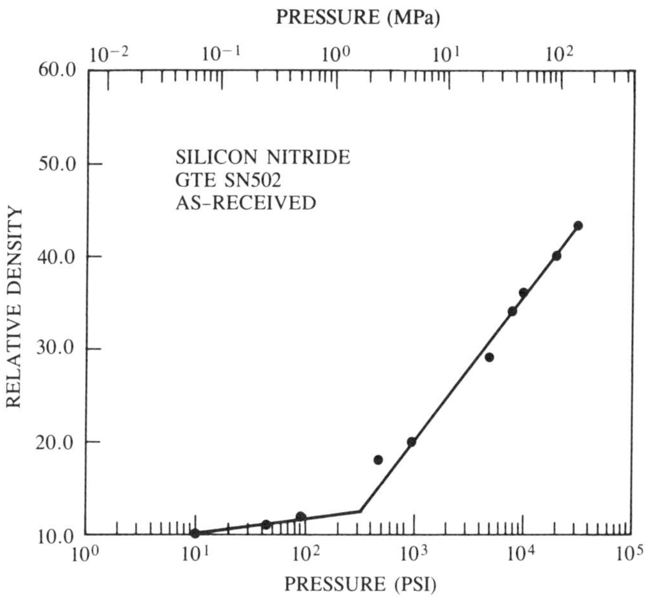

(d)

FIGURE 2-20 *contd*

A technique has been developed for determining the relative strength of agglomerates from powder compaction data (Niesz and Bennett 1968). The compaction data for several materials plotted as the logarithm of pressure versus percent of theoretical density are shown in Figure 2-20. The compaction curves for zirconia and as-received silicon (Figure 2-20a and b) show no break below 21 MPa. The break observed above this pressure can be attributed to the fracture of the individual particles, which leads to particle rearrangement or the breakdown of hard agglomerates. The breaks in the curves shown in Figure 2-20 (c and d) for the A16-SG alumina and the as-received silicon nitride indicate that their agglomerates start to break down at about 7 MPa and 3 MPa, respectively.

Relatively low breakdown pressure (below 20 MPa) suggests that the forces involved in forming these agglomerates are not strong enough to inhibit particle rearrangement during compaction. Therefore, these agglomerates can be broken down effectively by removing any moisture present and milling the powder with a milling aid such as stearic acid or triethanolamine. The curves shown in Figure 2-21 for the SN502 silicon nitride after dry milling for 24 hours with 1 wt% of a milling aid do not exhibit any

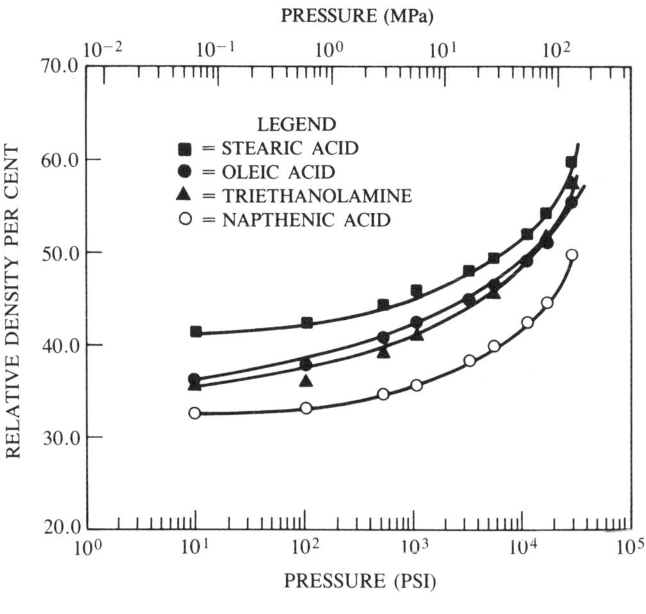

FIGURE 2-21 The compaction curves for dry-milled GTE SN502 silicon nitride powder (milled for 24 hours with 1% milling aid) do not indicate any breaks.

breaks between about 0.1 MPa and 7 MPa. This indicates that the soft agglomerates in this powder are effectively broken down by the small amount of milling aid.

MEASURING POWDER COMPRESSIBILITY

The following test provides information about the compressibility of a given ceramic powder. The green density that can be achieved is a function of the pressure used to compress the powder. In addition, this method can be used to determine the strength of agglomerates.

To measure the degree of compaction of a powder continually from 1 MPa to 172 MPa (172 MPa is the typical pressure used in compacting ceramic powders in a uniaxial die), pour the powder into a steel die 1.4 cm in diameter that has been lubricated with silicone, and level it. Mount the die assembly on an Instron Testing Machine, and compact the powder by applying a load at a rate of 0.13 cm/min. The Instron readout height, which indicates the volume of material in the die, is recorded at various pressures. The compact density is calculated as a percent of the theoretical density (that is, the relative density) and plotted against the log of the applied pressure.

As an illustrative example, the dimensions and mass of a pellet of A16-SG alumina pressed to 172 MPa were measured in the laboratory as: mass $(m) = 2.5$ g; width $(d) = 1.4$ cm; height $(h) = 0.74$ cm; theoretical density = 3.89 g/cm^3.

$$\text{Compact density at 172 MPa} = m\frac{\pi h d^2}{4} \tag{2-7}$$

The compact density can be determined from the Instron readout at various pressures. The height at any given time is given by:

$$h = h_f + \Delta h \tag{2-8}$$

where h_f is the final height of the pellet, and Δh is the change in the height of the pellet. The change in height, Δh, is calculated from the following relationship:

$$\Delta h = t\frac{L_n}{C_s} \tag{2-9}$$

where t is the time obtained from the Instron readout, L_n is the load rate fixed at 0.13 cm/min, and C_s is the chart speed, fixed at 5.0 cm/min.

BROAD POWDER PROCESSING GUIDELINES

To accurately define the characteristics of a powder and to prepare a powder that will retain the necessary flow properties during molding, will burn out without any distortion or cracks, and will produce complex shapes with close dimensional accuracy are challenging tasks. The effort is further complicated by the diversity of ceramic powders. Nevertheless, broad guidelines can be established.

- In an ideal situation, bimodal distribution will favor optimum packing.
- The packing efficiency can be improved with powders having wide particle sizes ranging from 1 μm to 44 μm.
- Milling can significantly improve the packing characteristics for coarse powders with a wide size distribution as well as for powders that tend to agglomerate. In addition, if sintering aids will be required for later processing, milling is the most practical method for achieving their homogeneous dispersion.
- The particle packing efficiency of ceramic powders with an upper size limit of 1 μm is influenced by the particle shape and the amount of agglomeration.
- Although spherical particles enhance packing efficiency, the part might not have sufficient strength after binder removal. On the other hand, equiaxed particles could provide optimum packing, thus resulting in a stronger part after binder removal.

TABLE 2-11. Powder characterization techniques.

Characterization	Technique	Standard
Particle size distribution (Adair et al. 1983)	Sieve analysis	ASTM B214-66
	Microscopy	
	Sedimentation	
	Light scattering	
	Electrical sensing	
	Gas adsorption	
	Sampling	ASTM B215-60
Particle shape and morphology	Microscopy	
	Scanning electron microscopy (SEM)	
Agglomeration number	Calculated	
Agglomerate strength	Presented separately	
Powder density	Apparent	ASTM B212-71
	Tap	ASTM B527-71
Part density	Green/sintered	ASTM C373-72
Surface area	BET	

- Agglomerates in powders with upper size limits of 1 μm can be reduced by the initial removal of moisture followed by premilling of the powder with a processing aid.

Table 2-11 lists the techniques that can be used to characterize a ceramic powder.

References

Adair, J. H., A. J. Roese, and L. G. McCoy. 1983. Particle Size Analysis of Ceramic Powders, Proc. Symposium on Processing for Improved Productivity, Am. Ceram. Soc., April 25-27.

Andreasen, A. H. M. 1930. Uber die Beziehung Zwischen Kornabstufung und Zwischenraum in Produkten aus losen Kornern (mit einigen Experimenten), *Kolloid Zeitschriff*, 50, 217-228.

Ayer, J. E. and F. E. Soppet 1965. Vibratory Compaction: 1—Compaction of Spherical Shapes, *J. Am. Ceram. Soc.* 49(4), 207-210.

Bailey, D. F. 1966. Molding of Ceramic Composition, U.S. Patent, 3,285,873.

Boctar, M. K., D. C. Freshwater, and B. Scarlett. 1965. The Effect of Particle Size Distribution on the Permeability of Filter Cakes, *Trans. Inst. Chem. Engrs.*, 43, T228-232.

Farris, R. J. 1968. Prediction of the Viscosity of Multimodal Suspensions from Unimodal Viscosity Data, *Trans. Soc. Rheol.*, 12, 281-301.

Funk, J. E. and D. R. Dinger. 1986. Modelling for Particulate Packing and Chemistry for CWM Rheology, Proc. Joint Direct Utilization AR&TD Contractors Review Meeting, PETC/METC, Pittsburgh, PA. 127-150.

Funk, J. E. and D. R. Dinger. 1992. Particle Packing, Part 1. Fundamentals of Particle Packing Monodispersed Spheres, Part 2. Review of Particle Packing of Polydisperse Particle System, *Interceram*, 41(1), 41(3), 10-14, 176-179.

Furnas, C. C. 1928. Relations between Specific Volume, Voids, and Size Compositions in Systems of Broken Solids of Mixed Sizes, U.S. Bur. of Mines Rep. Invest., No. 2894.

Matthews, B. A. and C. T. Rhodes. 1970. *J. Coll. Int. Sci.* 32, 332.

McGeary, R. K. 1961. Mechanical Packing of Spherical Particles, *J. Am. Ceram. Soc.*, 44(10), 513-522.

Messing, G. L. and G. Y. Onoda, Jr. 1978. Inhomogeneity-Packing Density Relationships in Binary Powders—Experimental Studies, *J. Amer. Ceram. Soc.*, 61(5), 363-366.

Milewski, J. V. 1974. Identification of the Maximum Packing Conditions in the Bimodal Packing of Fibers and Spheres, 29th Annual Tec. Conf. SPI Reinforced Plastics Division.

Milewski, J. V. 1978. Packing Concepts on the Utilization of Filler and Reinforcement Combination, in *Handbook of Fillers and Reinforcement for Plastics*, Van Nostrand Reinhold Co., New York.

Mutsuddy, B. C., 1983. Influence of Powder Characteristics on the Rheology of Ceramic Injection Molding Mixtures, *Fabrication Sci.*, 3, *Proc. Brit. Ceram. Soc.*, 117-137.

NATO Advanced Study Inst. 1977. *The Scientific Basis of Flocculation*, Christ's College, Cambridge, England.

Niesz, D. E. and R. B. Bennett. 1968. Structure and Properties of Agglomerates, in *Ceramic Processing before Firing*, ed. G. Y. Onoda, Jr. and L. L. Hench, John Wiley & Sons, New York, 61–73.

Onoda, G. Y. Jr. 1977. Green Body Characteristics and Their Relationship to Finished Microstructure, in *Ceramic Microstructures '76*, ed. R. M. Fulrath and J. A. Pask, Westview Press, Boulder, CO.

Onoda, G. Y. Jr. 1982. Physical Parameters for Characterizing Agglomerates, presented at the 84th Annual Meeting of the American Ceramic Soc. Cincinnati, OH.

Patankar, A. N. and G. Mondal. 1980. The Packing of Solid Particles, A Review, *Trans. Indian Ceram. Soc.*, 39(4), 109–119.

Rumpf, H. and H. Schubert. 1978. Adhesion Forces in Agglomeration Processes, in *Ceramic Processing before Firing*, ed. G. Y. Onoda and L. L. Hench, John Wiley & Sons, New York.

Schwartzwalder, K. 1938. Refractory Body and Method of Making Same, U.S. Patent 2,122,960.

Smith, W. O., P. D. Foote, and P. F. Busang. 1929. Packing of Homogeneous Spheres, *Phys. Rev.*, 34(11), 1271–1274.

Sohn, H. Y. and C. Moreland. 1968. The Effect of Particle Size Distribution on Packing Density, *Canad. J. Chem. Eng.* 46(6), 162–67.

Tadros, H. F. 1980. Physical Stability of Suspension Concentrates, *Coll. Int. Sci.*, 12, 141.

Tremper, R. T. and R. S. Gordon. 1978. Agglomeration Effects on the Sintering of Alumina Powders Prepared by Autoclaving Aluminum Metal, in *Ceramic Processing before Firing*, ed. G. Y. Onoda and L. L. Hench, John Wiley & Sons, New York, 153.

Westman, A. E. R. and H. R. Hugill. 1930. The Packing of Particles, *J. Am. Ceram. Soc.*, 13(10), 767–779.

Whalen T. J., J. E. Noakes, and L. L. Terner. 1978. Progress on Injection Molded Reaction-Bonded Silicon Carbide, in *Ceramic for High Performance Applications—II*, ed. J. J. Burke, E. H. Lenoe, and R. N. Katz, Brook Hill Publishing Co., Chestnut Hill, MA, 179–189.

3

Binder formulation

In general, moldable ceramic formulations are composed of more than one ingredient, and each ingredient has a characteristic influence on both molding and binder removal. Although it is difficult to identify clearly the role of individual binder ingredients in a complex mixture, it is important to delineate the basic functions. In this discussion, a binder formulation is considered as a system consisting of at least three components: a primary binder that is the major component of the total binder formulation, a plasticizing aid that functions as a flow modifier, and a processing aid that functions as a wetting agent and provides sufficient bonding between the ceramic powder and the primary binder. (The processing aid usually is composed of more than one chemical additive.)

In this chapter, the various materials available as binder components are described, and selection criteria are related to the general characteristics desired in the molding mix. Procedures are indicated for verifying that the specific materials will function appropriately as binders, plasticizers, and processing aids in the formulations. Various binder systems are formulated and characterized with respect to optimum binder content, proportions of plasticizers, and processing aids. Their effects on viscosity and flow behavior when combined with ceramic powders in a molding mix also are discussed.

BASIC SYSTEM INGREDIENTS

Selection Criteria for Primary Binders

The key to achieving a void-free, stress-free, and uniformly dense molded shape is the selection of primary binders that will give correct flow properties to a ceramic binder combination. For the best results, a composition must be chosen that will have a stable flow consistent with the shape, size, and dimensional stability of the molded part.

Selection Guidelines

To aid in the selection process some guidelines have been established (Strivens 1963):

- Binders that are hygroscopic or volatile under the mixing and molding conditions should not be considered, since they create defects in the molded parts during binder removal.
- Binders that migrate excessively during molding should not be considered, since they can cause an adverse skin effect, delamination, and crazing on the molded parts.
- Binders with low yield points (the minimum stress at which continuous flow can occur) can be advantageous during molding because they offer less resistance to flow. However, the absence of a yield point can create problems at the binder removal stage due to distortion of the parts.
- A binder should give a viscous melt on heating above its glass transition temperature. The high viscosity is essential to maintain the ceramic powder in a fine dispersion and to produce the friction and shear necessary to achieve adequate mixing. A high viscosity also is essential to produce back pressure during molding, but the viscosity should not be so high that the flow is restricted through the narrow channels of the die.
- A binder should be selected that can be used to adjust the time-dependent yield stress so that the curing time is long enough to allow for the stages of mixing, de-airing, and molding to be completed, and also to allow excess molding material to be reused.
- The binder should not induce anisotropic strains in the molded parts, or excessive shrinkage or expansion. The system should solidify from its plastic state at a temperature above the mold temperature and, after curing, should be removable without damage to the molded parts.
- In selecting a binder, consideration should be given to the process used to remove the binder in the subsequent step.

An optimal binder should:

- Have good flow properties.
- Be stable under mixing and molding conditions.
- Have good compatibility with the other components of the mix.
- Impart adequate strength to the molded parts after binder removal.
- Leave a minimal residue after binder removal.
- Have a long shelf life.
- Be low in cost.

Thermoplastic resins, thermoset resins, waxes, organometallics, metal alkoxides, and water-soluble organics are all classes of binders that can be used successfully to formulate mixes for injection molding.

Rheological Characterization

The flow of a plastic mass into a mold cavity is governed by its rheological behavior. The science of rheology is concerned with the flow or deformation of materials under the influence of applied stresses, and such flow or deformation can occur in two possible ways with respect to time. It can happen instantaneously, or it can occur in a more complex way over an extended period of time. Furthermore, the flow or deformation can be permanent, or it can be reversed on removal of the applied force. In addition, the reaction of the material to the applied force depends on its molecular structure. It is believed that the primary effect of shear is to break down molecular interactions arising from entanglements of the polymer chain. Because these entanglements are a function of the size of the polymer molecule and the number of molecules, both the molecular weight and the molecular weight distribution influence the rheology of polymeric binders.

Before the appropriate polymeric binder can be selected for formulating a ceramic powder–binder mix, the mix should be characterized rheologically to evaluate its flow behavior during molding. The factors used to characterize the rheological behavior of such mixes are shear and temperature–dependent viscosity plus melt elasticity. These factors provide important information for determining the optimal processing conditions, the appropriate process equipment, and the chemical composition of the binder.

The flow characteristics of viscous fluids, shown in Figure 3-1, can be described in the following terms:

- Newtonian flow, in which the viscosity is constant and independent of increasing shear rate.
- St. Venant flow, in which the viscosity is inversely proportional to the shear rate.
- Bingham flow, in which the material does not flow until the shear stress reaches a critical limit. The nonlinearity of Bingham flow has been attributed to physical changes in the network structure and the alignment of the flowing material within the network structure as the stresses increase or decrease. The flow does not start until the network structure no longer can support the stress. This finite stress is known as the yield point.
- Pseudoplastic flow, in which the viscosity decreases continuously with shear rate.

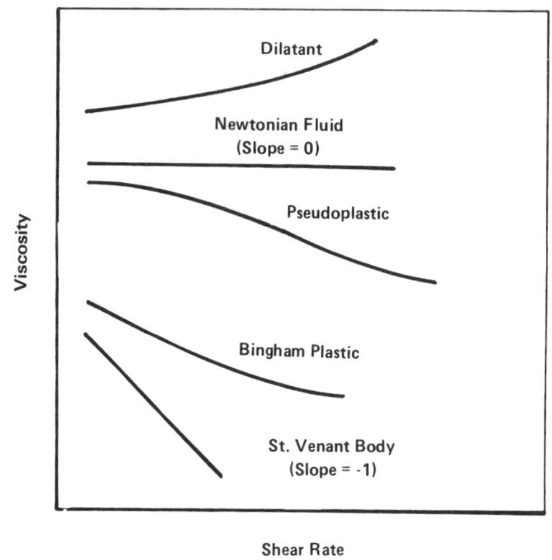

FIGURE 3-1. The rheological characteristics of viscous fluids.

- Dilatant flow, in which the viscosity increases with shear rate.
- Thixotropic flow, in which the apparent viscosity decreases with an increase in the time that the fluid is subject to shear.

In general, the rheological or flow behavior of molding mixes ranges between Newtonian, pseudoplastic, and dilatant flow. To more clearly define these types of flow behavior, the apparent viscosity of these mixes is plotted as a function of the applied shear rate. (The apparent viscosity is the ratio of the shear stress to the shear rate at a given shear rate.)

The following relationship, which correlates the shear stress, τ, to the shear rate, $\dot{\gamma}$, data for pseudoplastic and dilatant fluids, is known as the Ostwald–de Waele or power law equation:

$$\tau = K\dot{\gamma}^n \tag{3-1}$$

where K is the consistency coefficient, which is a constant related to the flow characteristics of a real fluid, and n is the power law exponent. The amount that the exponent, n, deviates from unity provides a relatively quantitative measure of how much the flow behavior of the real fluid differs from that of a Newtonian fluid. The exponent, n, has been defined as the flow behavior index of a fluid, and, like the coefficient of viscosity, can be considered a true property of the fluid.

The mathematical significance of n can be more clearly seen from the logarithmic form of the above power law equation:

$$\log_{10}\tau = \log_{10}K + n \log_{10}\dot{\gamma} \tag{3-2}$$

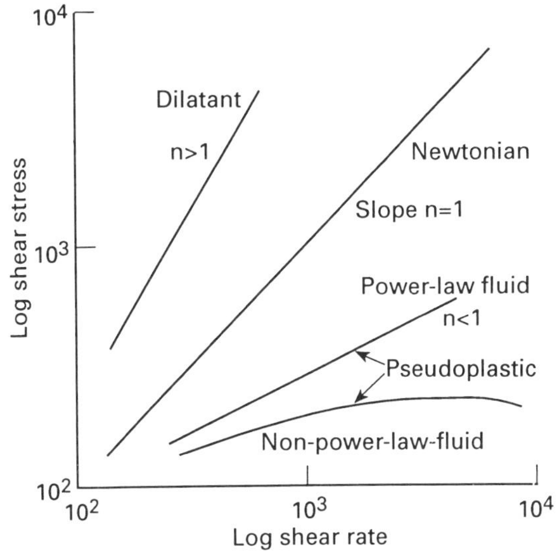

FIGURE 3-2 The significance of the power law index, n.

This is the equation of a straight line whose slope has the value n. As shown in Figure 3-2, a log–log plot of shear stress as a function of shear rate for a Newtonian fluid is a straight line with a slope of 1, for a dilatant fluid it is a straight line with a slope greater than 1, and for a pseudoplastic fluid it is a straight line with a slope less than 1.

Only fluid behavior that obeys the power law is represented by a straight line on this type of plot. Fluid behavior that does not obey the power law is represented by a curve. Although such fluids still can act like pseudoplastic or dilatant fluids, their behavior is more complex than can be described by the power law.

The Measurement of Viscosity

Two types of rheometer can be used to measure the apparent viscosity of polymeric binders and ceramic powder mixes—capillary and parallel plate rheometers. In a capillary rheometer (shown in Figure 3-3) the mix is forced through a stainless steel capillary tube by the movement of a plunger activated by a superimposed pressure. The mix is extruded at either a constant stress or a constant strain, usually the latter. The shear stress is expressed as:

$$\tau, \text{Pa} = \frac{Pr}{2L} = \frac{Fr}{2\pi R^2 L} \tag{3-3}$$

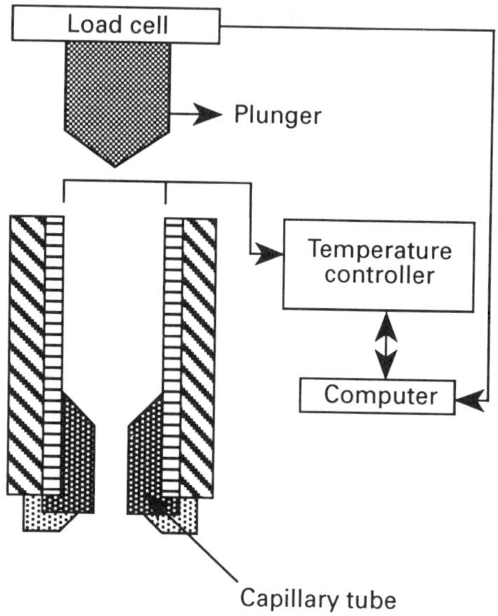

FIGURE 3-3. Schematic diagram of a capillary rheometer.

The shear rate is calculated from the following relationship:

$$\dot{\gamma},\ \mathrm{sec}^{-1} = \frac{4Q}{\pi r^3} = \frac{4V}{\pi r^3 t} \tag{3-4}$$

The apparent viscosity, η_o, thus can be expressed as the ratio of the shear stress to the shear rate:

$$\eta_o,\ \mathrm{Pa}\cdot\mathrm{sec} = \frac{P\pi r^4}{8LQ} = \frac{Fr^4 t}{8R^2 LV} \tag{3-5}$$

In equations (3-3) through (3-5) P is the pressure exerted by the ram, measured in Pascals (Pa); F is the force on the ram, measured in Newtons (N); r is the radius of the capillary and L its length, both measured in meters (m); R is the radius of the barrel, also measured in meters; Q is the flow rate measured in cubic meters per second (m^3/sec); V is the volume of extruded material measured in cubic meters (m^3); and t is the extrusion time, measured in seconds (sec). See Appendix 3-A for a discussion of various theoretical approaches that have been proposed to predict the viscosity of particulate-filled polymers.

Because the plunger moves at a constant speed, the crosshead speed is a direct measure of the shear rate. The above expression for shear rate is based on Newtonian flow behavior. For ceramic powder–binder mixes that do not show Newtonian flow a correction must be applied to calculate the shear rate at the capillary wall (Rabinowitsch 1929). This correction, called the Rabinowitsch correction, is calculated from the following expression:

$$\dot{\gamma}_w = \dot{\gamma}_a \frac{3n + 1}{4n} \tag{3-6}$$

where n is the slope in the plot of log τ_w/log $\dot{\gamma}_a$, the subscript w in $\dot{\gamma}_w$ refers to the so-called true wall shear rate, and $\dot{\gamma}_a$ is the apparent shear rate without slip. The true wall shear rate $\dot{\gamma}_w$, is then calculated from the values of $\dot{\gamma}_a$ and n.

Suspension slips have been studied since the early 1930s, and can be categorized as true slips or apparent slips. True slips occur when a suspension does not adhere to the walls of a tube, so that the material slides through the tube with little or no deformation. Apparent slips occur when there is a local variation in concentration, such as that caused by a layer of a low-viscosity fluid between the suspension and the tube wall. The effect of the apparent slip can be disregarded once its velocity is known. Equation (3-7) relates the velocity of the slip to the shear rate with and without the slip (Mooney 1931):

$$\dot{\gamma}_s = \dot{\gamma}_a + \frac{8V_s}{D} \tag{3-7}$$

Where $\dot{\gamma}_s$ is the apparent shear rate with the slip (1/sec), $\dot{\gamma}_a$ is the apparent shear rate without the slip, V_s is the velocity of the slip at the tube wall (cm/sec), and D is the tube's diameter (cm).

A viscous melt flowing into a capillary requires a finite length of time to reach equilibrium, so the effective length of the capillary and its actual length are not equal. There also are energy losses due to the elasticity of the melt. Both of these entrance effects can have a significant negative impact. One approach that has been proposed to correct the shear stress for entrance effects is given by equation (3-8) (Bagley 1957):

$$\Delta P = 4\tau_w\left[\frac{L}{D} + \frac{L}{D_e}\right] \tag{3-8}$$

where ΔP is the pressure drop, measured in pascals (Pa); τ_w is the shear stress at the capillary tube wall, also measured in Pa; L is the length of the capillary tube, measured in m; D is the diameter of the capillary tube, measured in m; and the term $(L/D)_e$, which represents the value of the entrance effect, is dimensionless. This term is assumed to depend only on the material's flow rate, but the use of a capillary with a large length-to-diameter ratio (for example, $L/D \approx 40$) eliminates the need for correcting the shear stress for entrance effects.

The Characteristics of Thermoplastic Polymers

Thermoplastic polymers, unlike thermosetting polymers, do not undergo any chemical reactions that develop crosslinks during molding or post-molding. Therefore, their hardening is reversible; they can be softened once they are hard by heating and then rehardened by cooling.

Thermoplastic polymers can be semicrystalline (for example, polyethylene, polypropylene, nylon) or glassy (for example, polystyrene, polyvinyl chloride). A semicrystalline polymer becomes a viscous liquid when melted, whereas a glassy polymer becomes a leathery or viscous doughlike material when heated above its glass temperature, T_g.

Some properties of commercial[1] thermoplastics that are useful for developing binder formulations for ceramic injection molding are given in Table 3-1.[2]

Waxes

Waxes have been used in ceramic forming since the beginning of the ceramic industry. However, their use as a primary binder is limited because of their lack of plasticity and their narrow melting range. Inadequate plasticity greatly affects the fluidity of the mix during the molding cycle, and a narrow melting range can result in an uncontrollable cooling cycle, which causes an excessive concentration of stress in the molded parts. The narrow melting range can also cause problems such as slumping and distortion

[1]The commercial binders discussed are ones familiar to the authors. The evaluations of these binders are based on the experience of the authors and are provided for illustrative purposes only. The mention of these products should not be construed as a warranty of their quality or performance, nor should it be inferred that other binders do not possess equivalent performance characteristics.

[2]The data in Tables 3-1, 3-2, 3-5, 3-6, and 3-8 are obtained largely from the manufacturers or from the trade literature. Because most of these products are commercial, the physical constants of a specific plastic, wax, plasticizer, or processing aid can vary, depending on the source.

TABLE 3-1. **Properties of thermoplastic resins.**

Properties	High melt index polyethylene	Low melt index polyethylene	Polystyrene	EVA
Grade or trade name	NA-250	MN 711-20	DP-210	AC-400[a]
Density (g/cm^3)	0.93	0.914	1.04	0.928
Melt index[b] (g/sec)	250	22	30	Beyond measurement range
Specific heat (cal/°C/g)	0.55	0.55	0.32	0.46
Thermal conductivity (W/m·K)	0.334	0.377	0.138	—
Thermal stability in air	Oxidation at 80°C	Same as NA-250	Start to break down above 250°C	Oxidation ca. 200°C
Compatibility	Fatty acids, waxes, and polymeric esters	Same as NA-250	Phthalates, adipates, oleates, and stearates	Waxes and fatty acids
Solubility parameter $(cal/cm)^{1/2}$	8.0	—	—	—

a. Comonomer ratio 86:14 wt%.
b. The melt index indicates the capacity of a polymer to flow and fill a mold. The melt index number is the number of grams of polymer extruded from a small orifice (about 1.27 mm in diameter) in 10 minutes at 100°C and 2160 grams load. In general, the viscosity is reduced as the melt index increases. Thus a high melt index indicates a low viscosity, so that, in effect, the melt index is the inverse of the viscosity.

during the binder removal process. However, these shortcomings of waxes can be eliminated by the addition of modifying agents that extend their melting range and increase their plasticity.

Waxes vary markedly in their chemical and physical properties, depending on whether they are vegetable, animal, or mineral in origin. Natural waxes that have been treated chemically to alter their properties are called modified waxes. Synthetic waxes are made by a chemical reaction, and blended waxes are physical mixtures of various waxes or of waxes plus polymers or other compounds. Four classes of waxes are discussed here:

- Paraffin waxes
- Microcrystalline waxes
- Synthetic hydrocarbon waxes
- Oxidized polyethylene waxes

Paraffin and microcrystalline waxes are derived from petroleum. The paraffin or straight-chain hydrocarbon waxes are extracted from the high-boiling fractions during the petroleum refining process, and are called straight-chain hydrocarbons because the carbon atoms in their molecular structures are linked together in succession without branching or forming side-chains. Paraffin waxes are marketed in various grades, based mainly on their melting points. However, because a wax's melting point reflects the melting points of its constituents, it is not necessarily indicative of its quality or chemical composition. The characteristics of a wax, along with its melting point, that must be considered in determining its quality include its density, compatibility, and wettability. Of the four grades of paraffin waxes commercially available, the refined grade is the one used for ceramic injection molding.

Microcrystalline waxes consist of mixtures of extremely small crystals, less than 1 μm in size. They are composed of hydrocarbons including straight-chained, branch-chained, monocyclic, and polycyclic compounds. The optimum ratio of each type of hydrocarbon structure present in a particular wax is controlled during the refining process to impart desired properties such as adhesive and cohesive strength, flexibility, and a high melting point. However, these properties do not bear a simple relation to the "molecular" weights, which are actually an average of the molecular weights of the constituent hydrocarbons.

Microcrystalline waxes differ from refined paraffin waxes in the molecular size and structure of their constituents. They are tougher, more flexible, and have higher tensile strengths and melting points than paraffin waxes. They bind solvents and oils much better than paraffin waxes and thus prevent sweating-out during the molding process. When paraffin waxes are viewed with a microscope, large, well-formed crystals are seen, when crystallized both from the melted wax and from solvents. Microcrystalline waxes form small irregular crystals from the melted wax, but do not form well-defined crystals of any size from solvents.

Synthetic hydrocarbon waxes are made by a well-known process for producing hydrocarbon fuels (the Fischer-Tropsch process). In this process, low-grade coal is burned in the presence of oxygen and steam to produce hydrocarbons and carbon monoxide, which then react in the presence of various catalysts to yield a wide variety of hydrocarbons. These are hard, high-melting hydrocarbon waxes, essentially a mixture of straight-chain (paraffin) hydrocarbons with an average molecular weight of 750. These waxes are similar to conventional paraffin waxes but with longer carbon chains. They have better high-temperature properties than the conventional paraffins and behave more like microcrystalline waxes, particularly with respect to their cohesive strength and flexibility.

TABLE 3-2. **Properties of some commercial waxes.**

	Paraffin	Microcrystalline polyethylene		
Properties	Shell-120	Okerin 1865Q	BASF-A	Paraflint H1-N1
Melting pt. (°C)	75	108	102	
Av. mol. wt	367	500	6000	750
Oil content	0.1	0.7	—	0.5 max.
Density (g/cm³)	0.75	0.9	0.92	0.945
Isopropyl ether	soluble	soluble	soluble	soluble
Vapor pressure mm Hg at 300°C	130	180	360	—

Oxidized polyethylene waxes are made from solid polyethylene. Because of their relatively low molecular weights (about 10,000), partly paraffinic structure, and high melting points, they fill the gap between the Fischer-Tropsch and paraffin waxes and the low-to-intermediate molecular weight polyolefin waxes. Their combination of well-defined high melting points with low viscosity and their compatibility makes them potentially good primary binders for moldable ceramic formulations. Some relevant properties of commercial waxes are given in Table 3-2.

The Characteristics of Thermoset Resins

Thermosetting resins are polymeric materials that are irreversibly hardened (or cured) by heating. With the exception of elastomers (rubbers), thermosets tend to be brittle. The freedom of movement of their constituent polymer chains when under stress is greatly limited by cross-linking between the chains, which creates a three-dimensional network that reduces the plasticity of the material. Because the rigidity and the hardness of thermosets are related to the amount of cross-linking they contain, their flexibility can be controlled either by reducing the number of cross-linking sites or by controlling the number of cross-links.

Catalysts, promoters, inhibitors, and stabilizers are added to thermosets to control the curing process. Catalysts initiate curing; promoters accelerate the catalytic action, thus increasing the curing rate; inhibitors, on the other hand, are used to delay or to control the curing rate during molding; and stabilizers delay curing at room temperature but not at molding temperatures.

Thermosets that are being used for ceramic injection molding include:

- *Phenol–formaldehyde*: Liquid or solid phenol–formaldehyde resins can be broadly classified as either one-stage or two-stage resins. A one-stage resin known as resol is formed if the molar ratio of formaldehyde to phenol is 1:1 or greater (that is, there is a molar excess of formaldehyde) and an alkaline catalyst such as ammonia or caustic soda is used. The resol molecules contain reactive hydroxymethyl groups ($-CH_2OH$), and on heating they condense, forming larger molecules without the addition of a curing agent.

  A two-stage resin, called a novolac, is formed if a molar excess of phenol and an acid catalyst are used. Novolac molecules do not contain reactive hydroxymethyl groups and thus need additives to be cross-linked. Usually these are compounds that can liberate additional formaldehyde.

  Furfuraldehyde can be used in place of the formaldehyde in the formation of a one-stage resin. It has been found that both the aldehyde group and the furan ring take part in the reaction, providing the cross-links in the polymeric chains.

  Phenol–formaldehyde resins cure very rapidly at high temperatures and have good flow characteristics. One of their chief uses is to modify conventional phenolic resins to enable higher loadings of the ceramic powder.
- *Epoxies*: The backbone chain of epoxy polymers contains both oxygen atoms and six-membered aromatic (or benzene) rings. Epoxies usually are supplied as two components to be mixed and cured. Single-component epoxies are also available. These polymers are cured by the absorption of oxygen without the production of either water or volatiles.
- *Polyesters*: Unlike other thermosets, polyesters polymerize rapidly at room temperature and pressure. They are a large family of polymers synthesized from saturated and unsaturated organic acids and alcohols and cross-linked by styrene, acrylic, or other monomers in the presence of a suitable catalyst. The polyesters can be cured rapidly with a small quantity of catalyst (less than 1.0 wt%), usually a peroxide such as methyl ethyl ketone peroxide, and the reaction is exothermic (occurs with the evolution of heat). An almost limitless number of polyester formulations are possible, and they can be tailored to suit a particular ceramic powder and molding conditions.
- *Allyls*: The most widely used allyls are diallyl phthalate and diallyl isophthalate. These are cured using peroxide catalysts. Similar to polyesters, there are a large number of monomers that can be polymerized to produce allyl thermosets.

The Characteristics of Organometallic Polymers

Organometallic polymers of interest in injection molding are broadly classified into four groups: metal alkoxides, silanes, siloxanes, and silazanes. Because polycarbosilane has been used in injection molding silicon carbide (SiC), it is the organometallic silane discussed in detail here. Alkoxides, which are likely candidates as primary binders for injection molding oxides and mixed oxides in aqueous media, are discussed in more general terms.

One way to prepare polycarbosilane is to dechlorinate dimethyldichlorosilane with metallic lithium. This step is followed by recrystallization and sublimation to obtain pure dodecamethyl-cyclohexasilane. After heat treating the reaction product in an autoclave at 400°C for 48 hours, solid polycarbosilane is formed. Removing the low molecular weight portion by acetone fractionation yields polycarbosilane with an average molecular weight of 1000 to 2000. In this typical molecular formula for polycarbosilane Me represents a methyl group ($-CH_3$):

$$[(Me_3-Si)_{0.5}(CH_2-\underset{|}{CH}-\underset{|}{\overset{\overset{Me}{|}}{Si}})_{1.0}\,(Si-Me_2)_{1.0}]_n$$

The successful use of polycarbosilane as a primary binder in injection molding depends on rheological behavior, thermal and hydrolytic stability, and the yield of silicon carbide after pyrolysis. These factors are influenced by the chemical and physical characteristics of the particular polycarbosilane used. The chemical nature of the precursor compounds and the processing parameters determine the chemical structure and average molecular weight of a particular polycarbosilane polymer. Polycarbosilanes require the addition of a small quantity (0.5–5 wt%) of a catalyst, such as a peroxide, to be cured (cross-linked) during or following the molding operation. A major drawback in the use of polycarbosilanes as binders is that they are readily hydrolyzed.

More recently, a family of siloxane polymers ($(R_2 - SiO)_n$, where R can be an aliphatic or aromatic hydrocarbon group) has been used successfully for injection molding silicon carbide (SiC) parts. These polymers are more thermally stable than polycarbosilanes, and most importantly do not readily absorb moisture so are not easily hydrolyzed.

The Characteristics of Water-Soluble Primary Binders

Water-soluble primary binders have a long history of use in ceramic forming (McNamara and Comeforo 1945; Treischal and Emrich 1946; Onoda

1978). However, their application to injection molding is relatively recent (Rivers 1978; Fanelli and Silvers 1988).

The formability of injection molded shapes with water-soluble binders such as methyl, ethyl, or hydroxypropyl/hydroxyethyl celluloses is dictated by gelation properties that are controlled by gel temperature, gel strength, and the degree of water retention. In turn, these properties depend on the extent of substitution by methoxy, ethoxy, or hydroxypropyl groups on the cellulose molecule, the substitution of hydroxypropyl groups of the cellulose on a mole-for-mole basis, and the molecular weight and the molecular weight distribution of celluloses. However, the role of these factors with agar and its derivatives, such as agarose, has not been elucidated. Thus most of the discussion here is confined to either the methyl or the ethyl celluloses.

Figure 3-4 shows the flow curves for hydroxyethyl cellulose for a solution concentration of 10 kg/m^3, as a function of the shear rates at 25°C and 70°C. In both cases, the upward trend is almost reversed on the return of the curve. At all temperatures, the solution is pseudoplastic and has no yield stress. The degree of pseudoplasticity of the cellulose binder solutions (at concentrations of 5.0 kg/m^3 and 15.0 kg/m^3) can be expressed as a power law index, as shown in Figures 3-5 and 3-6. Frequently, the effect of

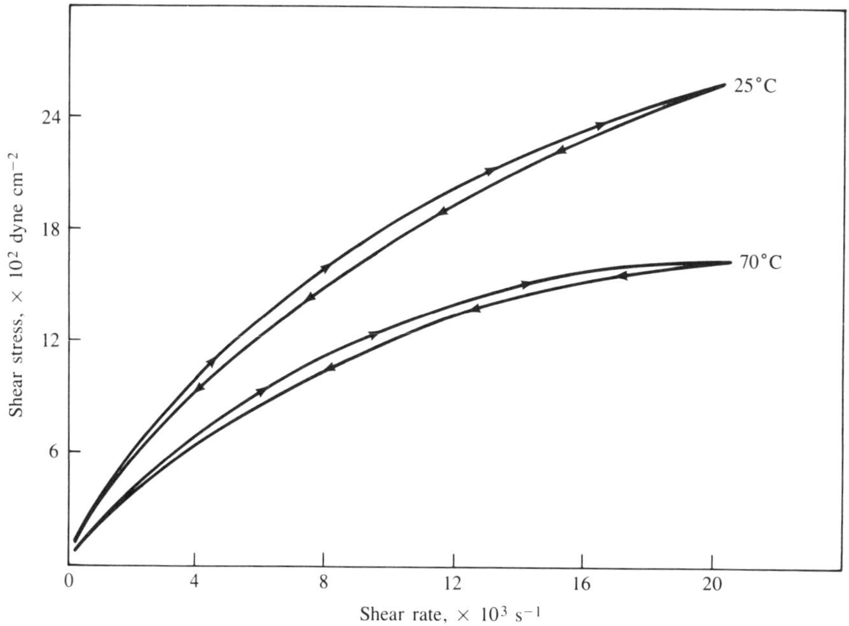

FIGURE 3-4. Flow curves for hydroxyethyl cellulose as a function of the shear rate.

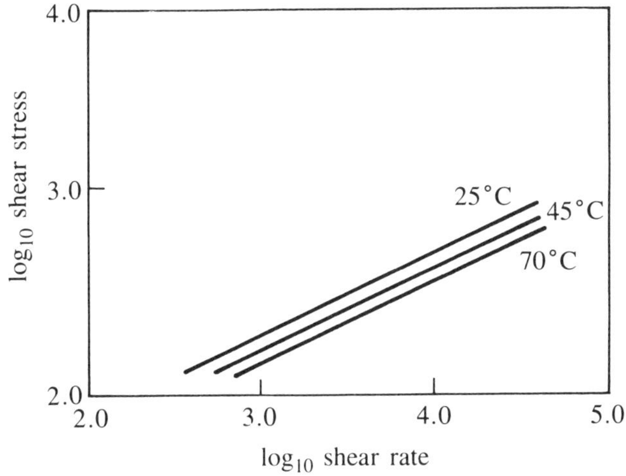

FIGURE 3-5. The degree of pseudoplasticity of cellulose expressed as a power law index, at a solution concentration of $c = 5\ \text{kg/m}^3$.

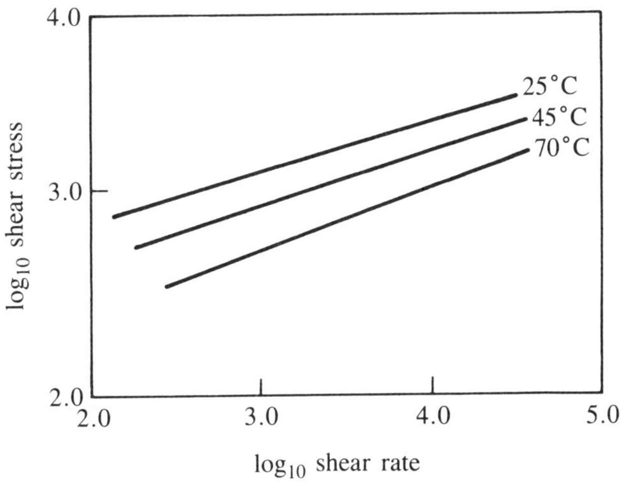

FIGURE 3-6. The degree of pseudoplasticity of cellulose expressed as a power law index, at a solution concentration of $c = 15\ \text{kg/m}^3$.

temperature on the consistency of the binder solution follows the Arrhenius relation; that is, the rate of change of the consistency of the binder solution is exponentially dependent on the temperature:

$$K = A_o \exp(-E/RT) \tag{3-9}$$

where K is the consistency coefficient, A_o is a constant, E is the activation energy for the process, R is the gas constant, and T is the absolute temperature. Figure 3-7 is based on this expression. Although the lines in this figure appear parallel, their slopes differ by 15%, indicating that the activation energy depends on the temperature.

The power law index, n, and the K values of aqueous solutions of methylcellulose (given in Table 3-3) also are considerably influenced by the molecular weight distribution obtained by blending celluloses with both high and low molecular weights.

Gelation of cellulose or hydroxypropyl cellulose solutions is caused by the hydrophobic interaction between cellulose polymers containing different amounts of methoxy or ethoxy substituent groups (Sarkar 1979). In

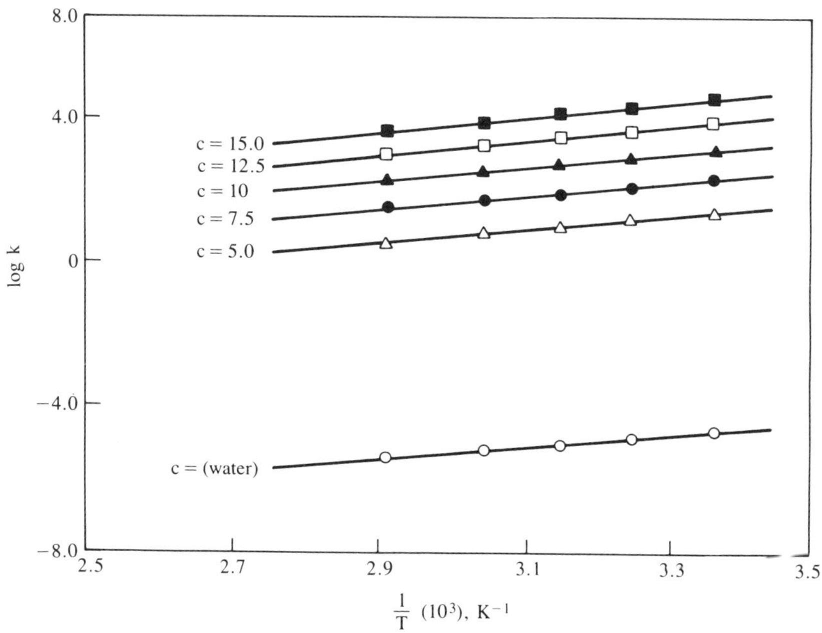

FIGURE 3-7. The activation energy and the consistency coefficient depend on the temperature and the concentration, respectively.

TABLE 3-3. Power law constants for 2% methylcellulose[a] blends of high and low molecular weight fractions.*

Viscosity of 2% solution (mPa·sec)		Wt. fraction of low mol. wt. component in blend	Viscosity of blend[b] (mPa·sec)	K^c	n^d
Low mol. wt. fraction	High mol. wt. fraction				
1500		1.0	1393	3176	0.719
500	7000	0.535	1353	3260	0.700
500	30,000	0.728	1369	3619	0.669
125	30,000	0.588	1329	4479	0.586
21.4	30,000	0.498	1393	5385	0.539
12	30,000	0.475	1445	5752	0.521

a. Nominal viscosity 1500 mPa·sec at 20°C.
b. 18.8 sec^{-1} shear rate at 20°C.
c. Consistency coefficient.
d. Power law index.
*Sarkar and Greminger (1983).

solutions at lower temperatures, molecules are hydrated, so there is little polymer–polymer interaction other than simple entanglement. As the temperature is increased, the molecules gradually lose their water of hydration. This is reflected by a drop in viscosity, as shown in Figure 3-8. Eventually, when sufficient, but not complete, dehydration of the polymer occurs, a polymer–polymer association takes place, and the system approaches an infinite network structure. This is reflected in a sharp rise in viscosity. These gels are completely reversible; they are formed on heating, but liquefy to their original consistency on cooling (Sarkar and Greminger 1983).

The degree of substitution and the type of substitutents strongly affect the gelation temperature. In general, the higher the total degree of substitution, the lower the gelation temperature. Maintaining the same amount of methoxy or ethoxy substitution and increasing the hydroxypropyl substitution increases the gelation temperature to a certain extent. On the other hand, the gelation temperature is directly related to the concentration, but it is not particularly dependent on the average molecular weight, possibly because solutions that have the same average molecular weight consist of different fractions that have similar molecular weights.

The effect of concentration on the gel strength of aqueous methylcellulose solutions with three different viscosities—low (25 mPa·sec), medium (400 mPa·sec), and high (1500 mPa·sec)—is shown in Figure 3-9, which indicates that the gel strength increases with increasing concentration. For

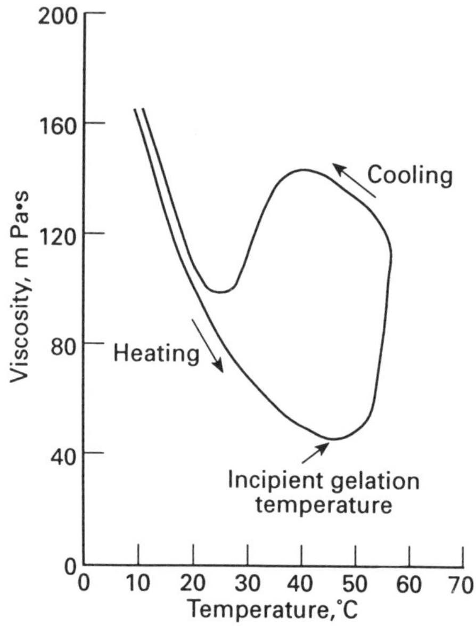

FIGURE 3-8. The relation of viscosity to polymer–polymer association with increasing dehydration of the polymer as the temperature is increased. After Sarkar and Greminger (1983).

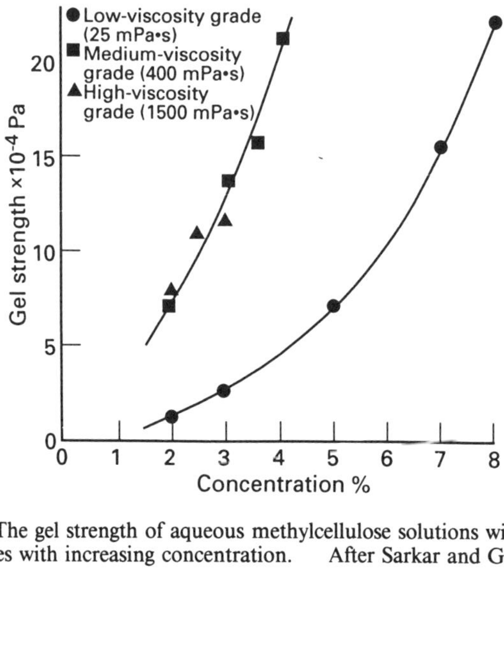

FIGURE 3-9. The gel strength of aqueous methylcellulose solutions with different viscosities increases with increasing concentration. After Sarkar and Greminger (1983).

the medium- and high-viscosity grades the relationship is the same. This indicates that the gel strength becomes independent of the molecular weight as the viscosity reaches 400 mPa·sec. Even though the gel strength of the low-viscosity grade is lower, it resembles the other solutions in its dependence on concentration (Sarkar and Greminger 1983).

Although small changes in the degree of substitution do not affect the gelation temperature significantly, the gel strength is extremely sensitive to such changes (Heyman 1935). For example, two 2% solutions of methylcellulose at 65°C with viscosities of about 4000 mPa·sec, one with a methoxy content of 28.3% and the other 29.6%, have gel strengths of 44,000 and 68,000 Pa, respectively. Similarly, a small quantity of hydroxypropyl substitution in methylcellulose can markedly increase its gel strength (Sarkar and Greminger 1983).

The effect of temperature on the gel strength of different concentrations of a methylcellulose solution with a viscosity of 400 mPa·sec is shown in Figure 3-10, where the gel strength increases with increasing temperature, reaching a maximum at about 70°C. As the temperature is increased beyond 70°C, the precipitation of methylcellulose in the form of discrete

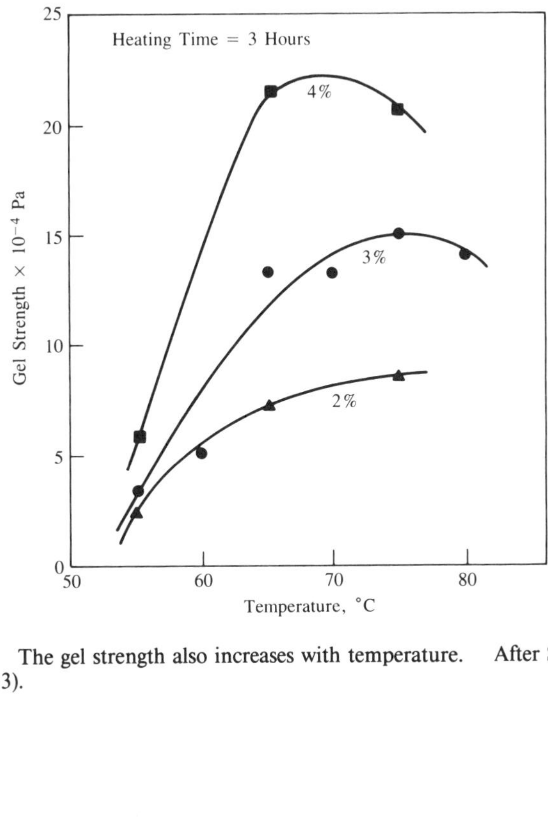

FIGURE 3-10 The gel strength also increases with temperature. After Sarkar and Greminger (1983).

particles begins to weaken the gel structure. This suggests that for injection molding the maximum benefit from the gel strength is obtained at 70°C.

Various additives can be used to alter the gelation properties. For example, most electrolytes, as well as substances such as sucrose and glycerol that have a high affinity for water, depress the gelation temperature because they dehydrate the polymer. Other compounds such as ethanol and propylene glycol act to elevate the gelation temperature. Any substance that depresses the gelation temperature can also increase the gel strength, and vice versa. Methylcelluloses can form complexes with hydroxybenzene compounds and also can interact with surfactant micelles to alter their viscosity and gel strength.

Cellulose-derived binders, agar, and agarose (derived from agar by additional purification) all belong to the chemical family of carbohydrates. Celluloses are composed of the carbohydrate glucose. Agar, which is derived from the red algae class of seaweed by a series of extraction and blanching operations, is composed of alternating units of two forms of the carbohydrate galactose, shown in Figure 3-11 (Fanelli et al. 1989). Its polymer chains form rigid bundles of twisted helixes (Arnott et al. 1974).

Methylcelluloses are soluble in cold water and gel in hot water. Gelling is due to hydrophobic, self-aggregation processes. Agar/agarose is soluble in hot water and gels on cooling. In agar gels, a cage network is formed by the aggregation of the double helical chains. The hydroxyl groups along the polymer chains interact strongly via hydrogen bonding with water molecules, which immobilizes water molecules within the cages of the gel network (Serwer 1983). The low-pressure injection molding of aqueous suspensions of a ceramic powder containing a cellulose polymer like agar and its derivative agarose as a binder, depends both on the polymer's rheology and on its thermo-gelating properties.

The apparent viscosity of an aqueous agar solution is shown in Figure 3-12 (Fanelli et al. 1989). When the agar is dissolved, the aqueous solution

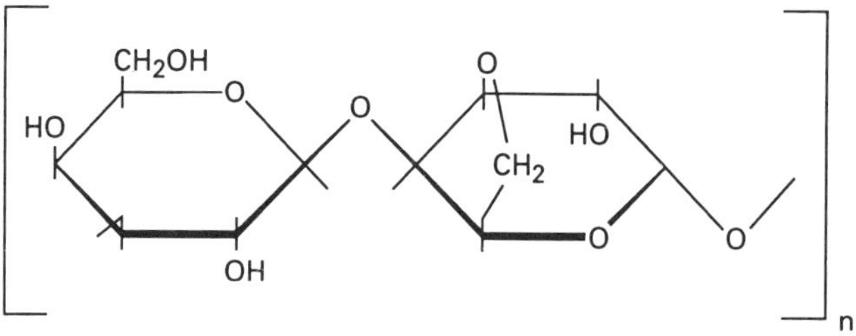

FIGURE 3-11. The chemical structure of agarose. After Fanelli et al. (1989).

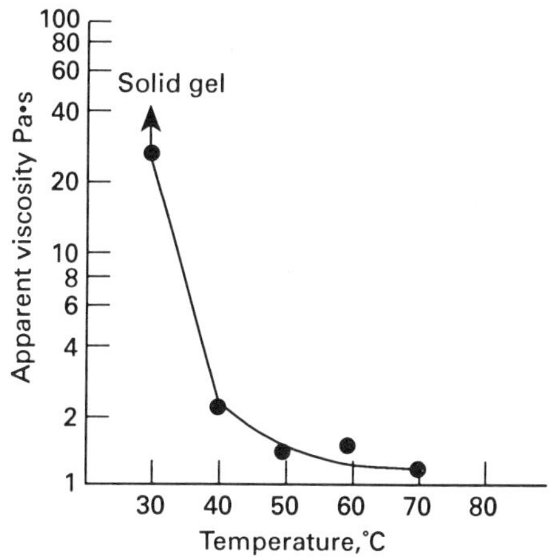

FIGURE 3-12. The apparent viscosity of a 2 wt% aqueous agar solution. After Fanelli et al. (1989).

maintains a low viscosity as it cools until the gel point is reached. At this temperature the solution gels to a rigid mass. The rheological behavior of a typical alumina–agar mix is shown in Figure 3-13 (Fanelli et al. 1989).

The characteristic strengths of the gels produced by various gelling agents differ widely. The gel strengths of some representative examples are compared in Table 3-4. Agar binder systems, which incorporate the two key properties of favorable rheology and high gel strength, are attracting interest for ceramic injection molding.

The Characteristics of Plasticizers

Plasticizers are stable, unreactive materials that are added to polymers to make them more flexible. In the mixture formed between a plasticizer and a polymer, molecules of the plasticizer are interspersed between molecules of the polymer, making the combined mass more flexible than the polymer alone. The key to the retention of the plasticizer in the mixture is to select a plasticizer that is compatible with the polymer. When the intermolecular forces between the polymer and plasticizer are equal or greater than the polymer's internal forces, the plasticizer will be retained.

The volatility of plasticizers should be low to assure their retention in the compound during processing and their effectiveness at the lowest possible concentration. No generic family of organic chemicals can be used as a

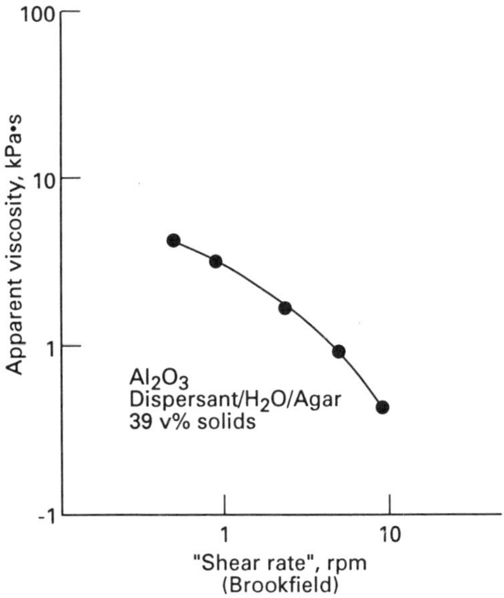

FIGURE 3-13. The viscosity of an alumina (Al_2O_3)–agar mix at 73°C ± 2°C. After Fanelli et al. (1989).

plasticizer for polyethylene. The conventional plasticizers, such as oleates and phthalates, are not compatible with polyethylenes. However, polyethylenes can be completely dissolved in hydrocarbon waxes or oils at temperatures above their melting points, and the resultant blends can be processed by conventional thermoplastic techniques. Ethylene–vinyl acetate (EVA) copolymers, which are mutually soluble with polyethylenes, sometimes are used to enhance their plasticity.

TABLE 3-4. Gel strengths of naturally occurring binders.

Gel[a]	Nominal strength (g/cm^2)
Agarose	2500
Agar	1500
Carageenan	800
Gelatin	30

a. 3 wt% aqueous solutions.

Plasticizers for polystyrenes, polypropylenes, and acrylics include phthalates, adipates, laurates, stearates, and oleates. The plasticity of EVA copolymers is strongly affected by the amount of acetate present in the polymer, which determines the extent of polymerization of the EVA copolymers and their consequent molecular structure. Waxes often are used with ethylene-vinyl acetates to modify their flow behavior. Figure 3-14 shows the compatibility between EVA, a plasticizer, and a processing aid.

For specific applications, waxes often are blended with a variety of plasticizing resins such as copolymers of ethylene and vinyl acetate and low molecular weight polyethylene and polypropylene. Selection criteria for plasticizers include compatibility with the wax, a marked improvement in the properties of the wax when in use, and the end use of the injection molded part.

Broadly classified as polyethylene homopolymers and copolymers, these plasticizers, which are compatible with most waxes and waxlike materials, are used in a wide variety of plastic applications as lubricants, flow modifiers, release aids, and dispersion aids. They also have the ability to wet and disperse inert materials such as pigments, carbon black, and fillers through-

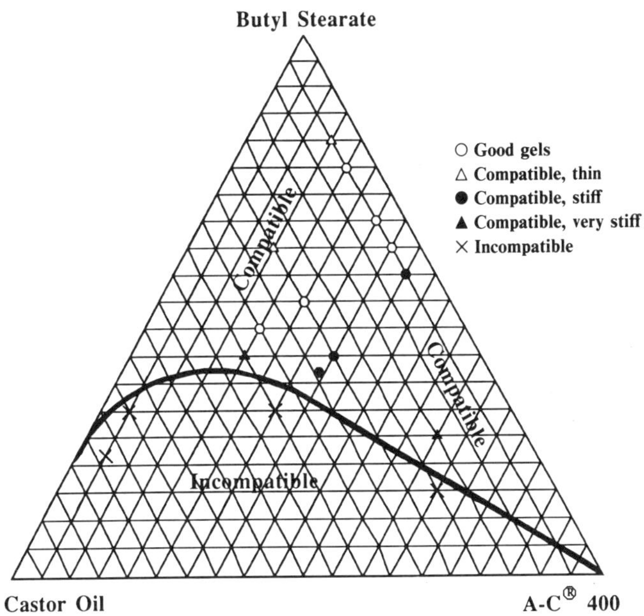

FIGURE 3-14. The compatibility between EVA, butyl stearate (a plasticizer), and castor oil (a processing aid).

TABLE 3-5. Properties of some plasticizers for wax-type binders.

	Homopolymers			Ethylene–vinyl acetate copolymers	Ethylene–acrylic acid copolymers
Properties	AC-6	AC-617	AC-1702	AC-400	AC-540
Mol. weight	2000	1500	1100	3500	3000
Density, (g/cm^3)	0.92	0.91	0.88	0.92	0.93
Melting pt. (°C)	106	102	85	95	108
Viscosity, (Pa · sec)	0.35	0.18	0.04	0.55	0.50
Comonomer ratio (wt%)	—	—	—	86:14	—

out a plastic or wax matrix. Homopolymers are produced in a variety of grades that differ in viscosity, softening point, hardness, density, and molecular weight. The homopolymers and the copolymers suitable for use as plasticizers for waxes are shown in Table 3-5.

The Characteristics of Processing Aids

Processing aids are additives that affect the performance of polymers and waxes during injection molding by modifying the cohesive forces between the polymer or wax and the ceramic filler. They reduce the viscosity of the mixture, improve the flow during molding, and ease the release of the injection molded part from the mold. Other important functions of processing aids are to uniformly distribute the binder components throughout the mix and to enhance the dispersion of the ceramic particles.

The stable dispersion of a ceramic powder in a mix is essential to minimize the formation of agglomerates, which cause flow instability and inhomogeneous microstructures. Agglomerated ceramic particles do not spontaneously disperse in the polymeric binder but are dispersed by shearing forces accompanying the mixing process. However, the dispersed particles start to reagglomerate as soon as the shearing ceases. Therefore, a processing aid is added that has the effect of hindering the particles from exerting mutual attraction.

The attraction between the particles results from long-range London–van der Waals forces, electrostatics, and hydrogen bonding. Other types of forces exist also, such as forces between molecules with permanent dipoles. These attractive forces can be minimized by two types of repulsive interactions, coulombic interaction between similarly charged particles and steric

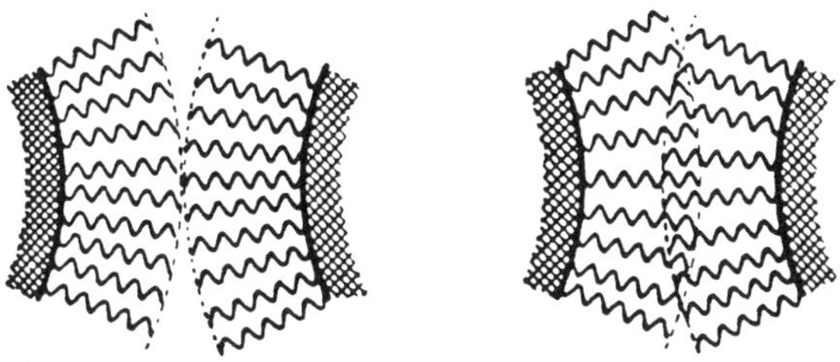

FIGURE 3-15. Steric stabilization by the adsorption of an organic fluid.

hindrance between long-chain molecules adsorbed on the surfaces of the particles.

The interaction potential is inversely dependent on particle separation, with its magnitude expressed by a constant (Hamaker 1937).[3] The stabilization of aqueous suspensions of particles is explained on the basis of coulombic repulsions between the charged surfaces of particles, by the D.L.V.O. theory.[4] It is not clear to what extent charged double layers are effective for stabilizing nonaqueous suspensions. An alternative approach to overcoming the attractive forces in organic fluids is to attach, or graft, polymer chains to the surfaces of the particles. By this means, steric hindrance, the so-called entropic repulsion (Napper and Netschey 1971), can act to prevent the particles from approaching one another more closely than the extended length of the polymer chains in the surrounding medium, as shown in Figure 3-15.

Dispersion stabilization via steric hindrance, however, requires a medium with low polarity (Napper 1983). The key point is that dispersions of fine particles in a molten polymeric binder or a solvent have a strong

---

[3]The Hamaker constant occurs in an equation for approximating the attractive forces between two spheres:

$$V_x = \frac{Aa}{12} \cdot \frac{1}{H_o}$$

where V_x is the potential energy, a is the radius of the spheres, H_o is the shortest distance between the surfaces of the spheres ($H_o = R \cdot 2_{a)}$, R is the distance between their centers, and A is the Hamaker constant, which can be calculated from refractive index data and has a value between 10^{-12} and 10^{-14} erg for various materials.

[4]The Derjaguin–Landau and Verwey–Overbeek (D.L.V.O.) theory is concerned with estimating the energy due to the overlap of electric double layers (generally repulsive) and London–van der Waals forces (generally attractive) in terms of interparticle distances and their summation to give the total interaction energy.

tendency to agglomerate unless they are stabilized by molecules that render the particle surfaces compatible with the molten binder or solvent.

When one of the components in a mix is a ceramic powder that is insoluble in the other component, their compatibility is defined as the capacity of the polymeric binder to hold the ceramic in fine dispersion without agglomeration. This is controlled by factors such as the surface energy associated with each component, the particle size and specific surface area of the ceramic powder, and the relative density and viscosity of the polymeric binder. If the interfacial energy of the binder and the ceramic is small, then their compatibility is greatly improved.

A solubility parameter, δ, has been defined as the square root of the cohesive energy density or the intermolecular energy (Hilderbrand and Scott 1950). The intermolecular energy is manifested as the energy of vaporization, E^v, or the internal pressure, p^i. Both of these are the sums of molecular interactions including ionic, hydrogen bonding, polar, induced dipolar, and dispersion interactions.

Compatibility can be accomplished by adding a small amount of one or more processing aids such as a surface active agent, a wetting agent, a coupling agent, or a combination of these agents. The molecules of the processing aids contain both polar and nonpolar moieties, usually in the form of a small polar head and a long nonpolar chain. The polar head is adsorbed on the particle surface, allowing the long nonpolar tail to become compatible with the polymer binder. When a sufficient amount of such processing aids is added to entirely cover the surfaces of all the individual particles, they are more likely to become completely dispersed.

Processing aids are derived from a variety of sources, including animal and vegetable fats and oils. The major fat-based processing aids fall mainly into four categories: fatty acids, fatty glyceryl esters, fatty bisamides, and fatty esters not based on glycerine. Table 3-6 gives some typical data on commercial processing aids.

Coupling agents used to improve the bonding of polymers to inorganic material surfaces are another type of processing aid. They are silanes, which are organofunctional silicones that are hybrids of silicon and organic substances related to polymers. Another role of coupling agents can be to form an interpenetrating polymer network that would act to further separate the particles. The generic composition of silane coupling agents is $R-Si(OR')_3$ (for example, $H_2C = CHSi(OCH_2CH_2OCH_3)_3$). These compounds, known as silane esters, can be thought of as esters of alkylsilicic acids. The way that a silane coupling agent is believed to function in a polymer filled with a ceramic powder is shown schematically in Figure 3-16.

A range of silane coupling agents is needed to match the functionality of the different polymers used as binders. Table 3-7 shows a line of commercially available silane coupling agents, their structures, and their suitability

TABLE 3-6. Properties of processing aids.

Physical properties	Processing aids: Oleic	Stearic	Oxidized menhadin
Form	Liquid	Powder	Liquid
Specific gravity (g/cm^3)	0.843	0.857	0.985
Melting point (°C)		56	
Boiling point (°C)		232	
Iodine number	89.9	2–4	105–115
Acid number		207	
Molecular weight	282	284	
Saponification number	190	210	210–220
Total saturated fatty acid (wt%)		90	15
Total unsaturated fatty acid (wt%)	100	10	85

for use with various polymers. The effect of two coupling agents (β-aminoethyl-γ-aminopropyl-trimethoxysilane and γ-glycidoxypropyl-methoxysilane) on the viscosity of a mix of silicon nitride and polyethylene is illustrated in Figure 3-17a (Lindquist et al. 1989). The methoxy silane (lowest curve) appears to be more effective in reducing the viscosity than the trimethoxysilane (middle curve).

Titanates are another family of coupling agents. Figure 3-17b illustrates the effect of a titanate and a combination of titanate–silane coupling agents on the viscosity of a mix of silicon nitride and polyethylene. Figure 3-17c indicates that the viscosity of this mix is decreased more by the addition of silanes and titanates than by a processing aid such as stearic acid.

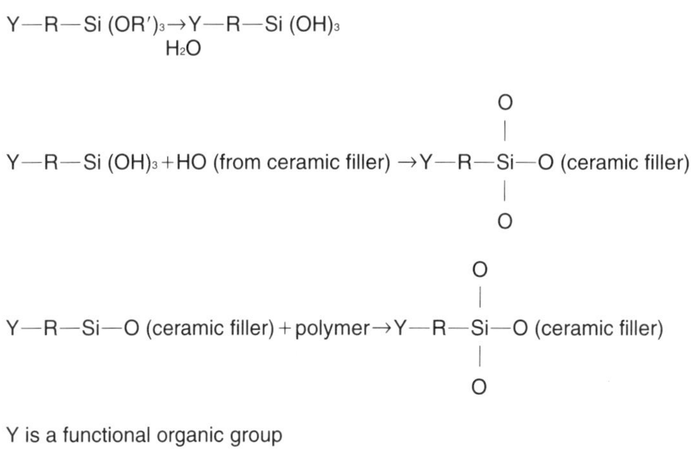

FIGURE 3-16. A schematic of the interaction between a silane coupling agent and a ceramic powder.

TABLE 3-7. Commercial silane coupling agents.

Type	Description	Structure	End-use polymer
A-187	γ-Glycidoxy-propyltrimethoxy-silane	$H_2C—CHCH_2OCH_2CH_2CH_2Si(OCH_3)_3$ (epoxide: H_2C and CH bridged by O)	Almost all polymers
A-172	Vinyl-tris (γ-methoxyethoxy) silane	$H_2C{=}CHSi(OCH_2OCH_3)_3$	Polyolefin
A-1120	β-Aminoethyl-γ-aminopropyl-trimethoxysilane	$H_2NCH_2CH_2NH(CH_2)_3Si(OCH_3)_3$	Epoxies, phenolics

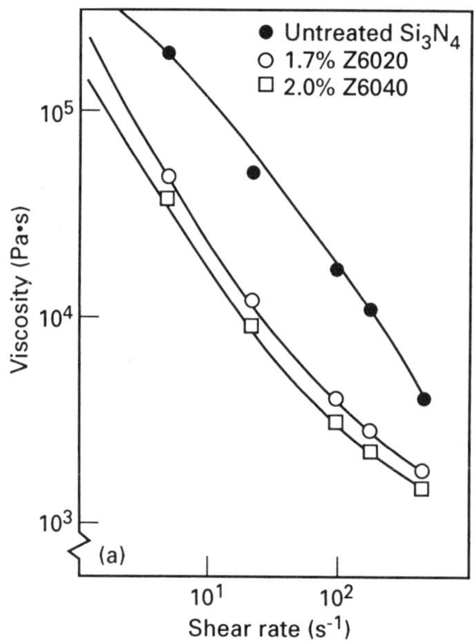

(a)

FIGURE 3-17. The effect of silane and titanate coupling agents on the viscosity of a mix containing silicon nitride (Si_3N_4) and a polyethylene binder. (a) The addition of a silane coupling agent decreases the viscosity. (b) A combination of a titanate and a silane coupling agent decreases the viscosity more than the titanate coupling agent alone. (c) The processing aid, stearic acid, does not decrease the viscosity as much as the silane and titanate coupling agents do.

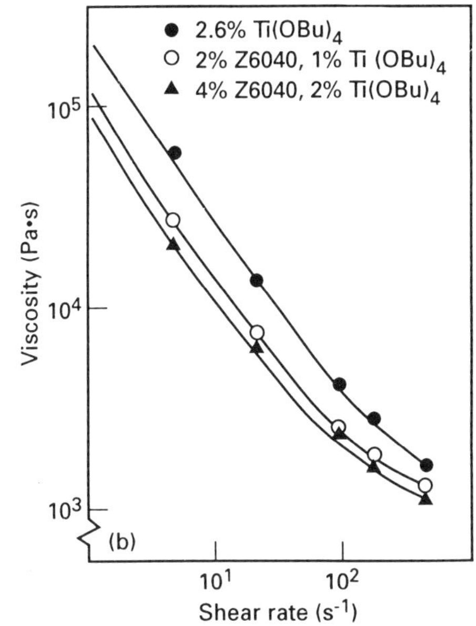

(b)

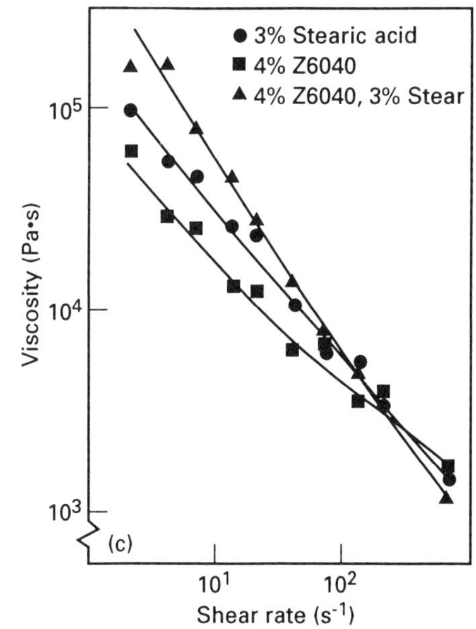

(c)

FIGURE 3-17 *contd*

The Rheology of Polymers as Primary Binders

The shear rates encountered in the gates and the mold range between 100 sec^{-1} and 1000 sec^{-1} but can occasionally reach 10,000 sec^{-1}, which is still considered suitable for injection molding. However, polymers with the lowest viscosity are preferable because the viscosity will increase rapidly with the addition of the ceramic powder. Therefore this discussion is limited to AC-400 EVA, NA-250 polyethylene, DP-210 polystyrene, and a low-molecular weight atactic polypropylene. The viscosities of these unfilled polymers are compared in Figure 3-18.

The viscosities of all these polymers are within the range that experience has shown is required for good flow during molding (less than 1000 Pa·sec; 10^4 poise). At a shear rate of 100 sec^{-1}, the viscosities of NA-250 polyethylene and atactic polypropylene are less than 100 Pa·sec, while at the same shear rate, the viscosities of DP-210 polystyrene and AC-400 EVA are above 100 Pa·sec.

The viscous flow of two polymers when filled with a ceramic powder is shown in Figure 3-19a and b. For example, at all concentrations of alumina (Al_2O_3) powder the flow is shown to be pseudoplastic at shear rates above 100 sec^{-1}. Figure 3-19 also shows that a higher loading of alumina powder can be achieved with NA-250 polyethylene than with DP-210 polystyrene.

Figure 3-20 illustrates the marked differences in the viscosities of polyethylene (NA-250) and EVA (AC-400) when filled with ceramic powders.

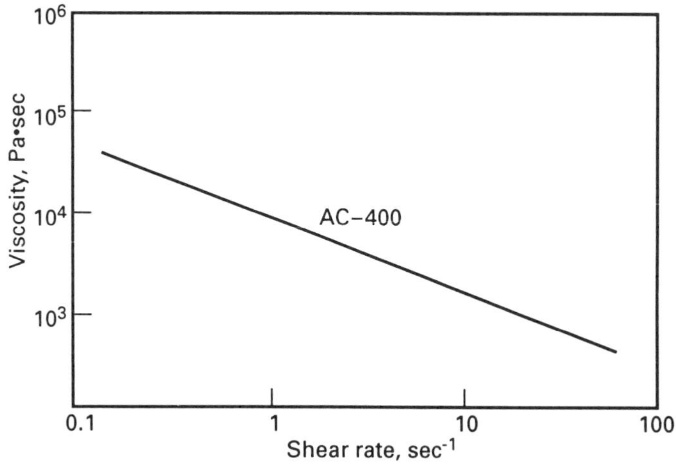

(a)

FIGURE 3-18. A comparison of the viscosities of four polymers with lower viscosities: (a) AC-400 EVA; (b) NA-250 polyethylene and DP-210 polystyrene; (c) atactic polypropylene at 160°C to 200°C. From Battelle Columbus, OH.

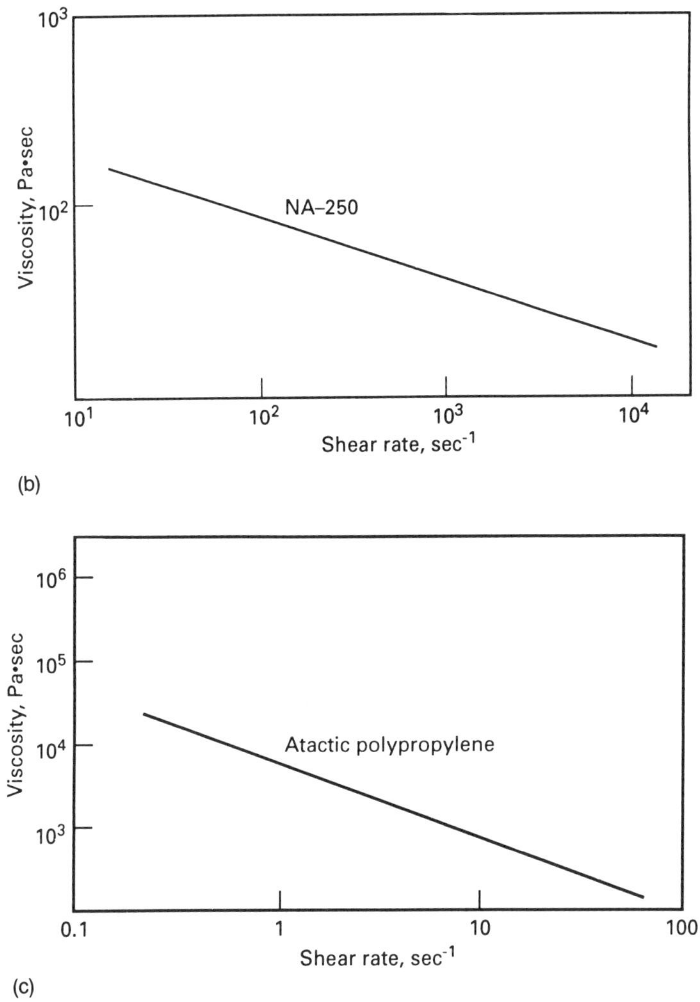

FIGURE 3-18 *contd*

At a loading of 60 vol% alumina powder and a shear rate of 100 sec^{-1}, the viscosity of AC-400 copolymer is an order of magnitude lower than that of NA-250 polyethylene. However, the viscosity of unfilled NA-250 polyethylene at this same shear rate is significantly lower than that of AC-400 copolymer. The difference in the viscosities between the three polymers containing the same alumina filler underscores the complex filler–polymer interactions that can be strongly influenced by factors such as wettability, adsorption, and bonding.

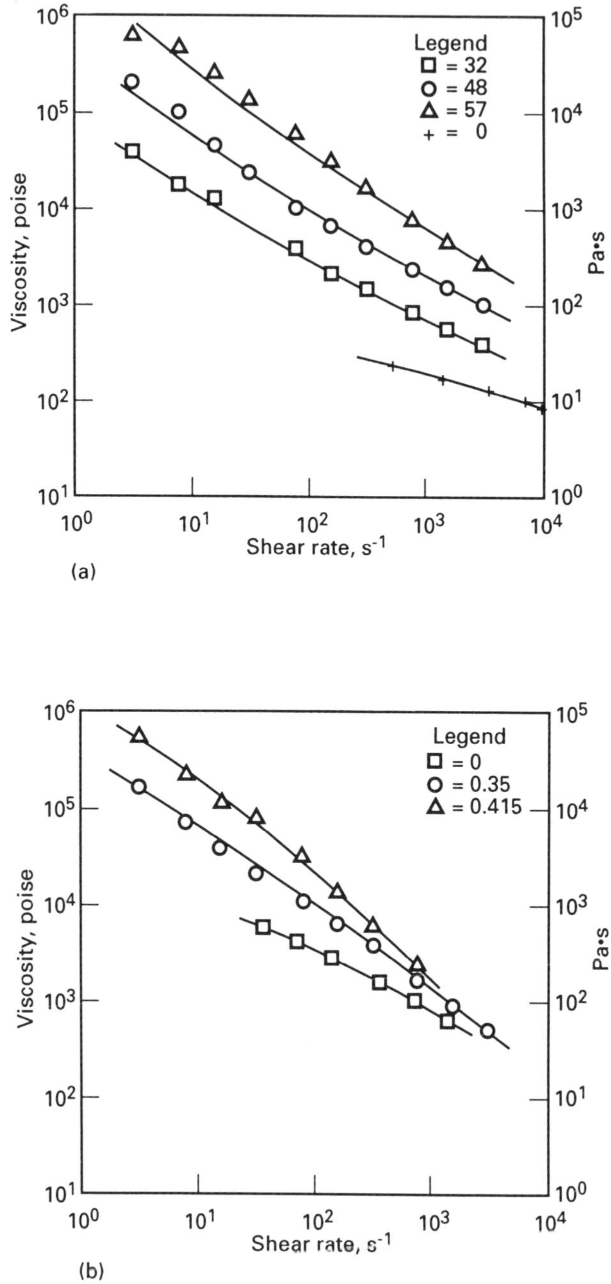

FIGURE 3-19. The viscosities as a function of shear rate of two polymers filled with alumina powder: (a) NA-250 polyethylene at 200°C; (b) DP-210 polystyrene at 200°C. From Battelle Columbus, OH.

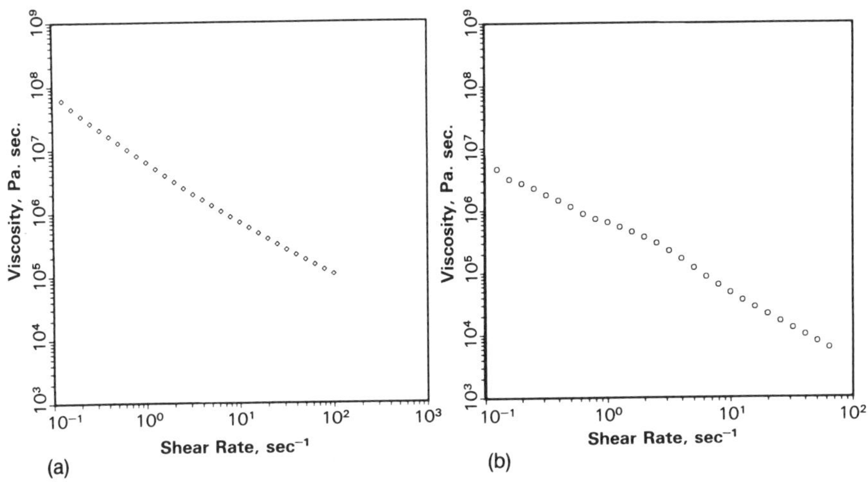

FIGURE 3-20. There is an order of magnitude difference in viscosity between (a) NA-250 polyethylene and (b) AC-400 EVA copolymer filled with 60 vol% alumina powder. From Battelle Columbus, OH.

The Rheology of Waxes as Primary Binders

The most effective approach for selecting a particular type of wax that can be used in a moldable composition is to determine the flow behavior of several waxes in the presence of a ceramic powder. Figure 3-21a shows the viscosity as a function of the shear rate for several synthetic waxes—BASF-A, N-45, and Paraflint H1-N1. As can be seen, the trend is for a decrease in viscosity with increasing shear rate. Even at a shear rate as low as 10 sec^{-1}, the viscosities of BASF-A and H1-N1 waxes are less than 10^3 Pa·sec. N-45 wax has a relatively higher viscosity, but if its viscosity were extrapolated to a shear rate of 10^3 sec^{-1}, this wax could also be potentially useful for molding. Although BASF-A and N-45 belong to the same family of waxes, their viscosities are markedly different because of differences in their specific characteristics, such as molecular weight and distribution, molecular structure, melting point, and the presence of impurities.

In Figures 3-21b and c, the viscosities of several microcrystalline waxes are plotted as a function of their shear rate. In Figure 3-21b, Be^2, and C1035 show a rapid decrease in viscosity with increasing shear rate. Figure 3-21c shows that the viscosity of Shell 400 and Mobil waxes decreases sharply at a shear rate of about 10 sec^{-1}, with a slight drift in viscosity observed for Shell 400. But it is not clear whether the drift is indicative of inherent instability

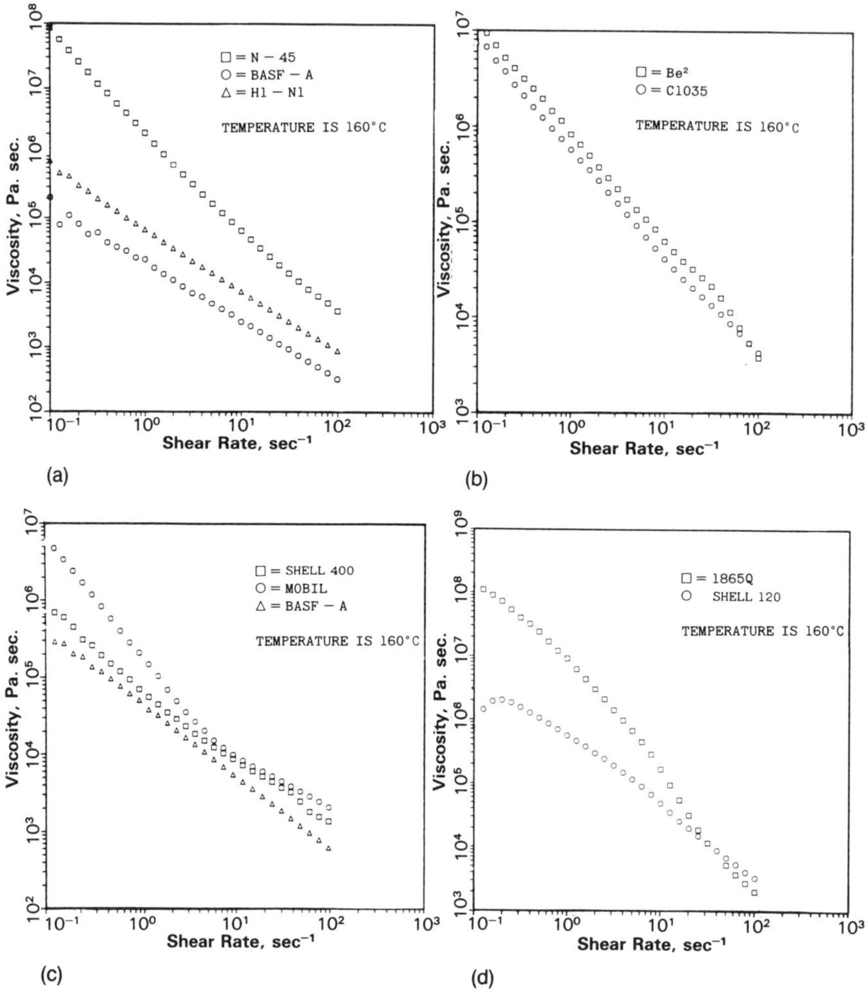

FIGURE 3-21. The viscosity as a function of the shear rate for different waxes: (a) BASF-A, N-45, and H1-N1 waxes with 58 vol% Al6 alumina; (b) Be² and C1035 microcrystalline waxes with 58 vol% Al6 alumina; (c) Shell 400, Mobil, and BASF-A microcrystalline waxes with 58 vol% Al6 alumina; (d) 1865Q and Shell 120 paraffin waxes with 58 vol% alumina. From Battelle Columbus, OH.

in this wax. Plots of the viscosity as a function of shear rate for two paraffin waxes are shown in Figure 3-21d. The viscosities of both these waxes decrease sharply with increasing shear rate.

In general, polymers and waxes that have different chemical structures can be expected to respond differently to deformation. A theory proposed to account for the effect of polymer chain branching relates the viscosity of a branched polymer to that of a linear polymer with the same molecular weight as follows (Bueche 1954):

$$\eta_b/\eta_c = g\ E(g) \tag{3-10}$$

where η_b is the viscosity of the branched polymer, η_c is the viscosity of the linear polymer, g is the ratio of the mean square radii of the branched and the linear polymers, and $E(g)$ is a factor related to the change in interaction or to the entanglements between polymer molecules, depending on the amount of branching or the reduction in the ratio g. It was predicted, from this theory, that for polymers of the same molecular weight the viscosity of branched polymers should be less than that of linear polymers.

It also has been observed that branched polymers tend to deviate from Newtonian flow behavior at lower shear rates than linear polymers with similar molecular weights (Kraus and Gruver 1965). Because waxes are likely to contain more long-chain branching than other polymers, the lower viscosity of waxes can be attributed to the presence of long-chain branching.

The viscosities of wax systems filled with ceramic powder at low shear rates (between $10^{0.1}$ and $10^1\ \text{sec}^{-1}$) indicate the presence of a network structure within the mix that could be due to interactions between the filler ceramic particles, possibly through active sites on their surfaces. The effect of this network structure on the viscosity of filled wax systems is dependent on the dynamic equilibrium between the breakdown and reformation of the structure of the filler particles. As the shear rate is increased, this equilibrium shifts toward a decreasingly network-like structure, until at relatively high shear rates (above 1.0 sec^{-1}) the network is essentially completely broken up. During this process, some of the wax can migrate to the surface, in which case the viscosity changes abruptly.

In Figure 3-22, the viscosity of pure polycarbosilane is plotted as a function of temperature. As can be seen, the viscosity of the polycarbosilane decreases initially and then increases. The initial decrease could be due to a lag before the reaction rate catches up with the temperature. The increase in viscosity could be caused by the presence of reactive groups on the polycarbosilane that condense with each other during the preparation for measurement. The conditions during measurement could stimulate the production of higher molecular weight species, which would increase the viscosity.

Figure 3-23 shows that the viscosities of polycarbosilane and silicon carbide (SiC) mixtures at three filler loadings—53 vol%, 40 vol%, and 30 vol%—increase with increasing filler loading. It should be noted that polycarbosilane mixes become very sensitive to flow behavior as the SiC content exceeds 40 vol%. However, it is not clearly understood whether this behavior is characteristic solely of the combination of polycarbosilane and SiC powder.

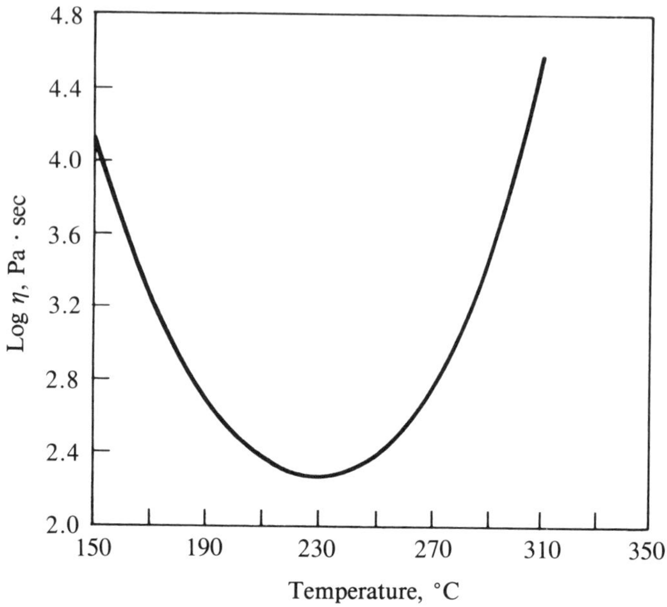

FIGURE 3-22. The viscosity of pure polycarbosilane as a function of temperature. From Battelle Columbus, OH.

The Power Law

The power law index (n) values for a number of ceramic powder and polymer mixes and ceramic powder and wax mixes, obtained from the curves of the logs of the shear rates as a function of the logs of the shear stresses, are given in Table 3-8. The results show that these polymers and waxes obey the power law and behave like pseudoplastic materials.

Bingham Flow

The power law index does not distinguish between pseudoplastic and Bingham flow. Mixes containing a high volume percent of ceramic filler (about 60 vol%) frequently show Bingham flow, the most familiar examples being the clay–water mixes used in the cold extrusion process. Mixes with Bingham flow behavior characteristically show a yield point.

The yield stress of filled systems such as polymer–ceramic mixes can be determined from the following relationship describing steady flow behavior (Casson 1959):

$$\tau^{\frac{1}{2}} = K_o + K_1 \dot{\gamma}^{\frac{1}{2}} \tag{3-11}$$

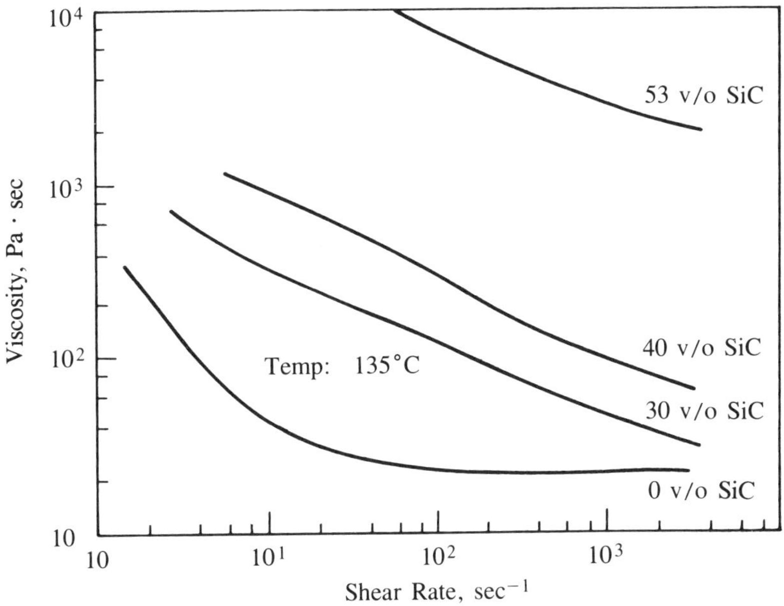

FIGURE 3-23. The viscosity of polycarbosilane at three different loadings of silicon carbide powder. From Battelle Columbus, OH.

where τ is the shearing stress, $\dot{\gamma}$ is the shear rate, and K_o and K_1 are constants related to the solid and liquid phases of the fluid mixture.

Equation (3-11) can be rewritten in a more representative form (Onogi et al. 1973):

$$\tau^{\frac{1}{2}} = K_o + K_1 (\dot{\gamma}\, \eta_a/\eta_o)^{\frac{1}{2}} \tag{3-12}$$

TABLE 3-8. Power law exponent for silicon carbide mixes.

		Power law exponent (n)			
		Filler loading			
Polymer type	Grade	0 v/o	36 v/o	53 v/o	61.5 v/o
Polycarbosilane	—	0.9			
Ethylene vinyl acetate	AC-400	0.44	—	0.48	—
Polyethylene	MN711-20	0.42	0.45	0.36	0.43
Oxidized polyethylene	BASF-A	0.48	—	0.53	—
Fischer Tropsch wax	H1-N1	0.54	—	0.28	—

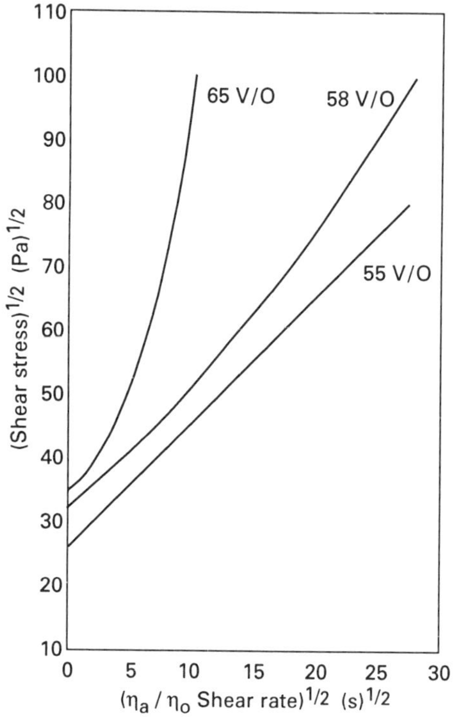

FIGURE 3-24. $\tau^{1/2}$ plotted against $(\dot{\gamma}\ \eta_a/\eta_o)^{1/2}$ for A16-SG alumina and polyethylene at three concentrations, to obtain the Bingham yield value.

where η_o is the zero shear viscosity and η_a is the apparent viscosity of the fluid. The plot of $\tau^{\frac{1}{2}}$ as a function of $(\dot{\gamma}\ \eta_a/\eta_o)^{\frac{1}{2}}$ produces a straight line whether or not the fluid is Newtonian. The yield stress, τ_y, is obtained from the intercept, K_o^2. Figure 3-24 shows the Bingham yield stress value (τ_y) for three concentrations of A16 alumina in polyethylene.

Thixotropic and Dilatant Flows

The flow behavior of a pseudoplastic fluid also can be time-dependent. Time-dependent flow behavior is characteristic of a fluid that is said to be thixotropic. If a thixotropic fluid is allowed to stand for a period of time and then is sheared at a constant rate, its apparent viscosity decreases with time until a balance is reached between structural breakdown and reformation. If the sheared fluid is allowed to stand, it eventually will regain its original structure. A thixotropic hysteresis loop for the fluid can be obtained by measuring the shear stress as the shear rate first is increased and then is decreased after a given time interval. Thixotropy could be an important consideration in injection molding mixes that contain water-soluble binders.

Occasionally, mixes of polymers or waxes with ceramic powders can cause the injection molding nozzle to clog or the screw to seize, because of dilatant flow of the mix. The phenomenon of dilatancy was identified as early as 1885, when it was observed that if a mixture of sand and water is confined in a balloon, it dilates if deformed (Reynolds 1885). In other words, the total volume of the sand–water mixture increases under pressure, in turn causing the level of liquid in the balloon to decrease. Besides dilating, such a mixture becomes more resistant to stirring as the rate of stirring is increased, until the pseudosolid material that is formed fractures.

When highly dilatant fluids fracture at high rates of shear, they are believed to manifest viscoelastic behavior. In that case, the reciprocal shear rate is exceeded by the relaxation time; so the suspension behaves like a solid—that is, when its shear strength is surpassed, it fractures.

The concept of a dilatant fluid can be generalized to all materials in which the apparent viscosity increases as the shear rate increases. To differentiate between volumetric dilation and no volumetric dilation, dilatancy also has been referred to as "inverted plasticity." Typical dilatant flow curves for two concentrations of A16-SG alumina and paraffin wax are shown in Figure 3-25.

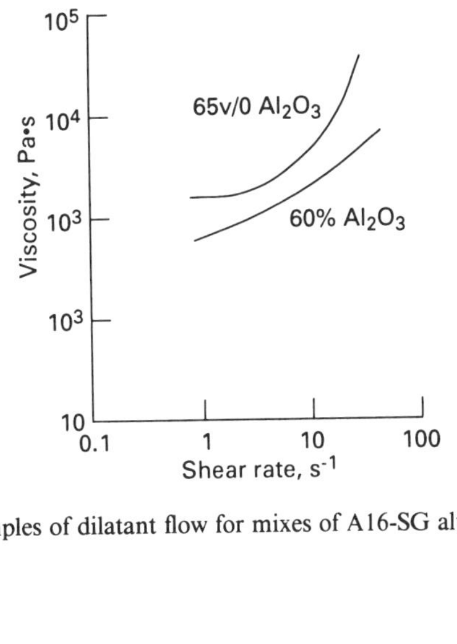

FIGURE 3-25. Examples of dilatant flow for mixes of A16-SG alumina and paraffin wax.

Hydrodynamic Viscosity

The viscosity of filled systems such as a suspension of ceramic particles in a viscous polymer or wax is largely influenced by particle–particle interactions. The viscosity of a suspension consists of its structural viscosity and its hydrodynamic viscosity. The structural viscosity represents the deviation from non-Newtonian flow as a function of shear rate; however, the hydrodynamic viscosity is independent of the shear rate and represents a limiting viscosity at high rates of shear. A detailed discussion of the theory of viscosity is given in Appendix 3-A.

Figure 3-26 illustrates the relative viscosities of optimal suspensions containing two particle sizes (bimodal). Figure 3-27 illustrates the calculated and measured relative viscosity data for a bimodal suspension system in which the ratio of one particle size to the other is approximately zero. In other words, one filler fraction contains very coarse particles compared with the other ($R_{12} \approx 0$). Figure 3-27 also illustrates the measured viscosity data for ratios of four different sizes. The 1:1 ratio corresponds to a monomodal

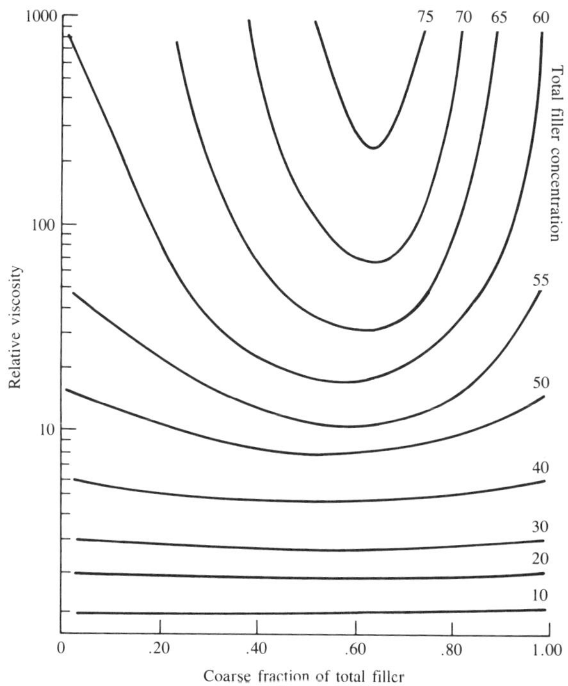

FIGURE 3-26. The relative viscosities of optimal suspensions of a bimodal powder. After Farris (1968).

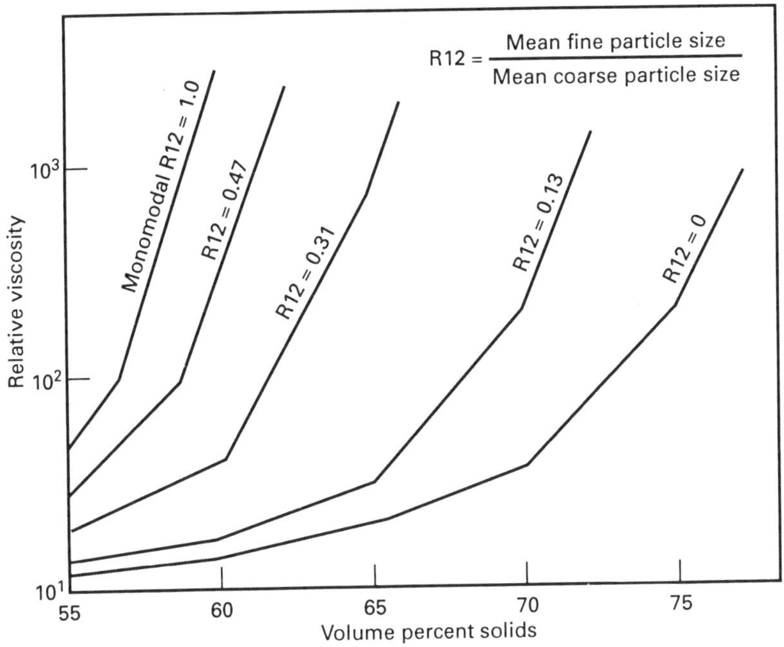

FIGURE 3-27. The relative viscosities as a function of the volume percent of solids in suspensions of monomodal and bimodal ceramic powders.

system, that is, a suspension containing only one particle size. Besides being an analytical tool, this approach provides a guideline for the change in the viscosity of the mix as a function of the filler (ceramic powder) concentrations. Therefore, the extent of the change in viscosity sets the upper limit for the filler concentration in a moldable mix.

In-Phase Shear Modulus

Some of the injection moldable fluids show a viscoelastic response to stress. A viscoelastic fluid is a viscous fluid with some of the properties of an elastic body. Such a viscous fluid returns to its original volume as the compressive stress is removed, that is, as the pressure on the material is reduced from the higher molding pressure to atmospheric pressure.

At moderate shear rates, the elastic recovery is extremely rapid and the relaxation time extremely short for fluids with low viscosities. Consequently, the elastic portion of the deformation is too small to measure. Such fluids are designated as viscous.

When viscoelastic fluids are stressed, some of the energy is stored elastically. This is accompanied by the deformation of various parts of the system into new nonequilibrium positions relative to each other. The rest of the energy is dissipated as heat, which is accompanied by various parts of the system flowing into new equilibrium positions relative to each other.

One approach to measuring the viscoelastic response is to assume that the material will respond as a Maxwell model, as shown in Figure 3-28. Then an apparent viscosity and an apparent modulus can be measured; these measurements depend on factors such as temperature, pressure, stress, geometry of deformation, and time.

Figure 3-29 shows that the stress response of viscous and elastic materials is different when they are subjected to sinusoidal strain. For a viscous material the stress response is one-half out of phase with the strain, and for a pure elastic material the stress response is in phase with the strain.

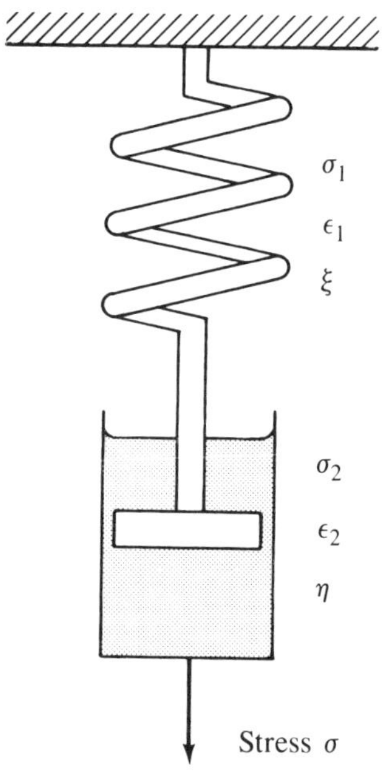

FIGURE 3-28. A Maxwell model as an illustration of the behavior of viscoelastic materials.

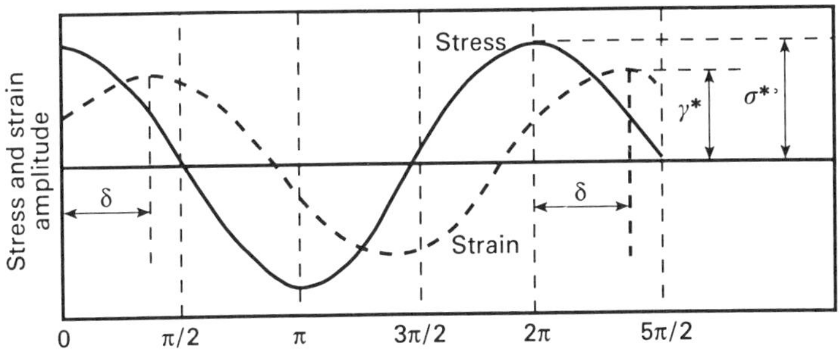

FIGURE 3-29. The stress response of viscous and elastic materials.

Because most polymeric materials are viscoelastic, the stress response shows a phase shift, and the derived dynamic viscosity (η^*), the storage (or elastic) modulus (G'), and the loss or viscosity (G″) are frequency-dependent:

$$\eta^* = \frac{\sigma^*}{\omega\gamma^*}\sin\,\delta \tag{3-13}$$

$$G' = \frac{\sigma^*}{\gamma^*}\cos\,\delta \tag{3-14a}$$

and

$$G'' = \frac{\sigma^*}{\gamma^*}\sin\,\delta \tag{3-14b}$$

where σ^* is the stress amplitude, γ^* is the strain amplitude, δ is the phase lag, and ω is the angular velocity. A standard approach for studying viscoelasticity in non-Newtonian fluids is to measure the amplitudes of the stresses and their phase shifts with respect to γ^*. These measurements do not change with time unless the fluid's behavior becomes thixotropic.

Measurements of the dynamic viscosity (η^*) and the dynamic shear modulus modulus (G') provide precise data. The data resulting from the calculation using the equations for the derived dynamic viscosity and the modulus can be used to characterize the rheology of ceramic formulations.

In measuring viscoelastic effects with an oscillating-plate method, the appropriate percent strain must be selected because the strain is proportional to the total angle through which the plate is turning. The larger the

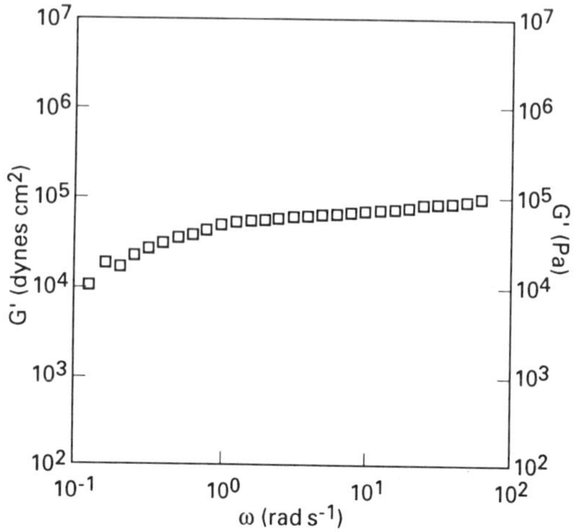

FIGURE 3-30. The viscosity and shear modulus of atactic polypropylene at 125°C. From Battelle Columbus, OH.

angle is through which the plate oscillates, the greater the strain. If the angle is too large, the fluid being measured can undergo internal cracking or segregation.

The dynamic shear modulus as a function of frequency (shear rate) is shown in Figure 3-30. The shear modulus of atactic polypropylene increases as the frequency increases, indicating that the faster the sample is deformed, the stiffer it becomes. However, as shown in Figure 3-31, the shear modulus of waxes appears to be independent of the shear rate. In other words, the stiffness of waxes does not increase as a function of the frequency (shear rate).

Figure 3-32 is a plot of the shear modulus as a function of the oscillation frequency for mixes with silicon carbide powder concentrations of 50, 55, and 63 vol%. At lower oscillation frequencies, as the solids loading increases, the rate of energy dissipation during shear flow increases. Figure 3-32 also indicates that the modulus becomes less sensitive to an increase in solids loading up to about 55 vol% with increasing frequency (rad/sec). However, in highly concentrated suspensions (above 63 vol%), there is particle–particle interaction analogous to the chain entanglements in polymer solutions. Consequently, the rate of deformation becomes an important determinant of flow behavior.

The fracture and/or segregation seen with highly filled suspensions at high shear rates can be considered a manifestation of viscoelastic behavior, where the relaxation time exceeds the reciprocal shear rate. As a result, the

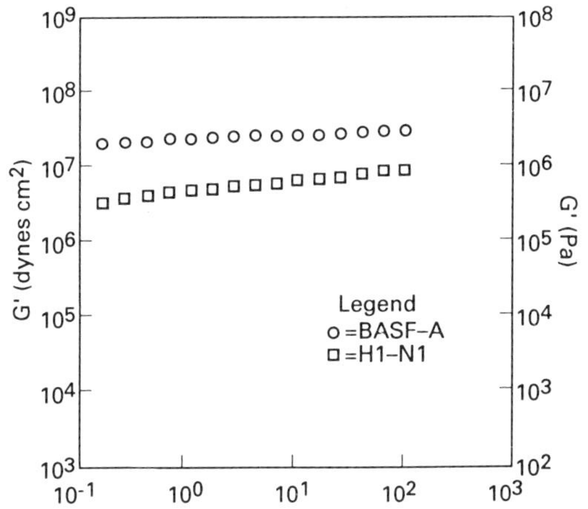

FIGURE 3-31. The shear modulus of BASF-A and H1-N1 waxes. From Battelle Columbus, OH.

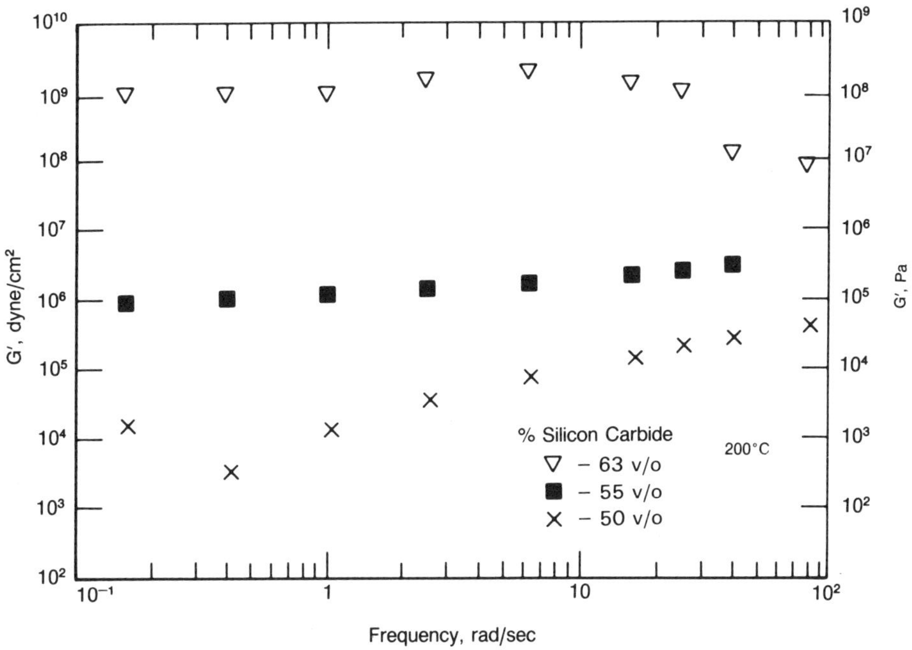

FIGURE 3-32. The modulus as a function of oscillation frequency for mixes with silicon carbide concentrations of 50, 55, and 63 vol%. From Battelle Columbus, OH.

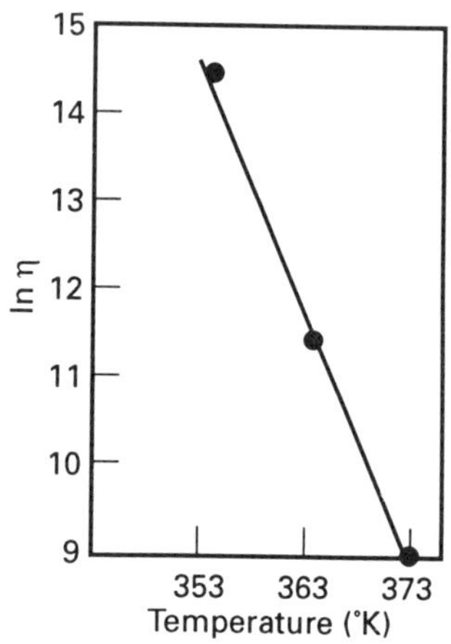

FIGURE 3-33. Log η as a function of temperature for Al_2O_3–EVA mixes at a constant shear rate.

suspension behaves like a solid, its shear strength is exceeded, and segregation occurs.

The Influence of Temperature on Fluid Flow

Temperature affects the flow behavior of fluids such as polymer melts, solutions, and filled suspensions. A decrease in viscosity occurs with both Newtonian and non-Newtonian fluids, primarily because of a decrease in intermolecular forces. A polymer chain with greater flexibility also can contribute to a reduction in viscosity. The relation between the viscosity and the temperature of a filled suspension can be expressed by a modification of the Eyring equation that allows for variations in the free volume of a fluid's structure:

$$\eta = \frac{hN_o}{V_m} \exp\left[\frac{E}{RT} + \frac{\gamma V_o}{V}\right] \tag{3-15}$$

where h is Planck's constant, N_o is Avogadro's number, V_m is the molar volume of the fluid phase, E is the activation energy needed to overcome the potential barrier between equilibrium positions, γ is a constant with a value between 0.5 to 1.0, V_o is the van der Waals volume of the molecule, and V is the average volume per molecule in the fluid.

In injection molding, it is customary to plot $\ln\eta$ as a function of T at a constant shear rate, rather than $\ln\eta$ as a function of $1/T$ at a constant force.

Figure 3-33 is an example of such a plot for alumina in ethylene–vinyl acetate copolymer. The linearity of this plot is indicative of the sensitivity of the process to a change in temperature. An estimate of the activation energy for viscous flow can be obtained from the slope.

The effect of temperature on particle–particle interaction must be taken into consideration with filled suspensions, in addition to the effect of temperature on intermolecular and intramolecular interaction in polymers. The extent of the temperature effect depends on the level of solids loading. If the temperature is high enough for the binder to migrate from the particle–particle interfaces, then the suspension could show dilatant rather than pseudoplastic behavior.

The effect of temperature on polymer rheology is similar to the effect of shear rate, but its effect on polymer–particle interaction is not predictable. Both a high temperature and a high shear rate weaken polymer–particle interaction. However, it should be kept in mind that the extent of the effect (reduction in the viscosity and in polymer–particle interaction) for a given change in the temperature or the shear rate cannot be predicted.

Wetting Characteristics

The dispersion of a filler in a polymer or wax involves three steps: wetting of the filler surfaces, separation of the filler particles, and stabilization of the dispersion. During the wetting process, the fluid must spread on the surface of the particle and then adhere to it. These steps are illustrated schematically in Figure 3-34. In the initial stage of the dispersion process, the

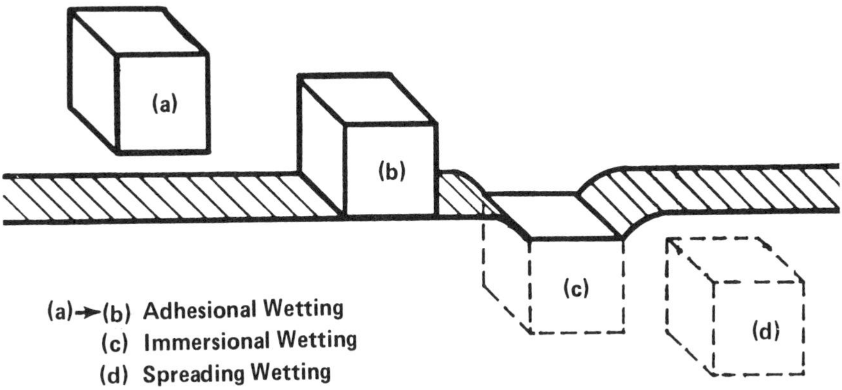

FIGURE 3-34. Schematic of the wetting process for a particle in a fluid.

polymer, plasticizer, and processing aid must wet the filler in order to displace air or absorb moisture from the surfaces and interstices of the agglomerates before the particles can disperse.[5]

The significance of the specific surface free energies or surface tensions arises because many measurable equilibrium surface quantities of macroscopic surfaces can be expressed in terms of their differences, such as the difference in their surface tensions. Thus the spreading pressure of a liquid on a solid, π_{sl}, and a vapor on a solid, π_e, can be expressed as:

$$\pi_{sl} = \gamma_s - \gamma_{sl} \tag{3-16}$$

$$\pi_e = \gamma_s - \gamma_{sv} \tag{3-17}$$

where γ_s is the surface tension of a solid, γ_{sl} is the surface tension of a liquid on a solid, and γ_{sv} is the surface tension of a vapor on a solid.

The contact angle at which a vapor, a liquid, and an inert solid meet can be expressed as:

$$\cos\theta = \frac{\pi_{sl} - \pi_e}{\gamma_{lv}} \tag{3-18}$$

Equation (3-18) can be expressed in terms of the three relevant surface tensions, which are measurable quantities:

$$\cos\theta = \frac{\gamma_{sv} - \gamma_{sl}}{\gamma_{lv}} \tag{3-19}$$

For stable equilibrium, $\gamma_{lv} > \theta$ and $\pi_e > \theta$, but π_{sl} can be either positive or negative. If $\cos\theta > 1$, then the solid is wetted completely by the liquid, which can spread freely over the surface. If $\cos\theta < -1$, the solid is not wetted. For intermediate values of $\cos\theta$ between +1 and −1, the cosine of the equilibrium contact angle is a direct measure of wettability.

Wetting involves a function called the adhesion tension, $\gamma\cos\theta$, where γ is the surface tension of the liquid, and θ is the contact angle of the liquid with the solid surface. A simple method has been devised to measure the adhesion tension for a number of processing aids in contact with powders, which is based on measuring the rate of unrestricted penetration of the processing aid into a column of A16-SG alumina powder in a small glass

[5]For a more detailed treatment of the wetting process, see Patton (1970) and Gaines (1972).

tube (Mutsuddy 1994). The adhesion tension, $\gamma \cos \theta$, then is estimated by using the following relation:

$$\frac{l^2}{t} = \left(\frac{r}{k^2}\right)\frac{\gamma \cos \theta}{2\eta} \tag{3-20}$$

where, for the advancing liquid, l is the penetration distance in time t, γ is the surface tension, θ is the contact angle, and η is the viscosity; r is the radius of the capillaries between the particles; and k is a tortuosity constant. In this equation, r/k^2 is a constant that depends on the packing for the column. By using standard packing, this constant can be eliminated from the calculation of adhesion tension. The calculated adhesion tensions for several selected processing aids and A16-SG alumina are listed in Table 3-9.

It has been observed that the higher the measured adhesion tension is, the more rapid the rate of dispersion during the mixing operation. However, this technique is not applicable with polymers that are solid at room temperature because a linear relation between the square of the penetration distance and the penetration time cannot be obtained.

The contact angles of high-melting polymers and waxes to ceramic substrates as a function of temperature and time have been measured by using a goniometer. Contact angles for some thermoplastic resins and waxes with alumina and silicon nitride substrates are listed in Table 3-10, and Figures 3-35 and 3-36 show the effect of temperature and time on the contact angles for various polymers (Saito et al. 1976; Dow et al. 1988).

Low molecular weight atactic polypropylene has the best wettability of the polymers and waxes. The paraffin wax and most of the microcrystalline waxes have low contact angles with both alumina and silicon nitride, whereas the contact angles of NA-250 polyethylene, CP-210 polystyrene, and BASF-A with these ceramics are relatively high. The contact angle of

TABLE 3-9. Calculated adhesion tension for selected processing aids and A16-SG alumina.

Processing aid	Adhesion tension, $\gamma \cos \theta$
Corn oil	22.0
Camphorated cottonseed oil	17.3
Peanut oil	12.4
Olive oil	8.5
Pine oil	7.6
Oleic acid	20.7
Menhaden oil	24.8

TABLE 3-10. Wetting behavior of thermoplastic resins and waxes with ceramics.*

Thermoplastics and waxes	Temperature, °C	Contact angle (θ), degrees Al_2O_3	Contact angle (θ), degrees Si_3N_4
Polyethylene, NA-250	200	≈33	≈36
Polystyrene D-210	200	≈40	—
EVAC AC-400	160	≈22	≈6
Atactic polypropylene	160	≈8	—
Paraffin Shell 120	160	≈15	<5
Microcrystalline 1865Q	160	≈14	<5
Microcrystalline Be-175	160	≈20	5–10
Microcrystalline C 1035	160	≈20	<5
Microcrystalline Shell 400	160	≈30	<5
Microcrystalline Mobil	160	≈20	<5
Polyethylene wax BASF-A	160	≈40	≈33
Paraflint H1–N1	160	≈32	≈18

*From Battelle Columbus, OH.

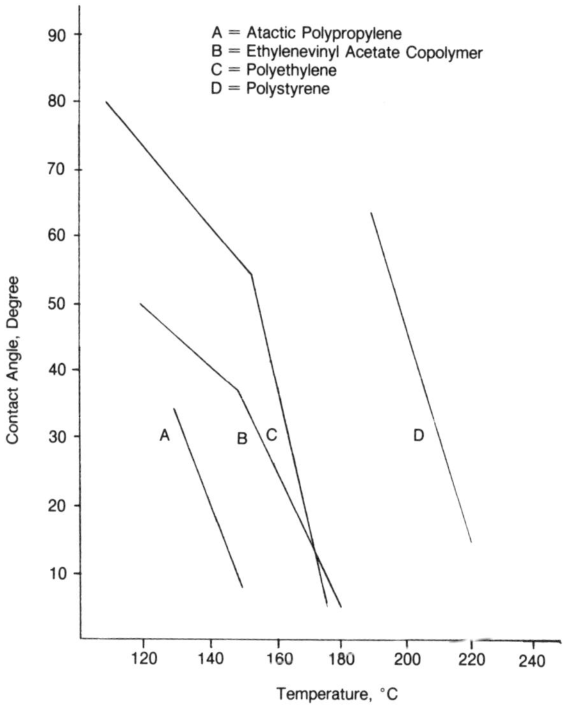

FIGURE 3-35. The contact angle of resin to an alumina substrate. After Saito et al. (1976).

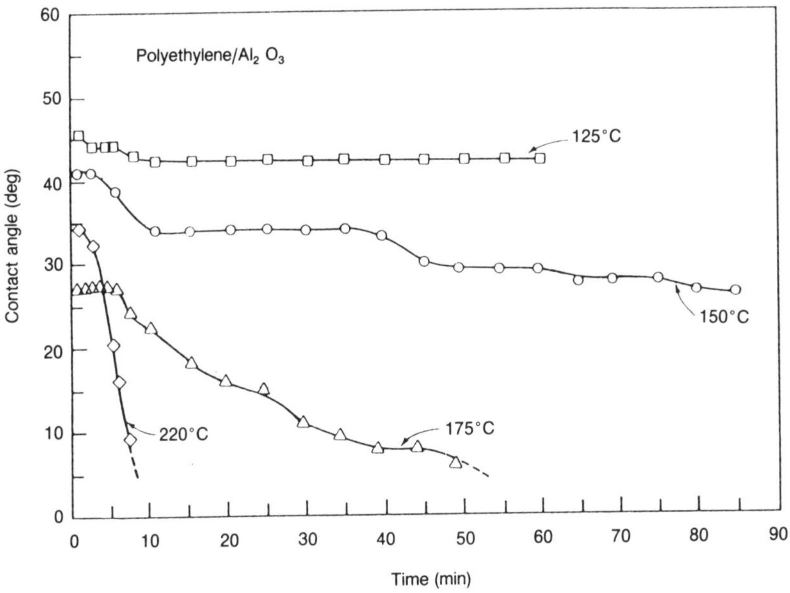

FIGURE 3-36. The contact angle of polyethylene to an alumina substrate as a function of time, at different temperatures. After Dow et al. (1988).

Paraflint H1-N1 wax with silicon nitride is somewhat lower than with alumina. The AC-400 copolymer as well as the waxes form consistently lower contact angles with silicon nitride than with alumina.

The implications of these differences in wetting could be related to some complex properties of the solid surfaces and the wetting agents. Such differences can be explained only if critical parameters such as the surface tension of the wetting agents, the surface energy of the solids, the impurities on the solid surfaces, and the surface morphologies of the solids are known. The measurement of the contact angle is useful as a selection procedure for the primary binder for injection molding because it gives an indication of the wetting behavior of the various polymers and waxes in relation to ceramic materials of interest.

In the final stage of the dispersion process, the dispersed particles are stabilized against flocculation. Although there is adequate information about the dispersion stability of suspensions at low particle concentrations (below 50 vol%) in low-viscosity liquids such as water and organic solvents, the information is limited for suspensions with particle concentrations above 50 vol%.

An unstabilized dispersion flocculates by the collision between primary particles, which reduces the number of particles in the system. The time required to reduce the number of particles to half their initial value, $t_{\frac{1}{2}}$, is given by (Verwey and Overbeak 1948):

$$t_{\frac{1}{2}} = \frac{3\eta}{4\,KTP_o} \tag{3-21}$$

where η is the viscosity of the fluid, K is the Boltzmann constant, T is the absolute temperature, and P_o is the initial number of particles in the system. At high particle concentrations (about 55 vol%), $t_{\frac{1}{2}}$ decreases rapidly.

The Selection Procedure for Plasticizers

Polyethylenes are completely miscible with hydrocarbon waxes or oils at temperatures above their melting points. Some information about synthetic plasticizers is given in Table 3-11. The viscosity curves for compositions containing 60 vol% alumina powder and 3 vol% of these synthetic plasticizers are shown in Figures 3-37 and 3-38. As can be seen, VP-109, Kemamide E, INT-33-UDKH, VP-900, and INT-EQ-6 have a significant effect on the mixing behavior of the polymer and alumina powder mix.

Table 3-12 shows the effect of the various plasticizers on the concentration of the alumina filler. When the plasticizer concentration is increased, the volume loading of the ceramic is increased stepwise from 57 vol% to the maximum amount that the binder phase can accommodate. Each of these plasticizers is capable of altering the wetting behavior of the polyethylene–alumina compositions to enable the achievement of maximum ceramic filler concentration. As Table 3-13 shows, as little as 3 vol% (1 wt%) of these specific plasticizers is sufficient to provide an optimum volume of alumina powder in a polyethylene base mix.

TABLE 3-11. Some information on synthetic plasticizers.

Grade	General description	Density, g/cm^3	Form
INT-33-UDKH	Polymeric condensation product of glycerides and organic acids	0.93	Powder
INT-EQ-6	Same as above	0.96	Fluid
INT-54	Same as above	0.89	Fluid
Kemamide E	Unsaturated fatty monoamide of oleic acid	0.81	Powder
VP-109	Unsaturated fatty acid	0.98	Fluid
VP-900	Polymeric ester	0.98	Fluid
VP-905	Polymeric ester	0.95	Fluid

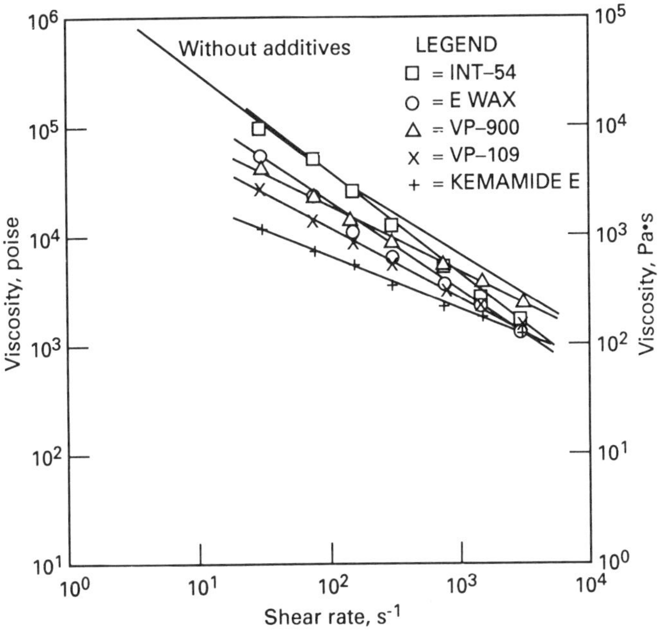

FIGURE 3-37. The effect of the synthetic plasticizers INT-54, E Wax, VP-900, VP-109, and Kemamide E, on the viscosity and the shear rate of a mixture containing 60 vol% alumina powder, 37 vol% NA-250 polyethylene, and 3 vol% plasticizer. From Battelle Columbus, OH.

Phthalates, adipates, oleates, and stearates are extensively used with polystyrene and other crystalline polymers to improve their fluidity. Other plasticizers, in addition to waxes, are compatible with EVA polymers.

Three grades of homopolymers have been examined for use as wax binders. Low-density polyethylenes possess a unique combination of properties: a high softening point and a low melt viscosity. Because of their low viscosity combined with good compatibility these resins can be readily dispersed in wax blends. Ethylene–acrylic copolymers (EAA) produce homogeneous wax blends. EAA copolymers are similar to ethylene homopolymers of comparable melt index and molecular-weight distribution, but the copolymers have better flowability at low temperatures (below 100°C). Two important properties of EAA copolymers are their wetting characteristic and their adhesion to different ceramic powders. The wettability of alumina and silicon nitride with these plasticizers is given in Table 3-14.

Both AC-617 and AC-1702 have contact angles (less than 20°) for alumina at which the wetting of the solid surface is likely to be extensive.

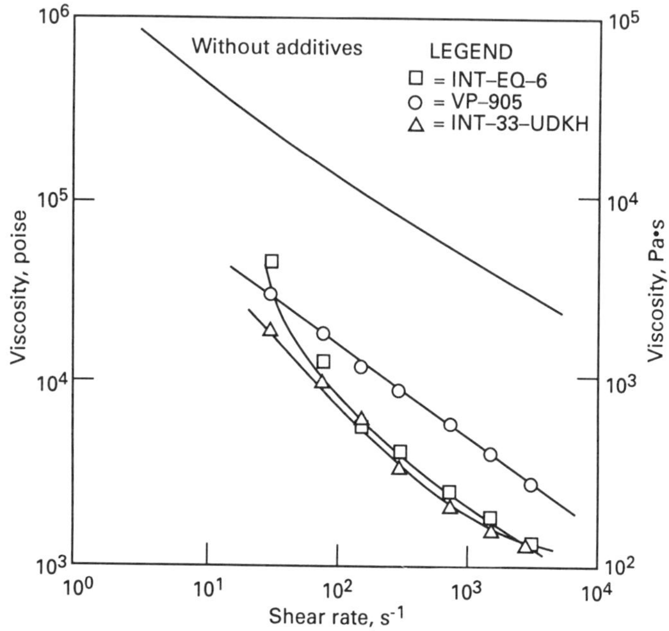

FIGURE 3-38. The effect of the synthetic plasticizers INT-EQ-6, VP-905, and INT-33-UDKH on the viscosity and the shear rate of the mixture described in the preceding figure. From Battelle Columbus, OH.

With the exception of AC-617, all of the plasticizers show good wettability with silicon nitride. In fact, the wettability with silicon nitride is better than that with alumina.

Investigation of the rheology of these plasticizers with several waxes in the presence of A16-SG alumina and milled silicon nitride illustrates their differences. The viscosity data shown in Figure 3-39a for different plasticizers with H1-N1 wax and alumina indicate that each plasticizer has a characteristic effect. For example, AC-400 (EVA) consistently maintains a higher viscosity over the entire range of shear rates. The homopolymer AC-6 and low-density polyethylene AC-617 have marked differences in their viscosities, and this is fairly consistent with the contact angle data presented earlier (see Table 3-14).

Figure 3-39b shows the viscosity data at 160°C for compositions of alumina and BASF-A wax with different plasticizers. The viscosities fall into three separate ranges. The mix composition containing EAA copolymer 540 shows a high viscosity with a nonlinear trend over the entire shear rate range; AC-6 homopolymer shows viscosity between the extreme mem-

TABLE 3-12. Effect of plasticizer aids on the maximum concentration of $A1_2O_3$ with NA-250 polyethylene.*

Concentration, fractions or parts, wt (vol)			Plasticizer	
Alumina	Polymer	Plasticizer	type	Work done[a]
0.842 (0.555)	0.137 (0.388)	0.021 (0.057)	VP-109	700
0.860 (0.591)	0.118 (0.349)	0.022 (0.060)	VP-109	1000
0.872 (0.614)	0.110 (0.333)	0.018 (0.053)	VP-109	2500
0.842 (0.553)	0.137 (0.387)	0.021 (0.060)	INT-33-UDKH	750
0.860 (0.589)	0.118 (0.348)	0.022 (0.063)	INT-33-UDKH	950
0.875 (0.619)	0.106 (0.322)	0.019 (0.059)	INT-33-UDKH	1750
0.842 (0.549)	0.137 (0.383)	0.021 (0.068)	Kemamide E	750
0.860 (0.588)	0.118 (0.340)	0.022 (0.073)	Kemamide E	700
0.881 (0.627)	0.101 (0.309)	0.018 (0.064)	Kemamide E	1875
0.842 (0.554)	0.137 (0.387)	0.021 (0.058)	VP 905	1900
0.851 (0.571)	0.128 (0.369)	0.021 (0.060)	VP 905	2700
0.859 (0.587)	0.121 (0.356)	0.020 (0.058)	VP 905	6500
0.842 (0.555)	0.137 (0.388)	0.021 (0.057)	VP 900	1700
0.851 (0.572)	0.128 (0.369)	0.021 (0.058)	VP 900	3750
0.860 (0.590)	0.120 (0.354)	0.020 (0.056)	VP 900	4500
0.842 (0.555)	0.137 (0.388)	0.021 (0.057)	INT-EQ-6	800
0.860 (0.590)	0.188 (0.349)	0.022 (0.061)	INT-EQ-6	5000
0.873 (0.618)	0.107 (0.325)	0.019 (0.057)	INT-EQ-6	1750

*From Battelle Columbus, OH.

a. See Chapter 4.

TABLE 3-13. Optimum plasticizer content of Al_2O_3 powder with NA-250 polyethylene.*

Concentration, fractions or parts, wt (vol)			Plasticizer	
Al_2O_3	Polymer	Plasticizer	type	Work done[a]
0.872 (0.614)	0.111 (0.330)	0.018 (0.053)	VP-109	2500
0.867 (0.604)	0.119 (0.331)	0.014 (0.039)	VP-109	2500
0.871 (0.612)	0.119 (0.359)	0.010 (0.028)	VP-109	2000
0.875 (0.619)	0.106 (0.322)	0.019 (0.059)	INT-33-UDKH	1750
0.874 (0.618)	0.113 (0.343)	0.013 (0.040)	INT-33-UDKH	2500
0.881 (0.632)	0.110 (0.340)	0.009 (0.033)	INT-33-UDKH	2600
0.881 (0.627)	0.101 (0.309)	0.018 (0.064)	Kemamide E	1875
0.880 (0.626)	0.108 (0.330)	0.013 (0.044)	Kemamide E	1875
0.881 (0.629)	0.110 (0.338)	0.009 (0.033)	Kemamide E	3800
0.873 (0.618)	0.107 (0.325)	0.019 (0.057)	INT-EQ-6	1750
0.873 (0.616)	0.114 (0.321)	0.013 (0.039)	INT-EQ-6	2450
0.874 (0.617)	0.117 (0.354)	0.010 (0.029)	INT-EQ-6	1800

*From Battelle Columbus, OH.

a. See Chapter 4.

TABLE 3-14. Wetting behavior of plasticizers with ceramics.*

Plasticizers	Contact angle (θ) with ceramics at 160°C, degrees: Alumina	Silicon nitride
AC-6 (homopolymer)	33–36	≈10
AC-617 (homopolymer)	≈15	≈50
AC-1702 (low-density polyethylene)	10–13	10–14
AC-540 (ethylene acrylic acid copolymer)	≈21	≈5
AC-400 (ethylene vinyl acetate copolymer)	≈22	≈6

*From Battelle Columbus, OH.

bers. AC-617 and low-density polyethylene AC-1702 homopolymers and EVA copolymer AC-400 have identical viscosities. In this instance, the contact angle data cannot be related to measured viscosity. A review of the contact angle data for the waxes (see Table 3-10) and for the plasticizers (see Table 3-14) would suggest some incompatibility in the polymer–polymer interaction, as well as powder–polymer interactions, for this particular wax (BASF-A), alumina, and plasticizer(s) mixes.

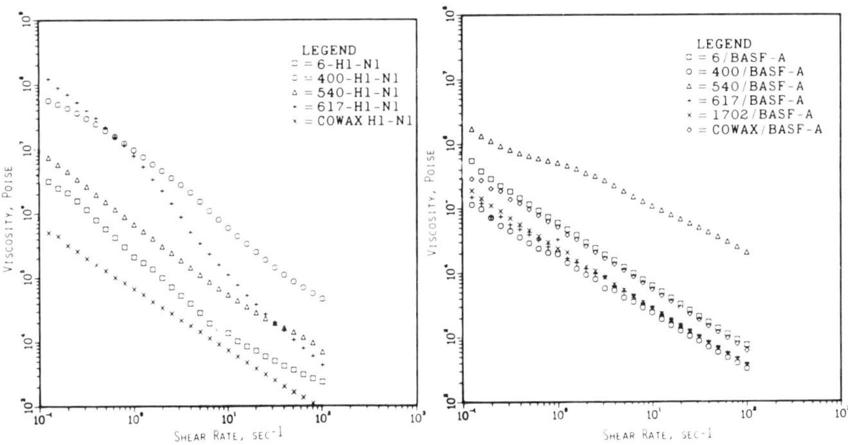

a. H1-N1 Wax with Surfaid 79 b. BASF-A Wax

FIGURE 3-39. The effect of various plasticizers on the viscosity of mixes of alumina powder and waxes. (a) H1-N1 wax; (b) BASF-A wax. From Battelle, Columbus, OH.

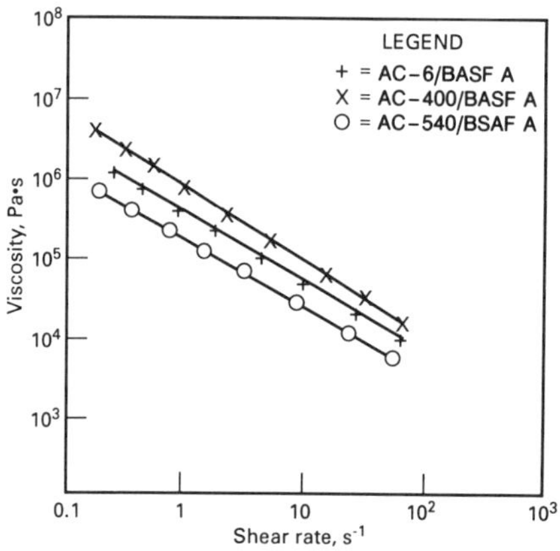

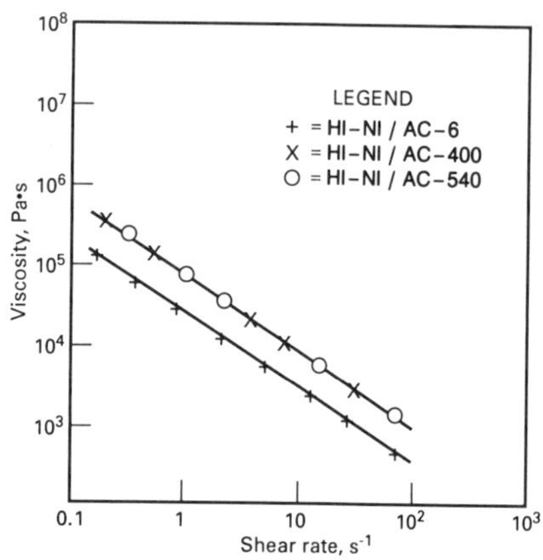

FIGURE 3-40. The effect of plasticizers on the viscosity of mixes of silicon nitride powder and waxes. (a) BASF-A wax; (b) H1-N1 wax. From Battelle Columbus, OH.

The viscosity curves for silicon nitride with BASF-A are shown in Figure 3-40a. The mix containing AC-400 (EVA) has a higher viscosity than with the other two plasticizers, and the mix containing AC-540 has the lowest viscosity. Viscosity–shear rate data for silicon nitride with the wax H1-N1 also are shown in Figure 3-40b. In this case, AC-6 (homopolymer) shows relatively lower viscosity than the other two plasticizers.

The purpose of the examples given above is to underscore the importance of gaining information on the chemical and physical criteria such as solubility parameter, compatibility, wetting, and powder surface characteristics before a combination (i.e., binder, plasticizer, and processing aid) of binder can be designed for injection molding a particular type of ceramic powder.

The Selection Procedure for Processing Aids

The most important stage of the dispersion process is stabilization of the dispersed particles to prevent flocculation. There are two basic mechanisms for achieving stabilization: one is electrical repulsion resulting from the interaction of the charged electrical double layers surrounding the particles, and the other is some form of steric hindrance, which is a constraint on chemical action due to the arrangement of atoms in the molecule. In this case, there is repulsion between the ceramic particles due to adsorbed polymer molecules on their surface that interact on the close approach of one particle to another. This repulsion prevents the particles from approaching close enough for flocculation to occur.

Several practical approaches are possible for selecting suitable processing (i.e., dispersing) agents that can be used with polymeric and/or wax type binders. An oil absorption test, a semiquantitative technique commonly used in the paint industry, can be used initially to select a processing aid. The oil absorption value is expressed as the number of parts of oil required to wet 100 parts of powder on a weight basis. A more convenient term is the volume percent required to achieve the desired end point.

The absorption data for A16-SG alumina and silicon with several oils are given in Table 3-15. A significant difference is seen in the absorption affinities of each powder. For alumina, the absorption affinity is strongly dependent on the oil type. The silicon-based powders are not particularly sensitive to the type of oil and require less oil to reach the end point than does the alumina.

To understand the reason why alumina powder is strongly dependent on the type of oil but silicon-based powder is not, one must consider their

TABLE 3-15. Oil absorption of ceramic powders.*

Oil		Volume fraction absorbed	
Type	Density, g/cm^3	Alumina A16-SG	Silicon
Oleic acid	0.89	0.37	0.46
VP-109	1.00	0.42	0.46
Menhaden	1.00	0.44	0.41
Cod-liver	0.92	0.54	—
Olive	0.91	0.54	—
Corn	0.93	0.56	0.47
Cottonseed	0.93	0.57	—
Soybean	0.92	0.58	—
Peanut	0.93	0.59	0.45
Pine	0.94	0.64	—
Transformer	0.89	0.67	0.47
Mineral	0.84	0.74	0.48

*From Battelle Columbus, OH.

surface chemistry. Assuming that the silicon particles are coated with a surface layer of silica, the net surface charge would be negative. Alumina, on the other hand, has a net positive surface charge.

The oils are composed of varying amounts of saturated and unsaturated fatty acids. The unsaturated fatty acids are negatively charged and control the absorption characteristics of the oil. With a negative charge, fatty acids can readily be absorbed on a positively charged surface, such as alumina powder, but not to any significant degree on a negatively charged surface such as silicon powder.

An alternative approach is to use the zeta potential technique, which measures the type and magnitude of the surface charge on a ceramic particle at a particular pH (Hunter 1981). In this technique a known electric potential is applied across an aqueous suspension of fine ceramic particles. Then, by using the direction and the speed of movement of the particles, the sign and the magnitude of the charge can be calculated. This technique measures the electrophoretic mobility (EPM) of the particles. The electrophoretic mobility data for different ceramic powders, silicon, silicon nitride, zirconia, and alumina have been recorded using a zeta meter, first on the as-received powders at various pH values and then after the powders have been coated with 1% oleic acid.

Figure 3-41a gives the EPM data of silicon particles with and without the oleic acid coating. The data show that the uncoated silicon particles have a point of zero mobility or zero point of charge (ZPC) at a pH of approximately 2.5. The same particles when coated with oleic acid show a small shift in the ZPC, to a pH of approximately 3.2. The rest of the curve is similar to the one obtained with uncoated silicon particles. These data indicate that silicon particles that are negatively charged over a wide pH range do not absorb a negatively charged processing aid such as oleic acid.

Similarly, Figures 3-41b and c gives the EPM data for silicon nitride and zirconia, both with and without the oleic acid coating. These powders are negatively charged over a pH range of approximately 2.0. This finding indicates that oleic acid has little or no effect on the behavior of their surface charges.

Figure 3-41d shows the electrophoretic mobility (EPM) data for A16-SG alumina with and without the oleic acid coating. As-received alumina powder has a zero point charge at a pH of 6.8, a value reported in the literature (Parffit 1969). When these particles are coated with oleic acid, their surface charge is reversed, and their zero point charge occurs at a pH of 2.5. These data indicate that oleic acid absorbs strongly on positively charged alumina particle surfaces, reversing the sign of the charge from positive to negative.

The above example of oleic acid can be generalized to evaluate the

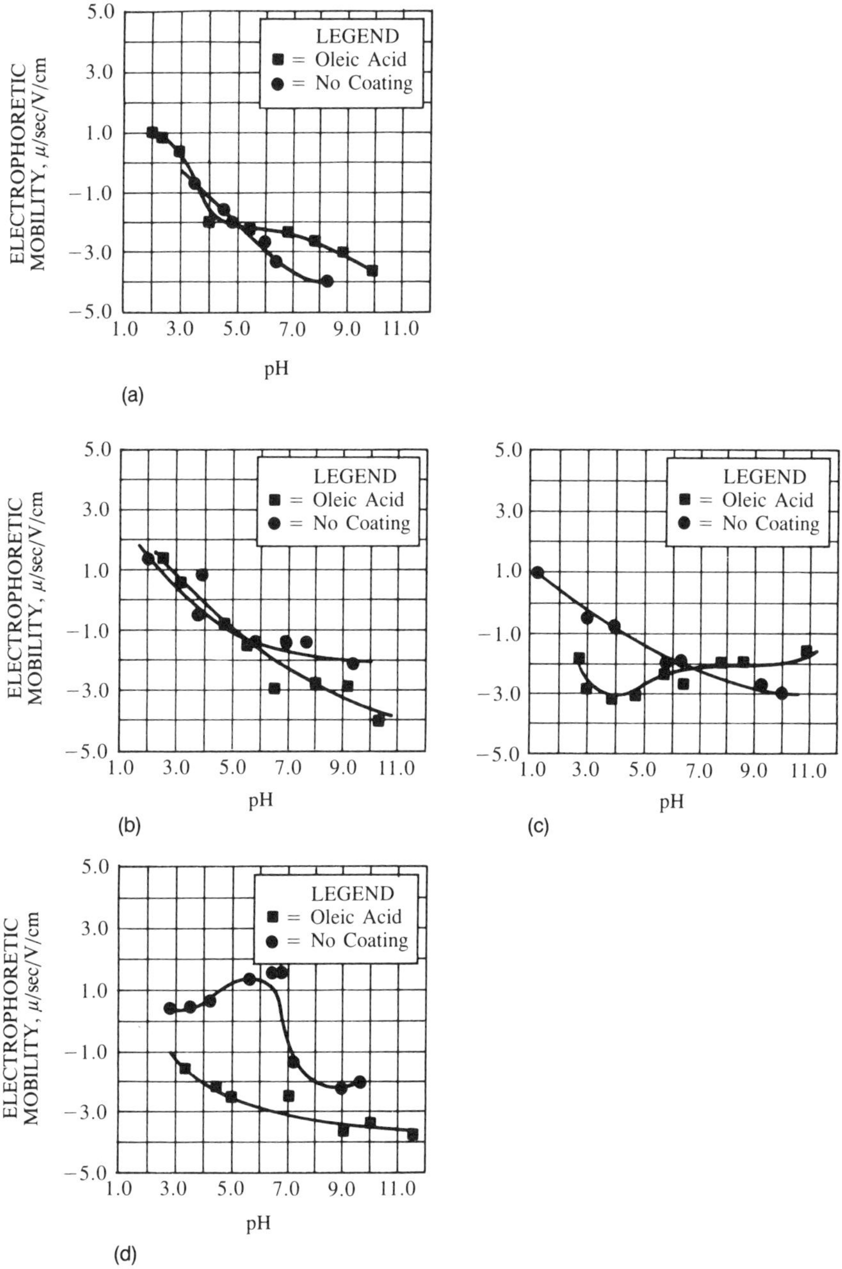

FIGURE 3-41. Electrophoretic mobility data for various ceramic powders uncoated and coated with oleic acid: (a) silicon; (b) silicon nitride; (c) zirconia; (d) alumina.

absorption characteristics of other processing aids with respect to particular ceramic powders. Both oil absorption and the zeta potential can be used to find a suitable processing aid for ceramic injection molding mix formulations. However, rheological evaluation still should be considered as the ultimate selection criterion.

THE OPTIMUM BINDER CONTENT AND FORMULATION OF BINDER SYSTEMS

Once the basic ingredients for a binder system have been ascertained, it is necessary to determine the total binder content for each ceramic powder and the relative proportion of each binder ingredient.

Optimum Binder Content

A useful guideline for determining the optimum binder content for a given powder is provided by rheological measurements. The results summarized in Table 3-16 are for the optimum binder content with the binder NA-250 polyethylene, but the data are not necessarily valid for any other type of binder. In addition to rheology, the critical powder volume concentration (CPVC) values obtained from the oil absorption test can be a useful method for determining the optimum binder content. (This is discussed in Chapter 4.) The results obtained from both methods are in good agreement, and they are consistent with the binder content for the formulations that can be molded successfully.

Mix Formulations

Numerous binder formulations for ceramic injection molding have been developed. Examples of some that are used commercially and therefore are readily available are given in Table 3-17 (parts a–e). This chapter has

TABLE 3-16. Optimum binder content determined from rheology data.

Ceramic	Grade or trade name	Binder	Vol.% of Binder	Wt.% of Binder
Al_2O_3	A16-SG	NA-250[a]	40.0	11.1
ZrO_2	Type C	NA-250	31.5	7.0
Si_3N_4[b]	Milled SN502	NA-250	42.0	17.5

a. Polyethylene.
b. With Y_2O_3 plus Al_2O_3 sintering aids.

TABLE 3-17a. Mix formulations for alumina, expressed in weight %.

After:	Thomas et al. (1988)	Tanaka et al. (1985)			A. Moteki (1960)
Al_2O_3	84.26	100	100	100	81.6
EVA[a]	3.59	4.23	4.47	4.73	—
PBMA[b]	4.40	9.86	8.94	7.88	—
Modified wax[c]	6.54	—	—	—	—
Phthalate:					
Dibutyl	1.21	1.61	1.70	1.72	—
Diethyl	—	—	—	—	2.0
Polystyrene	—	—	1.49	3.15	12.2
Stearic acid	—	3.16	3.34	3.53	4.1

a. EVA: ethylene vinyl acetate.
b. PBMA: polybutyl methacrylate.
c. Modified wax:
Microcrystalline, 60 wt%
Glycerol monostearate, 20 wt%
Dioctyl phthalate, 10 wt%
EVA 410, 10 wt%

TABLE 3-17b. Typical thermoset binder formulations.

Components	Volume %	Components	Volume %
Ceramic[a]	63.0	Silicon carbide[b]	47.0
Epoxy	7.3	Phenolfurfural-	
Coumarone-indene resin	3.2	phenolformaldehyde	47.0
Phenolformaldehyde	2.2	Graphite	5.0
Wax	24.3	Zinc stearate	1.0

a. Strivens (1963).
b. Whalen and Johnson (1981).

TABLE 3-17c. Claimed mix formulation for silicon nitride.

Components[a]	Weight %
Hydrocarbon wax	85–90
Epoxy resin containing hardener	5–10
(Bisphenol-A epichlorohydrin resin-containing hexahydrophthalic anhydride)	
Surfactant	3.6–5

a. French et al. (1984)

TABLE 3-17d. Commercial silicon nitride mix formulations.*

Si_3N_4	100 wt. parts	100 wt. parts
Acrylic resin (CB-1)	4.64	9.86
EVA[a]	6.94	4.23
Paraffin wax[b]	5.16	—
Dibutyl phthalate	2.33	—
Stearic acid	—	3.61

*Sanyo Chemical Industries.
a. EVA: ethylene vinyl acetate copolymer.
b. Paraffin wax: 60–70°C.

TABLE 3-17e. Typical binder formulation for silicon powder.*

	Weight %	Weight %
Isotactic polypropylene	7.68	5.79
Atactic polypropylene	7.68	5.79
Stearic acid	1.93	1.93
Microcrystalline wax	—	3.86
Silicon	82.71	82.63

*Edirisinghe and Evans (1987).

focused on the critical physical and chemical characteristics of the binder ingredients that should be carefully investigated in designing a binder formulation for ceramic injection molding.

APPENDIX 3-A VISCOSITY OF PARTICULATE-FILLED REINFORCED POLYMERS

Various theories have been proposed to predict the viscosity of particulate-filled reinforced polymers. Most of these theories are limited to a low concentration of filler and assume that the filled system remains a Newtonian fluid.[1] The first theoretical treatment was for very dilute systems of spherical particles (Einstein 1906, 1911, 1956):

$$\frac{\eta}{\eta_o} = 1 + 2.5V \tag{3A-1}$$

[1]In a Newtonian fluid, the viscosity is unaffected by changes in the shear rate. An example is water, in which the viscosity is altered only by changes in the pressure and the temperature.

where η is the viscosity of the suspension, η_o is the viscosity of the fluid, V is the total volume, and the constant 2.5 is the Einstein shape factor for a single sphere.

Subsequent treatments have attempted to extend this theory to concentrated suspensions and nonspherical particles.[2]

Another approach taken applied a general power-series equation in volume fractions (K_1 is the Einstein coefficient, 2.5):

$$\frac{\eta}{\eta_o} = 1 + K_1V + K_2V^2 + K_3V^3 + \cdots \tag{3A-2}$$

One assumption was that equation (3A-2) was a quadratic and that the relative viscosity, η/η_o, would become infinite when the volume fraction reached the value for cubic closed-packed spheres. The value of K_2 then would become 4.7, increasing slowly as the packing fraction decreased. An assumption then was made that after collision two particles, considered to be monosized, would separate along rectilinear paths (Vand 1948). This assumption led to the following equation in a glycerol solution up to a concentration of $V = 0.5$:

$$\frac{\eta}{\eta_o} = 1 + 2.5V + 7.349V^2 \tag{3A-3}$$

(In this equation the value of K_2 is about 7.35.) Subsequently, however, this assumption was shown to be incomplete, as it can be proved that the paths of recession of two colliding spheres are curvilinear and mirror images of their paths of approach (Mooney 1951; Manley and Mason 1952). At very high concentrations, a large number of terms would be required in equation (3A-2) in order to calculate η/η_o. The higher-power terms would have an increasing influence on the value obtained.

A useful approach considers an infinite power series in the general form of an exponential function:

$$\frac{\eta}{\eta_o} = \exp(2.5\ V) \tag{3A-4}$$

---

[2]These were reviewed by Lewis and Nielsen (1968) and Jeffrey and Acrivos (1976).

This equation has been generalized to satisfy all disperse systems by including a factor, h, that represents a complex function of particle interactions:

$$\frac{\eta}{\eta_o} = \exp\left(\frac{2.5\,hV}{1 - hV}\right) \tag{3A-5}$$

An identical equation for monodisperse spheres, which takes into account the crowding of particles in concentrated suspension, also has been developed (Mooney 1951):

$$\frac{\eta}{\eta_o} = \exp\left(\frac{2.5\,V}{1 - CV}\right) \tag{3A-6a}$$

or

$$\frac{\eta}{\eta_o} = \frac{2.5\,V}{1 - V/V_{max}} \tag{3A-6b}$$

A crowding factor, C, is defined in this equation as equal to $1/V_{max}$, where V_{max} is the volume fraction of ceramic powder in the mix that results in maximum packing.

Spheres in a cubic close-packed lattice ($V_{max} = 0.74$) are considered to exhibit infinite viscosity because of mechanical interlocking. Simple cubic packing ($V_{max} = \pi/6$) provides the maximum density that allows continuous movement. Thus the lower and upper limits of C are $1.35 < C < 1.91$. Equation (3A-5), based on experimental observations (Vand 1948), fits equation (3A-6) (Mooney 1951) if $C = 1.43$.

A modification of equation (3A-6) was derived that has been found to be experimentally valid (Krieger and Dougherty 1959):

$$\frac{\eta}{\eta_o} = (1 - CV)^{-2.5} \tag{3A-7}$$

The viscosity data for a mix of alumina (Al_2O_3 powder plus an ethylene vinyl acetate (EVA) copolymer binder are in good agreement with equation (3A-6). The relative viscosities at four powder loadings are given in Table 3A-1 for a ceramic powder filler composed of agglomerated, irregularly shaped particles with a mean particle size of 0.62 μm mixed with an EVA copolymer binder with an average molecular weight of 3500 and a melt flow index of 200.

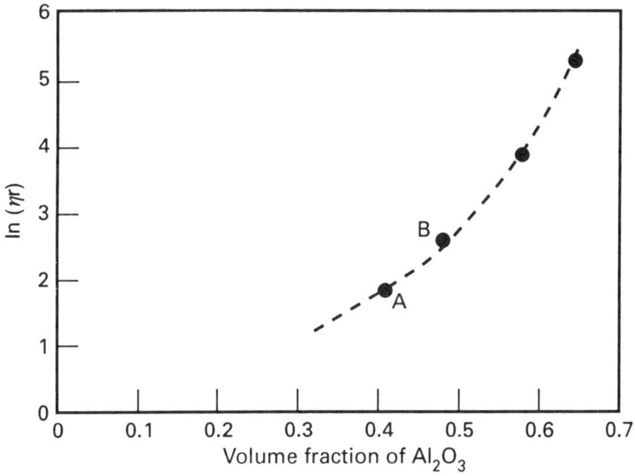

FIGURE 3A-1. A plot of equation (3A-8) using data from Table 3A-1 for a mix of alumina (Al_2O_3) powder and an EVA copolymer binder. After McColm and Clark (1988).

Equation (3A-6) also can be expressed in the more convenient form of a natural logarithm:

$$\ln \eta_r = \frac{2.5\ V}{1 - V/V_{\mathrm{max}}} \tag{3A-8}$$

Figure 3A-1, which is a plot of this equation, can be used to determine the slope at points A, B, and V_{max}. At point A:

$$7.5 = \frac{B}{1 - \dfrac{0.41}{V_{\mathrm{max}}}}$$

TABLE 3A-1. The relative viscosities of ceramic powder–binder mixes containing four different loadings of alumina plus the binder, ethylene vinylacetate (EVA) copolymer, determined at a shear rate of 100 sec^{-1}.*

Vol% Al_2O_3	Relative viscosity	$\ln \eta_r$
41.0	6.25	1.83
48.0	12.55	2.53
57.0	49.90	3.91
63.0	200.30	5.30

*After McColm and Clark (1988).

and at point B:

$$11.43 = \frac{B}{1 - \frac{0.48}{V_{max}}}$$

From the graph in Figure 3A-1 and the above relationships, the value of B is found to be 2.46, which closely fits the Einstein constant, 2.5. The V_{max} is 61.3%. This is attributed both to the lack of sphericity and to the size and number of agglomerates.

A good fit also has been found between experimental results and the following type of equation for V/V_{max} close to unity (Chong et al. 1971):

$$\frac{\eta}{\eta_o} = 1 + \left[1 + \frac{0.75\ V/V_{max}}{1 - (V/V_{max})}\right]^2 \tag{3A-9}$$

Equation (3A-6) was further modified for ellipsoidal particles (Brodnyan 1959):

$$\eta_o = \exp\left(\frac{(2.5\ V + 0.399\ (p - 1)^{1.48} V}{1 - CV}\right) \tag{3A-10}$$

where p is the axial ratio of the particles. This equation was found to fit experimental data up to $V = 0.55$. It was extended on an empirical basis to give the following relationship for acicular particles (Kitano et al. 1981):

$$\eta_o = 1 - \frac{V}{0.54 - 0.0125\ p} \tag{3A-11}$$

Equations (3A-1) through (3A-10) are valid for particles that have uniform diameter and are close to spherical in shape.

Current developments in ceramic powder processing are producing materials that closely meet these properties. The requirement for uniform shrinkage without discontinuous grain growth during sintering necessitates the use of powders with particles that are close to uniform in size. In this case, assuming that the particles are spherical, $V = 0.74$. If, however, the powder contains a range of particle sizes, the value of the crowding factor C in equation (3A-6a) changes.

Again, because of the formation of agglomerates, the relative viscosity

(η/η_0) for ultrafine particles can be affected significantly. An increase in the Einstein coefficient (K_1) is related to the presence of agglomerates. For an agglomerate, the ratio of the apparent, or hydrodynamic, volume to the true volume of the spheres that compose the agglomerate is the reciprocal of the volume packing fraction of the spheres that form the agglomerate, V_a. This is distinguished from the maximum volumetric packing fraction, $V_{\max}$.

Assuming model structures for the various sizes of agglomerates, the curve shown in Figure 3A-2 was plotted by using the following relation (Lewis and Nielsen 1968):

$$K_1 = 2.5/V_a = 2.5\frac{(V_s + V_l)}{V_s} \tag{3A-12}$$

where V_s is the total volume occupied by the spheres in the model structures of the agglomerates, and V_l is the volume of the immobilized liquid.

The immobilized liquid associated with an agglomerate is the liquid entrapped within the agglomerate plus the liquid contained at the edges of

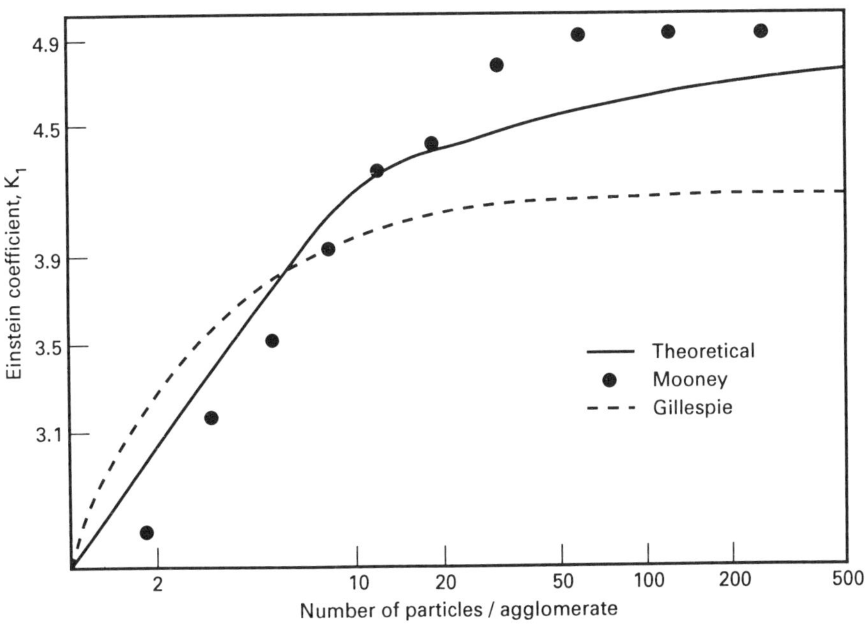

FIGURE 3A-2. The Einstein coefficient for a range of agglomerate sizes, obtained by fitting the Mooney equation to the viscosity data.

the agglomerate within an envelope formed by the outermost spheres composing the agglomerate. For the largest agglomerates (number of particles per agglomerate $n < 8$), all of the liquid essentially is contained within the structure. For the smallest agglomerates, the liquid is located primarily at the outer edges. For the rest of the agglomerates, the liquid is divided between the outer edges and the open volume at the center of the agglomerate.

For the largest agglomerates, V_a approaches a maximum value that depends on the type of packing. This value can vary from 0.637 for random, dense packing to a lower value for a more open structure—for example, 0.524 for cubic packing. The corresponding maximum values for K_1 are 3.92 and 4.77, respectively. The position of the theoretical curve for the smallest agglomerates above the points determined from the experimental results indicates that the liquid at the outer edges of an agglomerate probably is not entirely immobilized, as assumed by the model structures. Furthermore, if an expression other than the Mooney equation (3A-6) were used for the analysis, slightly different values of K_1 would be obtained from the experiment.

Theories concerning the viscosity of polydisperse systems suggest that for a bimodal system the fines combined with the liquid matrix could be treated essentially as a fluid through which the coarse particles flow (Farris 1968).

Expressed mathematically, the concentrations of the fines and the liquid volumes would be:

$$\theta_f = \frac{V_f}{(V_l + V_f)} \tag{3A-13}$$

where V_f is the volume of the fines, V_l is the volume of the liquid, and θ_f is the volume concentration of fines in the liquid.

When a filler is added to a liquid matrix, the viscosity increases, and the fluid stiffens. The stiffening factor depends only on the volume concentration and not on the particle size. Therefore, the stiffening factor is defined as (Farris 1968):

$$H(\theta_f) = \frac{\eta_f}{\eta_o} = \eta_r \tag{3A-14}$$

where $H(\theta_f)$ is the stiffening factor, η_f is the viscosity of the fluid and fines, η_o is the viscosity of the fluid alone, η_r is the relative viscosity, and $\theta_f = V_f/(V_l + V_f)$.

Adding the coarse filler to this already stiffened fluid stiffens it further by a factor of $H(\theta_c)$. This factor is the ratio of the new viscosity with both coarse and fine particles, η_c, to that of the fluid containing only fine particles, η_f:

$$H(\theta_c) = \frac{\eta_c}{\eta_f} = \frac{\eta_c}{\eta_0} \cdot \frac{1}{H(\theta_f)} = \frac{\eta_r}{H(\theta_f)} \tag{3A-15}$$

Therefore, the relative viscosity for the bimodal powder in the fluid matrix can be defined as:

$$\eta_r = H(\theta_f)\, H(\theta_c) \tag{3A-16}$$

where η_r is the relative viscosity for bimodal powder, $H(\theta_f)$ is the stiffening factor for fine particles, and $H(\theta_c)$ is the stiffening factor for coarse particles.

In equation (3A-16), θ_c is the concentration of coarse particles in the apparent liquid, which has a volume given by $(V_l + V_f)$. Thus:

$$\theta_c = \frac{V_c}{(V_l + V_f) + V_c} \tag{3A-17}$$

where V_c is the volume of the coarse particles.

When this reasoning is extended to more than two particle sizes, the equation for the relative viscosity becomes (Farris 1968):

$$\eta_r = H(\theta_1)\, H(\theta_2)\, H(\theta_3) \cdots H(\theta_n) \tag{3A-18}$$

References

Arnott, S., A. Fulmer, and W. Scott. 1974. The Agarose Double Helix and Its Formation in the Agarose Gel Structure, *J. Molec. Biol.*, 90, 269–284.

Bagley, E. B. 1957. End Corrections in the Capillary Flow of Polyethylene, *J. Appl. Phys.*, 28(5), 624–627.

Bordnyan, J. G. 1959. The Concentration Dependence of the Newtonian Viscosity of Prolate Ellipsoids, *Trans. Soc. Rheol.*, III, 61–68.

Bueche, F. 1954. *J. Chem. Phys.*, 22, 1570.

Casson, N. 1959. *Rheology of Disperse Systems*, ed. C. C. Mills, Pergamon Press, Elmsford, NY, 84.

Chang, I. B. and R. E. Beckwith, Jr. 1979. *Soc. Com. Chem.*, 6, 25–29.

Chong, J. S., E. B. Christiansen, and A. D. Baer. 1971. Flow of Viscous Fluid through a Circular Pipe, *J. Appl. Sci.*, 15, 369–375.

Dow, J. H., M. D. Sacks, and A. V. Shenoy. 1988. Dispersion of Ceramic Particles in Polymer Melts, *Ceram. Trans.*, 1(A), Ceram. Powder Sci., Am. Ceram. Soc. Inc., 380–388.

Edirisinghe, M. J. and J. R. G. Evans. 1987. Rheology of Ceramic Injection Molding Formulations, *Ceram. Trans. J.*, 86, 18.

Einstein, A. 1906/1911/1956. Investigation of Brownian Movement, *Ann. Physik.*, 19, 289 and 34, 591; Dover, New York.

Fanelli, A. J. and R. D. Silvers. 1988. Process for Injection Molding Ceramic Composition Employing an Agaroid Gel-Forming Material to Add Green Strength to a Preform, U.S. Patent, 4,734,237.

Fanelli, A. J., R. D. Silvers, W. S. Frei, J. V. Burlew, and G. B. Marsh. 1989. New Aqueous Injection Molding Process for Ceramic Powders, *J. Am. Ceram. Soc.*, 72(101) 1833–1836.

Farris, R. J. 1968. Prediction of the Viscosity of Multimodal Suspensions from Unimodal Viscosity Data, *Trans. Soc. Rheol.*, 12(2), 281–301.

French, K. W., J. T. Nair, and L. T. Turnbaugh. 1984. Composition for Injection Molding, U.S. Patent 4,456,713.

Gaines, G. L. Jr. 1972. Surface and Interfacial Tension of Polymer Liquids—A Review, *Poly. Eng. Sci.*, 12(1), 1–11.

Hamaker, H. C. 1937. *Physika*, 4, 1058.

Heyman, E. 1935. Studies of Sol–Gel Transformation. I. The Inverse Sol–Gel Transformation of Methylcellulose in Water, *Trans. Faraday Soc.*, 31, 846.

Hilderbrand, J. H. and R. L. Scott. 1950. *The Solubility of Non-electrolytes*, Reinhold, New York, 402, 431.

Hunter, R. J. 1981. *Zeta Potential in Colloid Science; Principles and Applications*, Academic Press, New York.

Jeffrey, D. J. and A. Acrivos. 1976. The Rheological Properties of Suspensions of Rigid Particles, *AIChE J.*, 22(3), 417–431.

Kitano, T., T. Kataoka, and T. Shirota. 1981. An Empirical Equation of the Relative Viscosity of Polymer Melts Filled with Inorganic Fillers, *Rheol. Acta*, 20, 107–209.

Kraus, G. and J. T. Gruver. 1965. *J. Polymer. Sci.*, A(3) 105.

Krieger, I. M. and T. J. Dougherty. 1959. A Mechanism for Non-Newtonian Flow in Suspensions of Rigid Spheres, *Trans. Soc. Rheol.*, III, 137–152.

Lewis, T. B. and L. E. Nielsen. 1968. Viscosity of Dispersed and Aggregated Suspensions of Spheres, *Trans. Soc. Rheol.*, 12(3), 1421–443.

Lindquist, K., E. Carlstrom, M. Perrson, and R. Carrlson. 1989. Organic Silanes and Titanates as Processing Additives for Injection Molding Ceramics, *J. Am. Ceram. Soc.*, 72(1), 99–103.

Manley, R. S. J. and S. G. Mason. 1952. Particle Motions in Sheared Suspensions II: Collision of Uniform Sphere, *J. Coll. Sci.*, 7, 354–369.

McNamara, E. P. and J. E. Comeforo. 1945. Classification of Natural Organic Binders, *Am. Ceram. Soc. Bull.*, 28(1), 25–31.

Mooney, M. 1931. Explicit Formulas for Slip and Fluidity, *J. Rheol.*, 2(2), 210–222.

Mooney, M. 1951. The Viscosity of a Concentrated Suspension of Spherical Particles, *J. Coll. Sci.*, 6, 162–170.

Moteki, A. 1959. Studies on the Factors Affecting the Workability (Fluidity) of Resins Used for the Injection Molding of Ceramics, *J. Ceram. Assoc. Japan*, 67(2), 387–399.

Mutsuddy, B. C. 1994. Measurement of Adhesion Tension of Liquid Processing Aids in Contact with Powders, *Am. Ceram. Soc. Bull.* (to be published).

Napper, D. H. 1983. *Polymeric Stabilization of Colloidal Dispersions*, Academic Press, New York.

Napper, D. H. and A. Netschey. 1971. Stearic Stabilization of Colloidal Particles, *J. Coll. Interface Sci.*, 37(3), 528–535.

Onoda, G. Y. 1978. The Rheology of Organic Binder Solutions, in *Ceramic Processing before Firing*, ed. G. Y. Onoda and L. L. Hench, John Wiley & Sons, New York, 235.

Onogi, S., M. Takayoshi, and Y. Warashina. Rheological Properties of Dispersions of Polymer Particles in Polymer Solutions, *Trans. Soc. Rheol.*, 17(1), 175–190.

Parfitt, G. D. 1969. *Dispersions of Powders in Liquids*, American Elsevier, New York, 316.

Patton, T. C. 1970. Reflections of a Paint Engineer on Paint Flow, Interface Physics and Pigment Dispersion, *J. Paint Tech.*, 42(551), 666–694.

Rabinowitsch, B. 1929. *Z. Physik. Chem.*, A(145), 1.

Rees, D. A. 1972. Polysaccharide Gels—A Molecular View, *Chem. and Ind.*, 630–636.

Reynolds, O. 1885. *Phil. Mag.*, 8, 20.

Rivers, D. 1978. Method of Injection Molding Powder Metal Parts, U.S. Patent, 4,113,480.

Saito, K. T. Tanaka and T. Hibino. 1976. Method of Producing a Ceramic Article by Injection Molding, U.S. Patent, 4,000,110, Dec. 28.

Sarkar, N. 1979. Thermal Gelation Properties of Methyl and Hydroxypropyl Methylcellulose, *J. Appl. Polym. Sci.*, 24(4), 1073.

Sarkar, N. and G. K. Greminger, Jr. 1983. Methylcellulose Polymers as Multifunctional Processing Aids in Ceramics, *Am. Ceram. Soc. Bull.*, 62(11), 1280–1284, 1288.

Serwer, P. 1983. Agarose Gels: Properties and Use for Electrophoresis, *Electrophoresis*, 4, 375–382.

Strivens, M. A. 1963. Injection Molding of Ceramic Insulating Materials, *Am. Ceram. Soc. Bull.*, 42(1) 13–19.

Tanaka, T. et al. 1985. Injection Molding of Alumina, *Yogyo-Kyokai-shi*, 93(9), 572–576.

Thomas, M. S. and J. R. G. Evans. 1988. Non-uniform Shrinkage in Ceramic Injection Moulding, *Brit. Ceram. Trans. J.*, 87, 22–26.

Treischel, C. C. and E. W. Emrich. 1946. Study of Several Groups of Organic Binders under Low-Pressure Extrusion, *J. Am. Ceram. Soc.*, 29(5), 129, 132.

Vand, V. 1948. Viscosity of Solutions and Suspensions, I and II, *Phys. Chem.*, 52, 227–299, 300–314.

Verwey, E. J. W. and J. Th. G. Overbeek 1948. *Theory of the Stability of Lyophobic Colloids*, Elsevier Science Publishers, New York.

Whalen, T. J. and C. F. Johnson 1981. Injection Molding of Ceramics, *Am. Ceram. Soc. Bull.*, 60, 216–220.

4

Mixing

The mixing method is the key to achieving a homogeneous mix—one free of agglomerates that contains the optimum content of filler powder and binder while still maintaining sufficient fluidity for molding. Although mixing is a critical process step for producing high-quality injection molded parts reliably, it generally is regarded as a somewhat simple operation. In part this attitude is due to the relatively crude mixing methods that can become effective if the operation is carried out over a sufficiently long period of time.

The design of an appropriate mixing system for injection molding includes:

- Optimizing the mixing conditions.
- Determining the factors that affect the performance of the mixing device.
- Identifying the causes of segregation in mixtures.
- Ascertaining when a homogeneous mix is obtained.

This chapter covers information needed for optimization of mixtures, the analysis of mixing processes[1] and the characterization of mixtures.

MIXING PROCESSES

Mixing involves the transport of the material in the mixture to produce the desired spatial arrangement of the individual components. Transport mech-

[1]Mixing processes are inherently difficult to analyze. The following references are suggested for further exploration of this subject: Frados (1976), Funt (1976), Middleman (1977), and Tadmor and Gogos (1979).

anisms, the primary mechanisms associated with any mixing process, can be grouped into three categories, as described below.

Diffusive Mixing

In general, if two materials are placed in a container, after a sufficiently long time has elapsed the molecules of both materials will commingle as a consequence of both the concentration gradient and random molecular motion. The result will be the formation of a uniform mixture on a submicron scale by molecular diffusion. The type of diffusion process usually encountered in practical mixing situations results from velocity gradients within the fluid. This diffusion process by itself, however, is too slow for the preparation of ceramic molding mixes.

Laminar Mixing

In nondiffusive mixing, the mixing components must be put into motion by some external agent. When the fluid viscosity is high, as in ceramic molding mixes, it usually is impossible to achieve turbulence. In this case, laminar mixing must be used although it operates at a lower rate.

The term "laminar" is applied to any operation that increases the randomness of the spatial distribution of the binder constituents or particles in a system without reducing their size. In laminar mixing, the binder constituents pass from a less to a more homogeneous state.

Dispersive Mixing

An important application of dispersive mixing is the incorporation of the ceramic powder into the molten organic binder. During dispersion the agglomerates of the constituent elements of the secondary component of the mixture are ruptured.

Whenever there is relative motion between a fluid and a particle, drag forces act on the particle's surface. If the fluid velocity is greater than the particle velocity, these drag forces tend to accelerate the particle. If the particle velocity is greater than the fluid velocity, they tend to retard its movement. If the particles are spherical and if the Reynolds number[2] for the flow about a particle is low, then according to Stokes' law, the drag force exerted on the particle is equal to $6\pi R\eta V$, where V is the relative velocity

---

[2]The Reynolds number

$$R = \frac{\rho V d}{\eta}$$

where ρ is the density of the fluid, V is the velocity of the fluid, d is the diameter of the tube, and η is the viscosity of the fluid.

between the fluid and the particle, R is the Reynolds number, η is the viscosity of the suspension, and 6π is a geometric factor for a very large diameter tube.

Shear stresses influence the mixing:

- High shear stresses promote dispersion in dispersive mixing.
- In any given system, there is some stress, defined as the critical stress, below which dispersion cannot occur. However, when the shear stress is just slightly larger than the critical stress, only the agglomerates with a favorable initial orientation will become dispersed.
- If the flow in the mixer is completely unidirectional, only the agglomerates with a favorable initial orientation will be ruptured. The others will be aligned with the flow, and no further dispersion will take place even with continued mixing. However, if the mixer causes the flow pattern continually to change direction, further dispersion can take place as the mixing continues. Eventually, the shear stresses created will position all the remaining agglomerates in an orientation favorable for rupture to occur.

The Effect of the Mixing Time

In a mixing operation, it is essential to know how long it will take to obtain a homogeneous mix, a period of time that has been termed a "degree of mixedness." Information about the rate of mixing will lead to an understanding of the fundamental mechanisms of mixing, to knowledge about the stability of the binder components, and to the ability to select the appropriate mixing device.

In the initial stages of mixing, the state of the binder components gradually changes from solid/liquid to semifluid, and ultimately to fluid. At this stage, the mixing mechanisms involve interdiffusion, mutual solubility of the binder components, and the interaction of the molten binder with the surfaces of the particles of the ceramic powder. Agglomerates form, both as a consequence of the change in interfacial tensions between the particles of the ceramic powder and the molten binder and as the entire mix gradually becomes a viscous mass. Subsequent mixing of this material is essentially dispersive (or shear).

Because mixing tends toward equilibrium, the rate (X) at which it attains this state can be expressed as an exponential function of time:

$$X = 1 - e^{ct} \tag{4-1}$$

where c is the rate constant, and t is the mixing time. As shown in Figure 4-1, X approaches its limiting value asymptotically. Plotting $-\log_e (1 - X)$

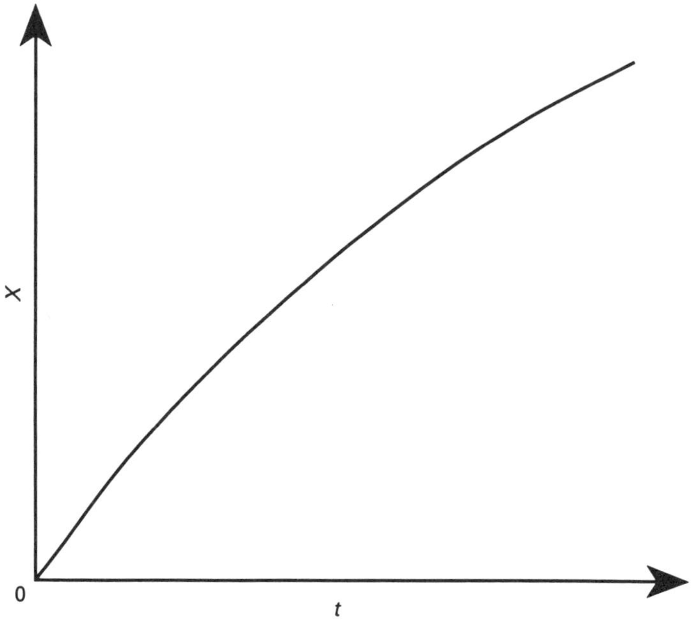

FIGURE 4-1 The rate of attainment of complete mixing.

(from equation 4-1) as a function of time produces a straight line with a slope equal to the rate constant, *c*. The value of this constant is one indication of a mixing machine's efficiency.

Using process tolerance capability as an indicator of mixing efficiency, the mixing operation's dependence on time can be demonstrated, as shown in Figure 4-2 (Billiet 1990). Insufficient mixing time (see Figure 4-2, Region I) adversely affects the process tolerance capability and there is an optimum mixing time that maximizes this capability. In other words, the tolerance that can be maintained with confidence reaches an optimum value, and prolonging the mixing beyond this period of time (see Figure 4-2, Region II) adversely affects the process tolerance capability.

Both temperature and speed (rpm) can strongly affect the mixing mechanism, as shown in Figures 4-3 and 4-4 (Dow et al. 1988). Also the oxidation in air with the mixing time for various binder components has been verified by using gel-permeation chromatography and infrared spectroscopy (Hunt et al. 1988; Takahashi et al. 1988).

The Mixing Technique

Commercial mixers generally are classified according to the mixing method or to the motion of the mixer.

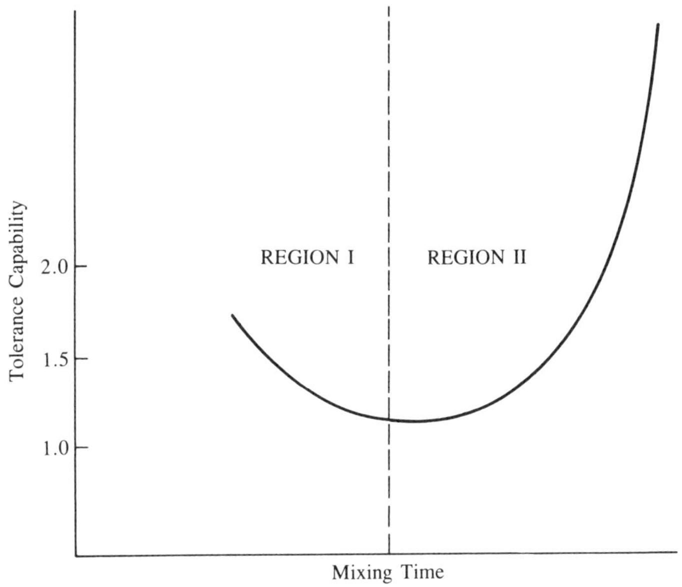

FIGURE 4-2 The dependence on time of the mixing operation. After Billiet (1990).

Three types of mixers conventionally are used to prepare ceramic injection molding mixes. Dispersive mixing is carried out primarily in Banbury mixers and roll mills. It is a batch operation although some mixing has been done on a continuous basis with specially designed extrusion equipment.

Internal mixers are batch machines for uniformly dispersing ceramic powder within the binder. A typical internal mixer, shown in Figure 4-5, consists of a figure-eight-shaped chamber that fits over two sigmoidal (S-shaped), counterrotating blades. The process of internal dispersion mixing can be demonstrated by means of a series of batch-mixing experiments involving the use of three mixing heads—sigmoidal, cam, and roller (shown in Figure 4-6)—in a torque rheometer machine.

The progress of the mixing can be followed in the torque rheometer (Figure 4-7a) to monitor the torque required to produce a uniform mix. A typical response is shown in Figure 4-7b. The mixing torque, which is proportional to the shear stress of the mix, is an indication of the work required to mix the ingredients; and typically it shows a rapid rise when the binder is added to the ceramic powder. The irregular peaks (in Figure 4-7b) indicate the melting and consolidation steps. Once the torque reaches a

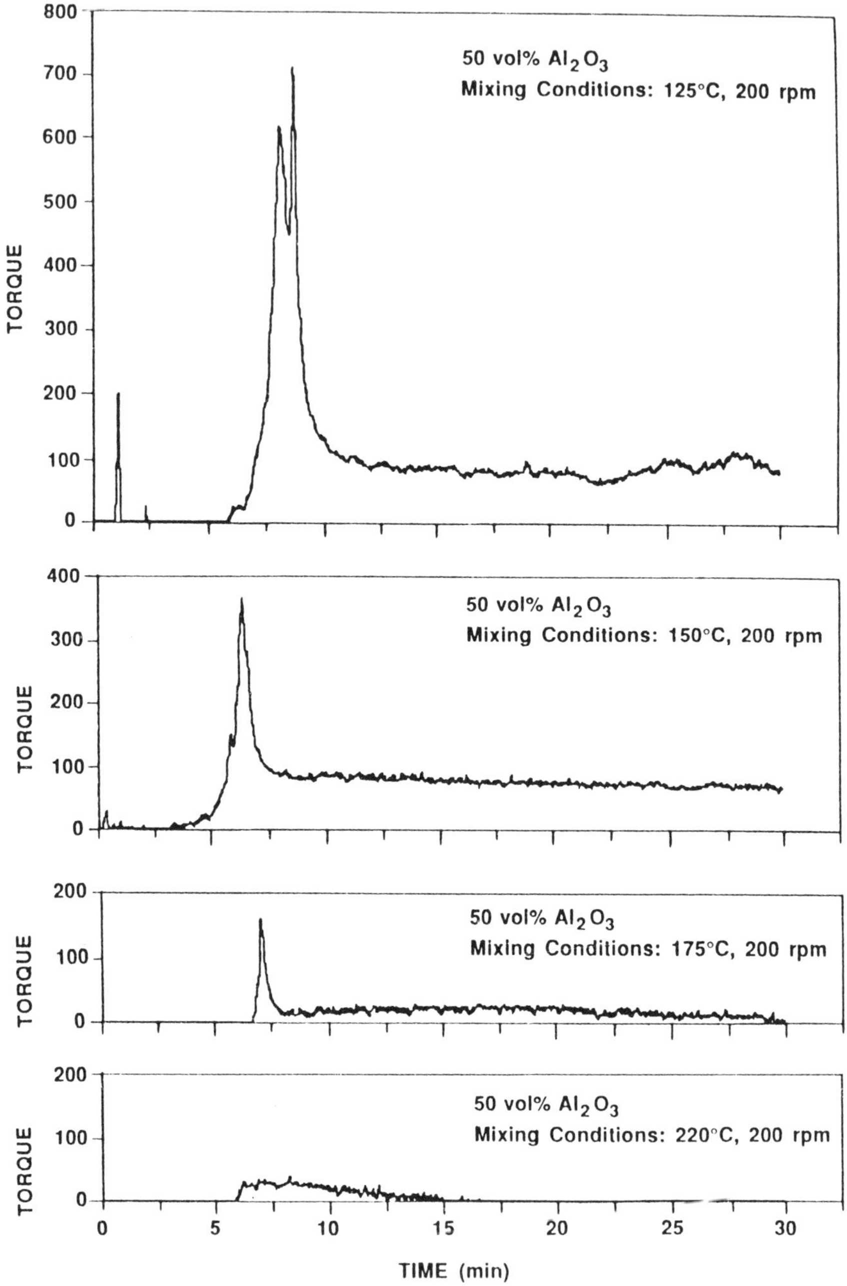

FIGURE 4-3 Torque as a function of mixing temperature. After Dow et al. (1988).

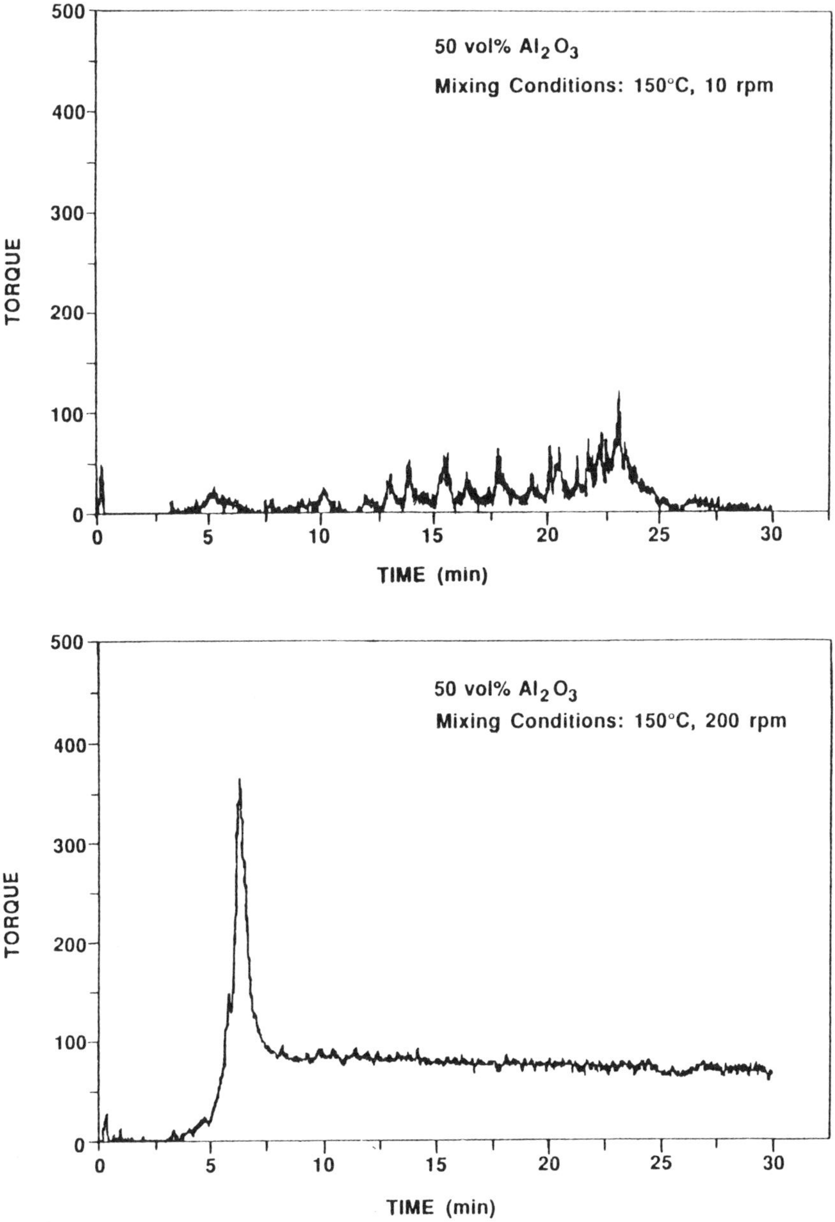

FIGURE 4-4 Torque as a function of mixing speed (rpm). After Dow et al. (1988).

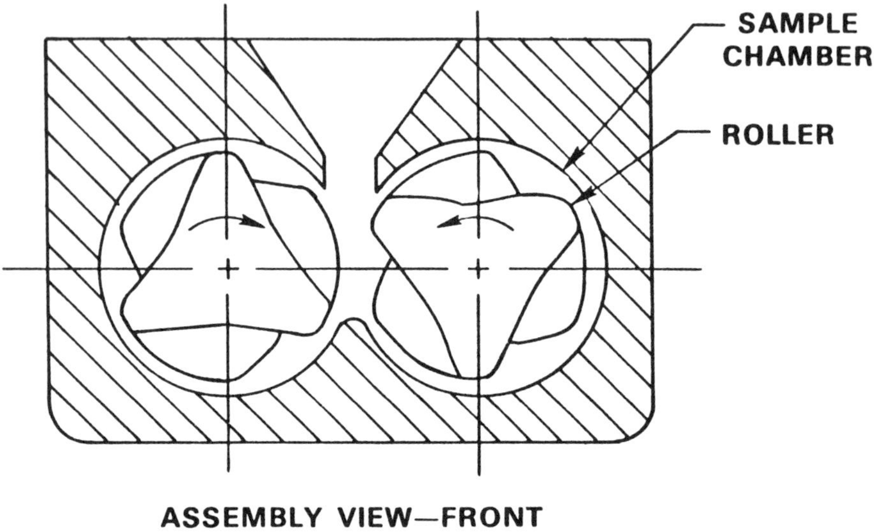

FIGURE 4-5 Schematic cross-sectional view of an internal mixer.

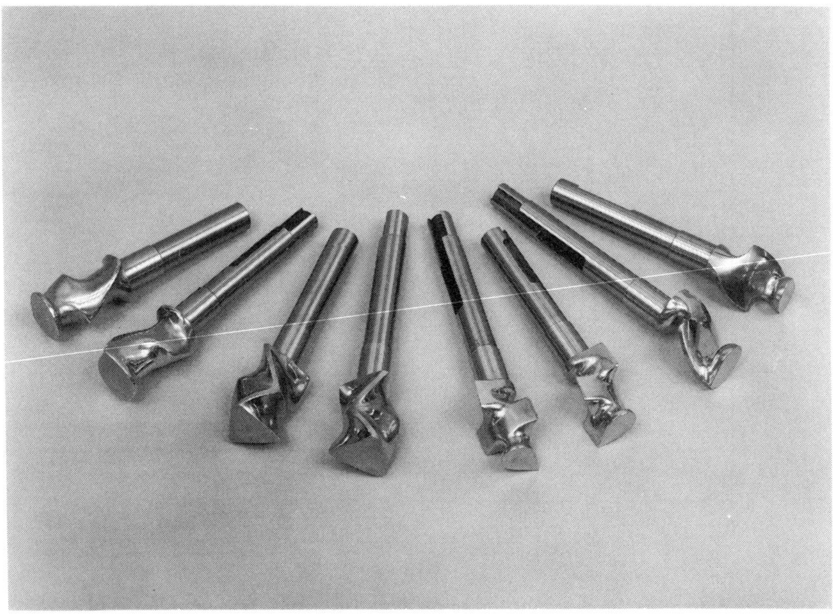

FIGURE 4-6 Interchangeable mixing heads. Courtesy of C. W. Brabender Instruments Inc.

FIGURE 4-7(a) Torque rheometer. Courtesy of C. W. Brabender Instruments Inc.

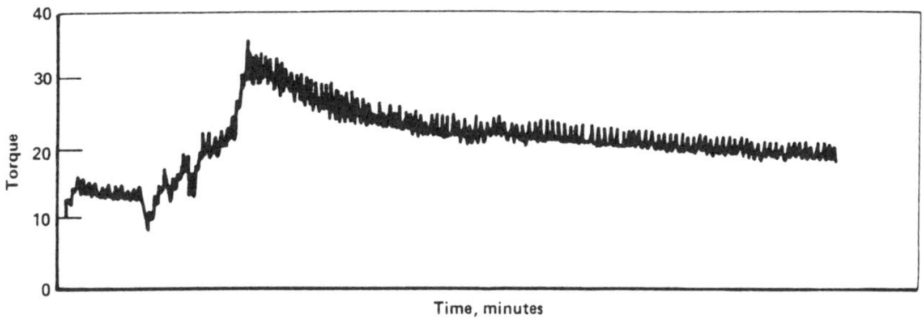

FIGURE 4-7(b) Typical torque response in mixing a ceramic powder and wax-type binder. From Battelle Columbus, OH.

steady state, which indicates that the viscosity of the mix is uniform, no additional mixing takes place. The mixing time is defined as the time required for a mix to reach this steady state.

The results obtained in using different mixing heads with a batch type mixer indicate that the breakdown of the agglomerates formed during mixing depends largely on the geometry of the mixing head rather than on the speed, time, or temperature. The best results are obtained with a roller mixing head. However, if ceramic powder premilled with a processing aid first is briefly mixed at a temperature where the entire mix is fluid, followed by mixing at a lower temperature where the shear action becomes effective, the number of agglomerates is markedly reduced. The improvement in deagglomeration achieved in this way is independent of the mixing head geometry.

The agglomerates in a mix also can be substantially reduced by altering the sequence followed in introducing various ingredients. Sequence selection is strongly influenced by the surface-charge characteristics of the powder and by its compatibility with different binder ingredients. Data on oil absorption, contact angle, and electrophoretic mobility are all used to determine a powder's surface charge and its compatibility with the binder. The mixer's modes of action can be interpreted according to the following mathematical analyses.

Dispersive mixing takes place in the flight clearances, which are labeled h_1 and h_2 in Figure 4-8. The relationship of the volume available for dispersive mixing to the dimensions of the cavity is given by:

$$V_2 = l_{\text{cav}} h_2 l_2 \tag{4-2}$$

where l_{cav} is the length of the cavity, h_2 is the clearance between the flight on each rotor and the cavity wall, and l_2 is the length of the rotor.

Laminar mixing takes place mainly in the channels between the rotor and the cavity wall. The relationship of the volume available for laminar mixing to the dimensions of the cavity is given by:

$$V_1 = \pi l_{\text{cav}} h_1 d_{\text{cav}} \tag{4-3}$$

where l_{cav} is the length of the cavity, h_1 is the clearance between the flight on each rotor and the cavity wall, and d_{cav} is the diameter of the cylindrical cavity.

Distributive mixing in an axial direction takes place partly in the channels, because the flights are not purely axial, and partly in any free volume remaining above the intermeshing region.

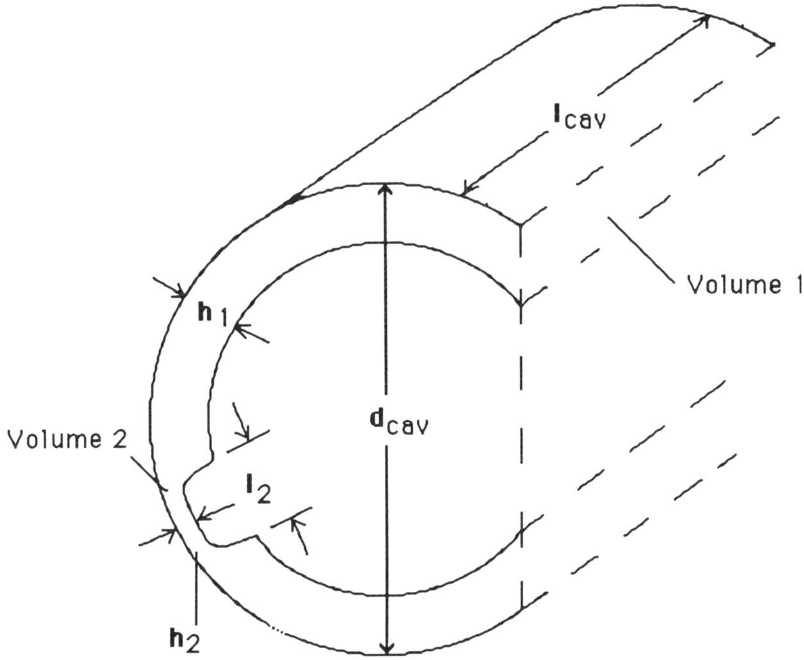

FIGURE 4-8 Idealized dimensions for an internal mixer. After J. R. A. Pearson (1985).

Depending on the volume actually loaded in the batch, the relationship of the free volume, V_0, to the dimensions of the rotors and the cavity is given by:

$$V_0 \leq l_{cav} d_2 \tag{4-4}$$

where V_0 is the free volume, l_{cav} is the length of the cavity, and d_2 is the diameter of the rotor.

The operating variables for a typical internal mixer are the ratio of the volume of the mix to the volume of the mixer, the sequence in which the ingredients are added and the rate at which they are added, the heat-exchanger-fluid temperature, the rotor speed, and the mixing time.

By using roll mixing, a very homogeneous mixture of macro-defect-free (MDF) cement[3] at high solids loading has been achieved (Birchall et al.

[3]MDF cements are hydraulic cement-water-soluble polymer composites with low water-to-cement ratios. They are dense chemically bonded materials that are macro-defect-free, that is, without the relatively large internal voids found in conventionally mixed pastes.

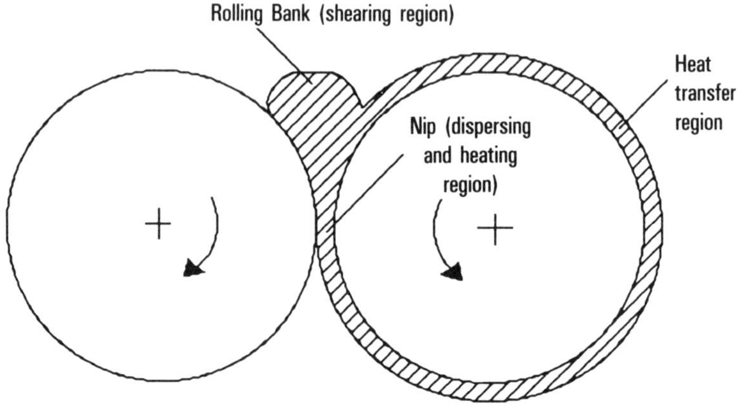

(a)

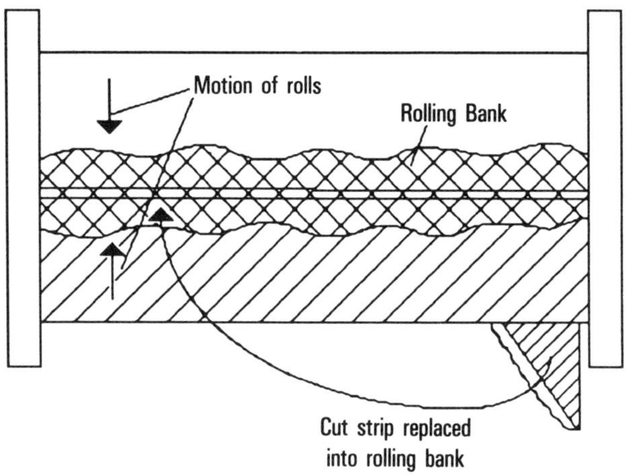

(b)

FIGURE 4-9 (a and b) Cross-sectional view of a twin-roll mill showing the large rolling bank. After J. R. A. Pearson (1985).

1981). A twin-roll mill is shown schematically in Figure 4-9. Roll mixing involves passing a mixture of ceramic powder and binder through a relatively wide nip (1 – 10 mm) to disperse and warm the mixture; slabs of material for subsequent molding are produced. The system consists of a parallel pair of rigid rotating cylindrical rolls. Relative to the diameter of these rolls, their length is large and their minimum separation very small.

Both single- and twin-screw extruders are used for mixing, but the latter, shown in Figure 4-10, are more universally used. Both types can be found in a wide variety of channel geometries. These extruders involve interruptions to the screw flight, reversals of helicity, barriers in the main channel, grooves in the barrel, and outsize clearances.

FIGURE 4-10 A twin-screw extruder. Courtesy of C. W. Brabender Instruments Inc.

A number of the geometries in a helically symmetric system are designed to ensure that none of the material contained in the center of the C-shaped regions (see Figure 4-11) will pass from the feed pocket to the nose end of the screw without being mixed. Interruptions to the screw flight promote distributive mixing in the channel cross-section, which effects an exchange of material among different positions along the channel length. Barriers in the main channel provide regions of dispersive mixing through which each element of the material must pass. This arrangement aids in achieving a homogeneous mix.[4] An additional advantage of continuous mixers is their lower residence time, which reduces the possibility of binder degradation.

The Agglomerate Problem

One of the most persistent problems encountered in achieving a well-dispersed, homogeneous mix is the presence of large numbers of agglomerates,

[4]For additional information about the flow characteristics of these types of extruders see: Irving and Saxton (1967), Uhl and Gray (1967), and Tadmor and Gogos (1979).

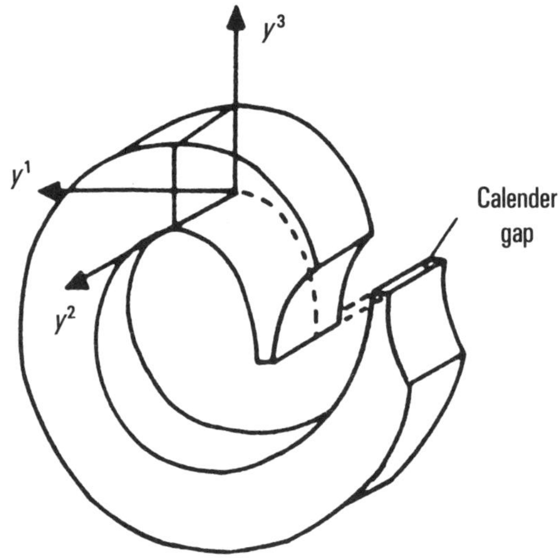

FIGURE 4-11 A C-shaped chamber represents a section of the helical channels on a screw. After J. R. A. Pearson (1985).

which normally are associated with any fine ceramic powder. As discussed in Chapter 2, electrostatic forces, van der Waals forces, and hydrogen bonding from adsorbed water all can lead to the agglomeration of fine particles. The first step in minimizing their formation is to remove as much moisture as possible from the powder before preparing the mix. Also it is important to premill the powder with a processing aid that causes the agglomerates to break down.

Certain fatty acids, such as oleic acid and stearic acid, and oils, such as menhaden oil, Kemamide E, or W-910 (discussed as processing aids in Chapter 3), have been found to be quite effective in breaking down agglomerates during milling. However, they are less effective if added during the mix-preparation step along with the ceramic powder and other binder ingredients.

Agglomerates also can form during the mixing process as different binder ingredients are introduced. They are formed both as a result of the change in interfacial tensions between the powder particles and the melted binder ingredients and during the gradual conversion of the mix into a viscous mass.

The maximum force that tends to divide a dumbell-shaped agglomerate in a fluid with viscosity η is given by (Tadmor and Gogos 1979):

$$F_{\max} = 3\pi\eta\dot{\gamma}r_1r_2 \tag{4-5}$$

where r_1 and r_2 are the radii of the two particles composing the agglomerate, and $\dot{\gamma}$ is the rate of change of shear experienced by the fluid that occurs when the dumbell's axis is at an angle of 45° with respect to the direction of the shear. Equation (4-5) indicates that the maximum force acting on a dumbell is proportional to the shear stress ($\eta\dot{\gamma}$) and the product r_1r_2. Thus, better mixing is achieved at high shear stresses.

The Effect of Error in Mixing

Error in mixing has been shown to critically affect feedstock composition, resulting in the loss of process tolerance capability, T. Assuming that no air has been entrapped in the feedstock, then the shrinkage factor, K, is related to the inverse cube root of the volumetric loading, P (Billiet 1990):

$$K = P^{-\frac{1}{3}} \tag{4-6}$$

K also is related directly to the dimensions of the mold cavity and inversely to the dimensions of the final injection molded part:

$$K = \frac{L_o}{L} \tag{4-7}$$

and:

$$P = \frac{V_d}{(V_d + V_c)} \tag{4-8}$$

where L_o is the dimension of the mold cavity; L is the corresponding dimension of the final product; V_d is the volume of the discrete phase of the feedstock, that is, the volume occupied by the powder; and V_c is the volume of the continuous phase, that is, the volume occupied by the organic binder.

The derivatives of L with respect to the volume of the continuous phase, V_c, and the volume of the discrete phase, V_d, provide the corresponding variation in dimension:

$$\frac{dL}{dV_c} = \frac{-LP}{3V_d} \quad \text{and} \quad \frac{dL}{dV_d} = \frac{LPV_c}{3V_d^2} \tag{4-9}$$

Replacing the volume, V, by weight, W, divided by density, ρ, gives:

$$\frac{dL}{L} = \frac{-P\rho W_c}{3V_d\rho_c} = \frac{PV_c\rho W_d}{3V_d^2\rho_d} \tag{4-10}$$

$$\frac{dW}{W} = \frac{\pm 3}{1-P} \cdot \frac{dL}{L} \tag{4-11}$$

$$\frac{dW}{W} = E_f \quad \text{and} \quad \frac{dL}{L} = \frac{T}{2} \tag{4-12}$$

$$| \ E_f \ | = \frac{3T}{2\,(1-P)} \tag{4-13}$$

where E_f is the maximum mixing error, i.e., that error in a mix composition that would result in the loss of the targeted process tolerance capability, T. This equation indicates the importance of taking particular care in weighing and handling the ingredients of the mix during the mixing stage.

ASSESSING THE QUALITY OF THE MIX

It is essential to achieve a uniform distribution of all the binder ingredients throughout the mix. In order to characterize the texture of the mix completely, it is necessary to specify the size, the shape, and the concentration of the minor components plus the location of each agglomerate. In practice, this is accomplished by using statistical methods. Two statistically defined quantities have been proposed that are analogous to the statistical theory of turbulence: the scale of segregation as a measure of the size of the agglomerates and/or the size of each binder ingredient, and the intensity of segregation as a measure of their concentration (Danckwerts 1952).

Both the scale and the intensity of segregation can be used as indicators of the homogeneity of a mix, either individually or combined. Frequently they are more useful in combination. At present, there are no instruments for making routine quantitative assessments of a mix's quality. For laminar or dispersive mixing, a mathematical method can be applied that gives a statistical measure of the maximum homogeneity in a mix by analyzing individual samples withdrawn from the mix. (See Appendix 4-A for an explanation of this statistical method.)

The measurement of mixedness is fraught with both experimental and sampling errors. Consequently, it often is not possible to verify adequately the validity of the theoretical criteria for the degree of mixedness. A computer program that uses a numerical technique, the Monte Carlo method, to simulate the mixing process has proved useful. The Monte Carlo method involves sampling theoretically or empirically derived statistical distributions to approximate real physical phenomena without referring to the actual physical systems.[5]

THE PREPARATION OF THE FEED MATERIAL

Preparation of the feed material for injection molding is the final step in the mixing process. Usually it is prepared by granulating the cold mix in a rapidly rotating knife granulator. Granulation produces a nonuniform material that contains a large fraction of fine particles less than 150 μm in diameter. The presence of such fine particles in the mix can cause bridging and packing at the feed end of the injection molding machine, which can become a serious problem with a plunger-type machine. In such machines, the discharge of the packed material from the barrel may necessitate complete dismantling of the machine. In addition, granular feed material tends to entrap a quantity of air that cannot be released inside the machine. Consequently, this entrapped air moves with the soft moldable material into the cavity, where it becomes locked in the molded product as voids.

In the plastics industry, it is customary to use uniformly sized pellets that are free of entrapped air, which are produced with a kneader-extruder. This machine combines the mixing action of a double-arm mixer with the convenience of an extrusion screw for discharging the viscous material. During operation, the mixing blades operate tangentially to each other in a horizontal trough. The blades rotate either toward or away from each other at differential speeds. The extrusion discharge screw, located in the cavity beneath the mixing blades, also rotates during both the mixing and the discharge cycles. During the mixing cycle, the screw continuously moves materials within reach of the mixing blades. Once this cycle is completed, the discharge screw is reversed to empty the machine. A commercial kneader-extruder is shown in Figure 4-12.

Figure 4-13 is a micrograph made to ascertain the extent of deagglomeration in a mix prepared in this type of kneader-extruder. It shows that the agglomeration was effectively reduced even without premilling of the alumina powder with a processing aid.

---

[5]See Tocher (1963) and Hammersley and Handscomd (1964) for a discussion of the Monte Carlo method.

FIGURE 4-12 A commercial kneader-extruder. Courtesy of Ross Machineries.

Contamination

At present no reliable data exist concerning the potentially serious problem of contamination of the mix by metals during the mixing process. The major contaminants are the transition metals—for example, iron, manganese, and nickel—that are commonly present in hardened steel and tool steel. Their presence can have a detrimental affect on sintering as well as on the properties of the finished ceramic product. Efforts are being made to minimize the problem of mixer wear by using improved alloys and by coating the mixing blades with hard refractory metal or ceramic coatings.

Loading the Ceramic Powder Mix

In formulating injection molding mixes, the objective is to attain the optimum binder content for molding. This optimum content is a compromise that balances maintaining the rheological conditions necessary for molding

FIGURE 4-13 A mix prepared in a kneader-extruder.

with minimizing the amount of binder. According to guidelines for determining binder concentrations, the binder content should be at least 2 vol% greater than the void volume (Bailey 1966). However, in order to create what is, in effect, a binder matrix in which the ceramic particles are suspended, the binder content can be as high as 105 to 115% of the void volume. Clearly, in seeking a practical method for determining the optimum binder content for a given powder, a wide range of possible binder concentrations is available for investigation.

The Critical Powder Volume Concentration

An empirical equation has been proposed that defines an effective liquid content, $\overline{V}_L$, in a mix that can be directly related to the critical powder volume concentration. This equation expresses the dependence of the relative viscosity, η_r, on the concentration of the solids in the mix, $\overline{V}_L$:

$$\eta_r = \alpha \overline{V}_L^{-\beta} \tag{4-14}$$

where the value of the constant α is 1 because the viscosity of the mixture is the same as that of the resin used in the mix when the effective liquid concentration is unity ($\overline{V}_L = 1$). The parameter β is dependent on the wettability between the resin and the powder.

It has been suggested that the maximum volume fraction of solids after pyrolysis (V^*_m) is a measure of the maximum packing efficiency, which is in agreement with the critical powder volume concentration (Wright et al. 1990). It has been further suggested that even if the thickness of the binder film between particles is as small as 10 molecular layers, viscous flow still is possible.

The most practical way to evaluate the critical powder volume concentration is a modified American Society for Testing and Materials oil absorption test, ASTM D-281-31, (Bierwagen 1972) using a torque rheometer. (See Appendix 4-B for an explanation of this test.) With this method the wetting characteristics of certain processing aids on ceramic powders also can be investigated. For the purposes of this discussion, the oil oleic acid has been selected.

Figure 4-14 shows changes in torque with incremental additions of oleic acid to various powders. As can be seen, as the oleic acid is added, the torque increases up to a maximum and then decreases. The optimum oleic acid content is the amount added at the maximum value of the torque. The critical particle volume concentration (CPVC), calculated from these experimental results, is summarized in Table 4-1. (The equation used for these calculations is given in Appendix 4-B.) Both the binder content used for preparing the mixes and the binder content calculated from the CPVC data for each powder show good agreement with these values for actual moldable compositions.

In addition to determining the CPVC, this oil absorption test can serve as a useful indicator of the absorption surface characteristics of the powder. This is shown in Figure 4-14 in which the torque shows only gradual increases up to 12 volume percent oleic acid, then with the addition of only a small amount of oleic acid the torque increases sharply. Such a sharp rise suggests complete wetting and saturation of the surfaces of the powder particles. The onset and completion of the sharp rise in the torque will vary with different powders (see Figure 4-15), depending on the available surface area and the characteristics of the powder surface.

The Effect of Ceramic Powder Loading on Shrinkage

Both green density and shrinkage after sintering of an injection molded part can be related to the volume fraction of ceramic powder in the mix as:

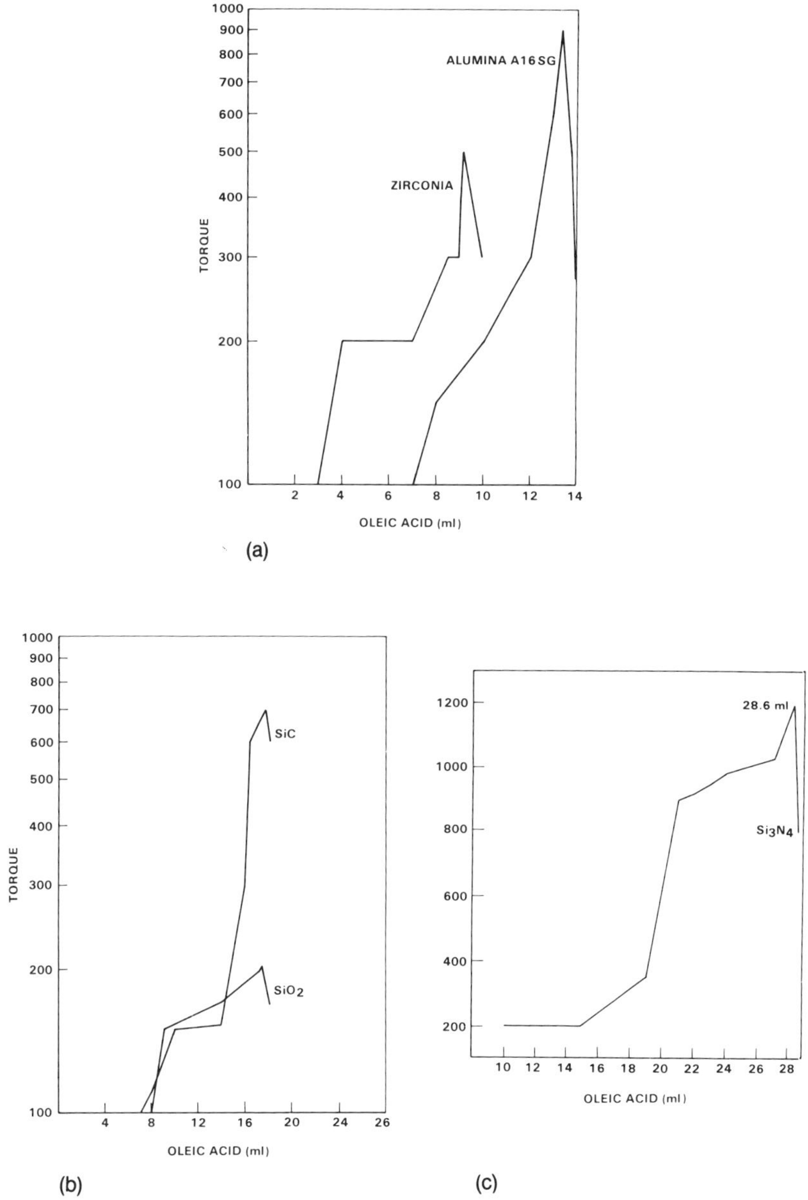

FIGURE 4-14 Changes in torque with incremental additions of oleic acid to various powders: (a) with A16 SG alumina and with zirconia; (b) with silicon carbide and with silica; (c) with silicon nitride. From Battelle Columbus, OH.

Table 4-1. Binder content for various powders.

Ceramic powder	CPVC	Binder content calculated from CPVC, wt%	Binder content from actual molding formulations, wt%
Al_2O_3 A16-SG	65.0	12.0	12.5–13.5
ZrO_2	71.6	6.5	7.0–8.0
SiO_2	56.6	22.5	20.0–21.0
Si_3N_4			
As-received	35.0	37.0	—
Milled	53.3	19.7	19.0–20.0
Si			
As-received	60.3	22.0	—
Milled 24 hr	68.8	16.3	15.0–15.5

From Battelle Columbus, OH.

$$\rho_G = \frac{V_p}{V_m} \cdot SG_p \tag{4-15}$$

where ρ_G is the green density of the molded parts measured in g/cm^3, V_p is the volume fraction of the powder measured in cm^3, V_m is the total volume fraction in the mix measured in cm^3, and SG_p is the specific gravity of the powder measured in g/cm^3.

The shrinkage during sintering, $\Delta L/L$, is related to the volume fraction of powder in the molding mix (Mangels and Trela 1984):

$$\frac{\Delta V}{V} = \frac{\rho_G}{\rho_{TD}} - \frac{\rho_S}{\rho_{TD}} = 3\left(\frac{\Delta L}{L}\right) - 3\left(\frac{\Delta L^2}{L}\right) + \frac{\Delta L^3}{L} \tag{4-16}$$

where $\Delta V/V$ is the shrinkage in volume, ρ_S is the sintered density, and ρ_{TD} is the theoretical density. The ratio of ρ_G/ρ_{TD} is equal to the volume fraction of solids, V_p/V_m (Mangels and Trela 1984).

Figure 4-16 illustrates the relation between the volume fraction of the powders and the sintering shrinkage, for various sintered densities.

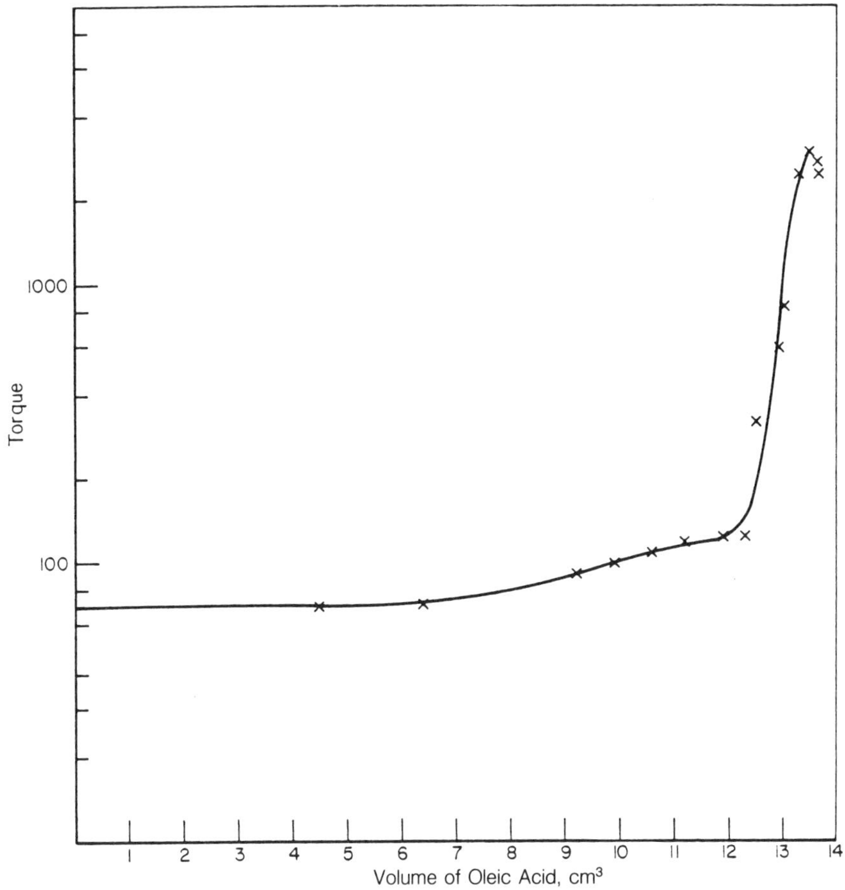

FIGURE 4-15 Torque tracing of the absorption of A16-SG alumina powder by oleic acid.

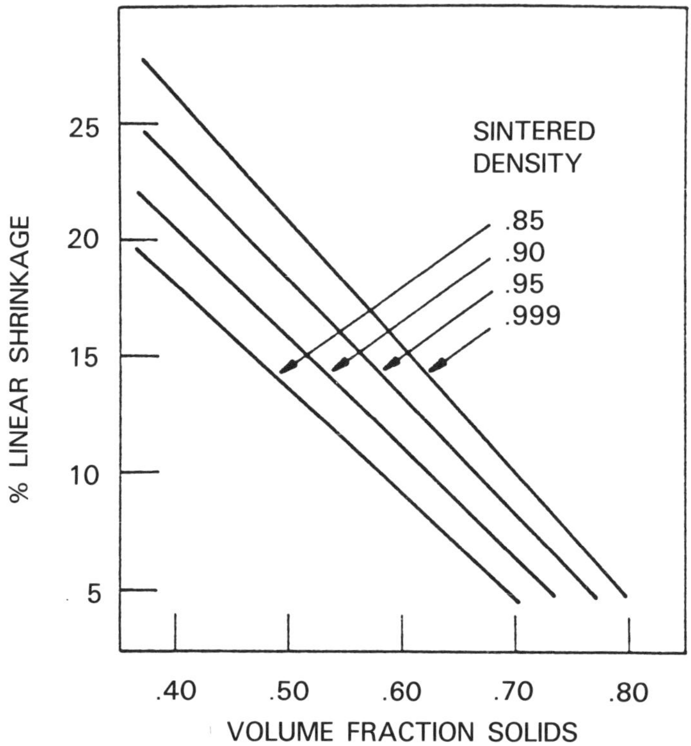

FIGURE 4-16 The effect of the volume fraction of powders on the sintering shrinkage of sintered ceramics. After Mangels and Trela (1984).

APPENDIX 4-A STATISTICAL METHOD FOR DETERMINING THE HOMOGENEITY OF A MIX

The degree of homogeneity attained can be determined in two ways:

- From the standard deviation, from the perfectly mixed material, of the compositions of the samples withdrawn after the specified homogenizing time.
- From the "degree of mixing," which can be expressed as the ratio of the actual standard deviation to the mean arithmetic value.

The arithmetic mean, a familiar statistical value, is calculated from the sum and the number of all the individual values according to the following equation:

$$M = \frac{a_1 + a_2 + a_3 \cdots a_n}{n} = \frac{\Sigma a}{n} = \frac{1}{n} \cdot \Sigma a \qquad \text{(4-A-1)}$$

where M is the mean arithmetic value, a is an individual value, and n is the total number of individual values.

The variance is found by summing the squares of all deviations from the mean multiplied by the number of the individual samples:

$$Q^2 = \frac{1}{n-1} \cdot \Sigma d^2 \qquad \text{(4-A-2)}$$

where Q is the variance, $\Sigma\, d^2$ is the sum of the squares of all deviations, and n is the total number of individual samples.

The standard deviation, which is the square root of the variance, is calculated as:

$$s = \sqrt{Q^2} \quad \text{or} \quad S = \sqrt{\frac{\Sigma d^2}{n-1}} \qquad \text{(4-A-3)}$$

where S is the standard deviation, Q is the variance, $\Sigma\, d^2$ is the sum of the squares of all deviations, and n is the number of individual samples.

The following relation is useful for judging the homogeneity of a mix:

$$H_m = \frac{S}{M} \cdot 100 \qquad \text{(4-A-4)}$$

where H_m is the homogeneity of the mix, S is the calculated standard deviation, and M is the mean arithmetic value.

APPENDIX 4-B METHOD FOR DETERMINING THE CERAMIC POWDER LOADING FOR VARIOUS BINDER SYSTEMS

When formulating mixes for injection molding, the objective is to obtain the optimum binder content for molding. A practical method has been developed for determining the optimum binder concentration for a given powder without having to characterize the powder. This method, a modification of the ASTM D-281-31 oil absorption test, eliminates the operator variance factor by using a torque rheometer.

In this modified method, the powder is mixed with an oil, which is added at a constant rate, and the changing torque is recorded (Bierwagen 1972). The volume at which the maximum torque occurs—the critical powder volume concentration (CPVC)—is calculated as follows:

$$\text{CPVC} = \frac{\text{Volume of powder}}{(\text{Volume of powder} + \text{Volume of oil})} \tag{4-B-1}$$

Methodology

The following procedure is followed:

- Attach a roller-type mixing head to the rheometer.
- Load the mixing chamber, at room temperature, with a premeasured volume of dry powder while the blades are rotating at 50 rpm. To avoid inhomogeneous mixing, the chamber must be filled completely with powder.
- Add the oleic acid dropwise from a burette.
- Note the rapid increase in torque with the initial addition of oil.
- Allow the torque to reach a steady state with mixing time.
- Continue to add oleic acid.
- Note the maximum torque.
- Record the total volume of oleic acid consumed.
- Calculate the CPVC from the volume of the ceramic powder and the volume of oleic acid added at maximum torque.

References

Bailey, D. F. 1966. Molding Ceramic Composition, U.S. Patent 3,285,873.

Bierwagen, G. P. 1972. CPVC Calculations, *J. Paint Tech*, 44(574) 46–55.

Billiet, R. L. 1990. Injection Molding of Advanced PM Materials in SE Asia, *Metal Powder Report*, 45(5), 326–332.

Birchall, J. D. A. J. Howard, and K. Kendall, 1981. Cementitious Product, Imperial Chemical Industries, U.K., Eur. Patent 0021682.

Danckwerts, P. W. 1952. The Definition and Measurement of some Characteristics in Mixtures, Apply. Sci. Res. 3,279.

Dow, J. H., M. D. Sacks, and A. V. Shenoy. 1988. Dispersion of Ceramic Particles in Polymer Melts, *Trans.* 1(A), *Ceram. Powder Sci., Am. Ceram. Soc.*, 380–388.

Frados, J. (ed.) 1976. *Plastics Engineering Handbook*, 4th edn., Van Nostrand Reinhold, New York.

Funt, J. M. 1976. *Mixing of Rubber*, R.A.P.R.A. Shawbury, U.K.

Hammersley, J. M. and D. C. Handscomd, 1964. *Monte Carlo Methods*, Methuen & Son, London.

Hunt, K. N., J. R. G. Evans, and J. Woodthorpe. 1988. The Influence of Mixing Route on the Properties of Ceramic Injection Molding Blends, *Brit. Ceram. Soc.*, 380–388.

Irving, H. F. and R. L. Saxton. 1967. Mixing of High Viscosity Materials, in *Mixing Theory and Practice II*, ed. V. W. Uhl and J. B. Gray, Academic Press, New York, Chapter 8.

Mangels, J. A. and W. Trela. 1984. Ceramic Components by Injection Molding, in *Advances in Ceramics*, ed. J. Mangels, Am. Ceram. Soc., Columbus, OH., 9, 220–233.

Middleman, S. 1977. *Fundamentals of Polymer Processing*, McGraw-Hill, New York.

Pearson, J. R. A. 1985. Mechanics of Polymer Processing, Elsevier Applied Science Publishers, 532.

Tadmor, Z. and C. G. Gogos, 1979. *Principles of Polymer Processing*, SPE/Wiley/ Interscience, New York.

Takahashi, M., S. Suzuki, H. Nitanada, and E. Areri. 1988. Mixing and Flow Characteristics in the Alumina Thermoplastic Resin System, *J. Am. Ceram. Soc.*, 71(12), 1093–1099.

Tocher, K. D. 1963. *The Art of Simulation*, English Univ. Press, London, England.

Uhl, V. M and Gray, J. B. (eds.) 1967. *Mixing: Theory and Practice II*, Academic Press, New York.

Wright, J. K., M. J. Edirisinghe, J. G. Zheng and J. R. G. Evans. 1990. Particle Packing in Ceramic Injection Molding. *J. Am. Ceram. Soc.*, 73(9), 2653–2658

5

The molding process and mold design

Injection molding is a relatively simple cyclic process in which a granular mix is heated until softened, and then forced into a mold where it cools and resolidifies to produce a part of the desired shape (Rubin 1973; Pearson 1985).

A complete cycle consists of the following operations:

- The mold is closed and a locking force is applied.
- The plunger moves forward, carrying a fresh charge of material into the softening zone of the cylinder and displacing previously softened material through the nozzle, which is in intimate contact with the mold. This softened material flows through the sprue into runners that terminate in gates leading to the mold cavity. As the cavity fills, the material in the cavity begins to cool and contract. During this time the pressure in the mold also increases. In the packing period, which includes the pressure buildup, material flows into the cavity at a relatively low rate.
- Pressure is maintained on the plunger for a period of time during which the material freezes at the gate and no flow occurs through the gate in either direction. The pressure in the cavity at the time when the gate freezes is called the sealing pressure. The ultimate mold shrinkage, the sink marks, and the release behavior are all determined by this sealing pressure.
- The plunger retracts.
- The mold opens. Some residual pressure remains in the cavity.
- The part is ejected from the mold.

From this basic description of the injection molding process, it is evident that the molding machine must include provisions for holding a supply of

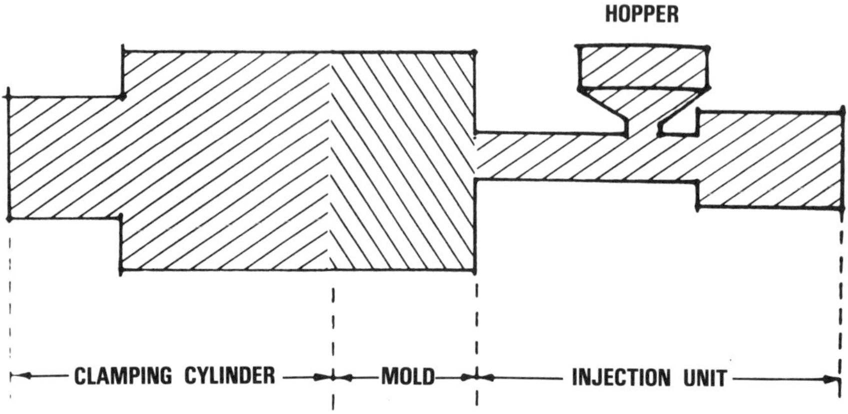

FIGURE 5-1 Basic elements of the injection molding process.

material (hopper), a method of forcing the material into a mold (injection unit), and a method of opening and closing the mold (clamping cylinder), as shown in Figure 5-1, as well as methods for controlling time, temperature, and pressure.

Once an appropriate molding machine is selected, the molding conditions must be determined, based on the machine, the material, and the part being molded. The size and the shape of the mold cavity are determined by the part to be molded. The mold design—the number of cavities per mold, their size, and their shape—plays a major role in the fabrication of usable, cost-effective injection molded ceramic parts.

This chapter describes machines available for injection molding, how they function, and how they are controlled. Molding conditions are established for several materials molded with a recriprocating screw machine. The parts of a mold are described, and the effects of their size, shape, and quality discussed.

MOLDING MACHINES

Many types and sizes of machines are available (McKelvey 1962; Whelan and Craft 1981), and schematics of two basic types are shown in Figure 5-2. A plunger-type injection molding machine (Figure 5-2a) differs from a screw-type machine (Figure 5-2b) in the way that the material is conveyed and plasticated. The plunger machine heats the material by conduction and convection only, whcreas in a screw machine part of the heat is created by the friction between the screw, the material, and the cylinder. In a plunger machine, the pressure loss in the cylinder can be significant; the pressure at

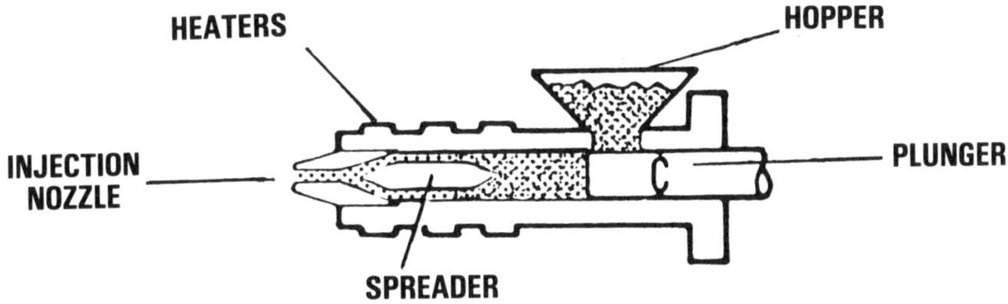

FIGURE 5-2a Schematic of a plunger-type machine.

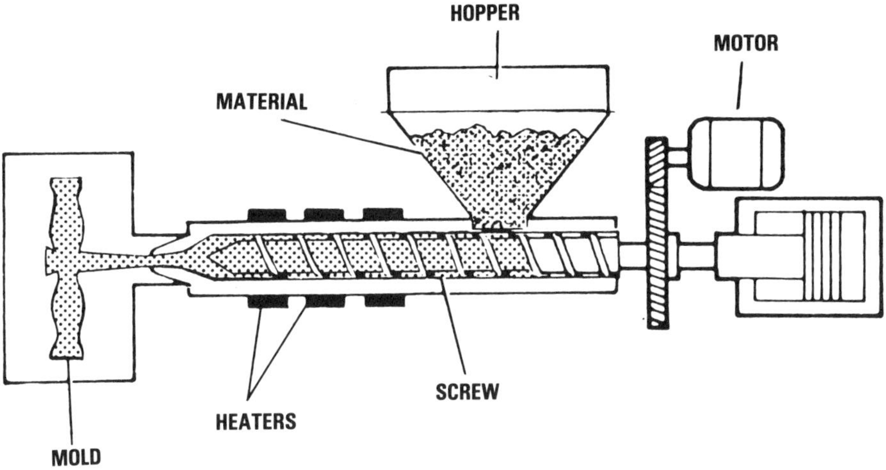

FIGURE 5-2b Schematic of a screw-type machine.

the end of the plunger is much higher than at the nozzle. In a screw machine the ram pressure is almost the same as the nozzle pressure. Figure 5-3 shows a 50-ton screw machine that has been used to injection mold ceramics.

Machine selection largely depends on the specific requirements of the product to be molded and the degree of versatility desired; but, regardless of the type, certain features are essential in an injection molding machine, as discussed below.

The Machine Feed System

The accuracy with which the amount of material needed to fill the mold is measured to the front of the screw or the plunger affects the quality of the product. Overfilling causes undue strains in the mold; and in toggle-clamp-

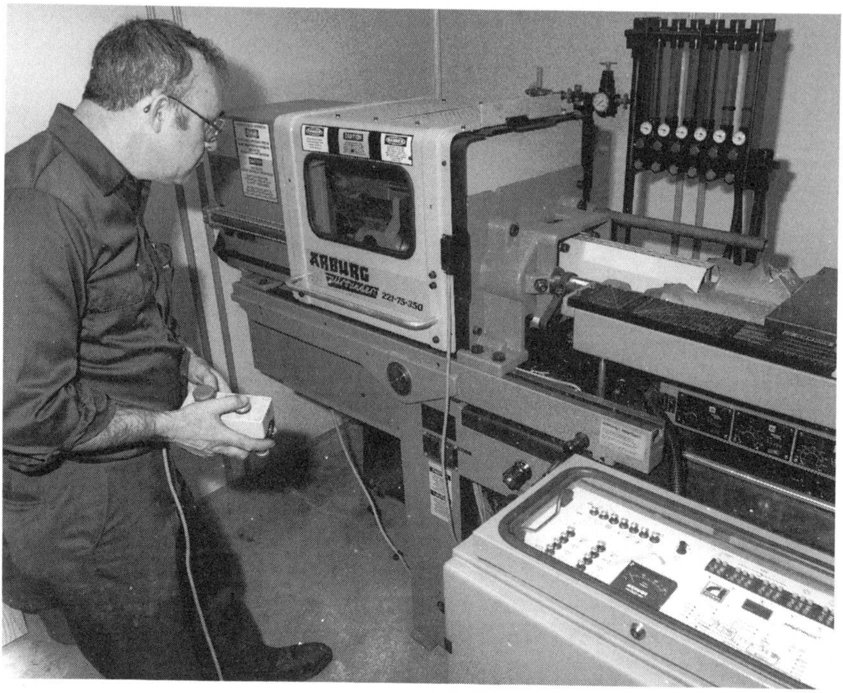

FIGURE 5-3 Arburg 221U/170P reciprocating screw machine. Courtesy Polymer Machinery Corp.

ing machines, overfilling can break the tie bar if the clamping force is too great and the overfilling is not noticed in time. An overfilled mold is difficult to eject, and the part usually is damaged.

Variation in the flowability, size, and shape of the ceramic granules causes inconsistent mold filling. This can be prevented by using a reliable feed system—either a volumetric metering unit or an automatic weigh feeder. A weigh feeder ensures that there is a constant cushion of material once the feed weight is adjusted to equal the weight of the part.

Injection Unit

The primary function of an injection unit is to heat the material in the cylinder uniformly and then to pump the softened material into the mold cavity. The injection unit must be able to perform the following operations:

- Move on its support to bring the nozzle into contact with the sprue bushing of the mold (in the clamping unit) or to retract it.

- Generate pressure between the nozzle and sprue bushing.
- Rotate the screw during the feed stage.
- Move the screw axially during the injection stage.
- Build up the holding pressure.

In addition to these functions, many modern machines provide one or more of the following capabilities:

- A mechanism for hydraulic retraction of the nonrotating screw.
- Simultaneous axial and rotational motion of the screw for injection with a rotating screw.
- Control of variable strokes, velocities, number of revolutions, and pressure.
- Easy assembly or disassembly, with swivel mount for rapidly changing the screw.

In a standard design, the injection unit is supported by the right-hand section of the substructure and the stationary platen, as shown in Figure 5-4. Small and medium-size machines often support the carriage on bars situated parallel to the main axis. The axis of the cylinder also is parallel to these bars, usually in the same axial plane. The bars can be furnished with pistons over which hydraulic cylinders move. These cylinders, which are

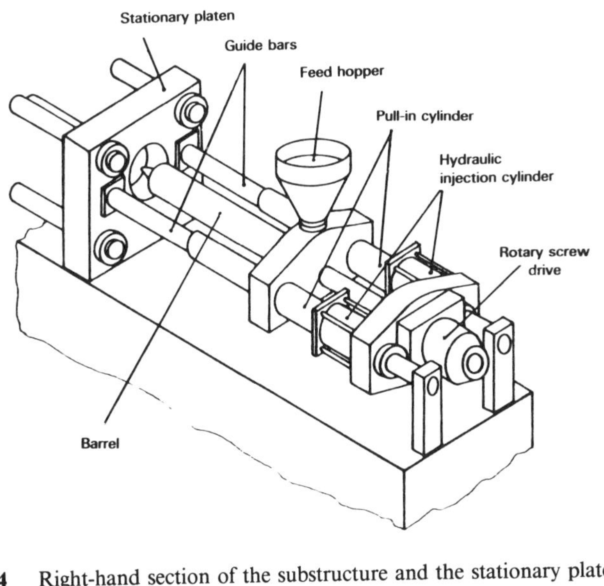

FIGURE 5-4 Right-hand section of the substructure and the stationary platen support the injection unit. Courtesy Carl Hanser Verlag.

connected to the carriage, provide the full-length stroke for the carriage's travel. Oil can be supplied through the bars.

It always must be possible to align the injection unit with the center of the sprue bushing. This normally coincides with the machine axis. The motion has to be smooth and steady, although frequently at a speed as high as 200 to 250 mm/sec. During carriage movement there must be no tilting, twisting, or other deviation from the center axis.

The heat of friction produced by the screw drive through the rotating screw to the material provides a considerable amount of the energy needed to plasticize the feed material. Consequently, a considerable quantity of energy is consumed during the feeding stage. Also the drive must be powerful enough to develop a high starting torque.

A rotating screw drive is characterized by the type of drive and its position. There are three types of drive.

- An electric motor with reduction.
- A hydraulic motor with reduction.
- A direct hydraulic drive.

An electromotive screw drive usually is used for large injection molding machines that have a clamping force of more than 200 metric tons. Hydraulic motors frequently are used to convert the supplied hydraulic power into mechanical power. These motors can be similar in design to hydraulic pumps and even operate like them in many cases. Rotating screw drives can use gear motors, rotary abutment motors, vane motors, or piston motors. The last two types are preferred.

Hydraulic motors have a number of advantages compared with electric motors:

- They provide variable speed, which is controlled with constant oil volumes. This is a simple way to obtain an adequate number of speeds.
- They offer a relatively wide linearity between the torque and the speed; that is, the speed stays directly proportional to the torque.
- The torque is limited by the pressure.
- The starting torque at onset is approximately the same as the operating torque.
- They provide a fast start and fast braking because of the small moment of inertia.
- A low weight-to-power ratio is provided, which favors acceleration of the moving mass during injection and braking.

To control noise during rotation of the screw the use of slow-running hydraulic motors as direct drives is suggested. Exact positioning of the screw during injection is very important for the quality of the moldings. This can be managed best when the mass of the movable parts is small; so a design that keeps the rotary drive stationary and connects it with the screw coupling by means of a hollow shaft is preferable. The direct hydraulic drive is light enough to be directly mounted on the screw shank. The screw's rotating drive is advantageous for the molding process. The torque and the speed of the rotating screw convey the material and impart the desired shear and homogenization. The torque, T_x, required for a screw of diameter D_x can be calculated by using the model law (Fischer 1976; Potente 1981):

$$T_x = T_o \,(D_x/D_o)^{2.7} \tag{5-1}$$

where T_o is the known torque value of a given screw diameter, D_o.

Maximum power can be suppled readily with either an electric or a hydraulic motor. However, supplying it over a range of speeds is difficult. To accomplish this either a gear drive or a more costly shift transmission can be used in conjunction with an electric motor. The stepwise adjustment of speed is necessary to reproduce processing parameters. This is essential to ensure consistency in the quality of the molded product. A hydraulic motor—one in which the rate of oil flow is in fixed amounts—is preferred. It provides an optimized speed for achieving reproducible molding conditions.

In a single-stage reciprocating screw machine, a properly designed screw unit produces a melt at a high output rate with minimum thermal degradation, that has a uniform viscosity and temperature. Therefore, it is important to select a suitable screw design. The basic screw design characteristics for pitch, helix angle, and length of zone sections, as shown in Figure 5-5, are adapted from those of conventional extruders. Screws usually are designed to have constant pitch and variable-root geometry.[1]

Current design practice divides this type of screw into three distinct zones beginning at the hopper end of the machine, as shown in Figure 5-6: the feed, the transition, and the metering zones.

The function of the feed section, which consists of five to nine flights with a relatively deep, uniform flight depth, is to deliver to the transition section the quantity of material required to keep the discharged end of the screw completely filled with softened material. In the transition section, the material is compressed, softened, and mixed. The flight depth in this section generally decreases uniformly from the end of the feed section to the beginning of the metering section.

---

[1]All terms specific to injection molding are defined in the Glossary.

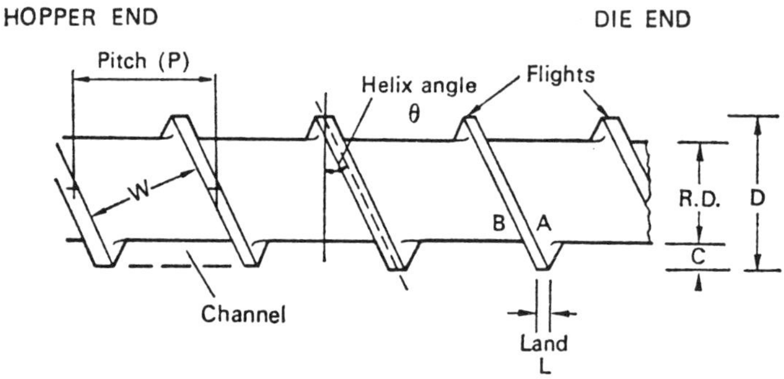

P is the pitch
D is the screw diameter
W is the channel width
C is the channel depth
L is the land width
R.D. is the root diameter
θ is the helix angle
A is the screw leading edge
B is the screw trailing edge

FIGURE 5-5 Basic screw design adapted from conventional extruders.

The metering section typically consists of four to six flights with a relatively shallow, uniform flight depth. The function of this section is to complete the softening and mixing in order to achieve maximum homogeneity, at the mix temperature, and then to pump the material into the mold cavity.

The compression ratio is the ratio of the volume of a metering section to the total volume of the screw. Compression ratios available in current commercial machines range from 1.5 to 4.5 to 1, with the general purpose

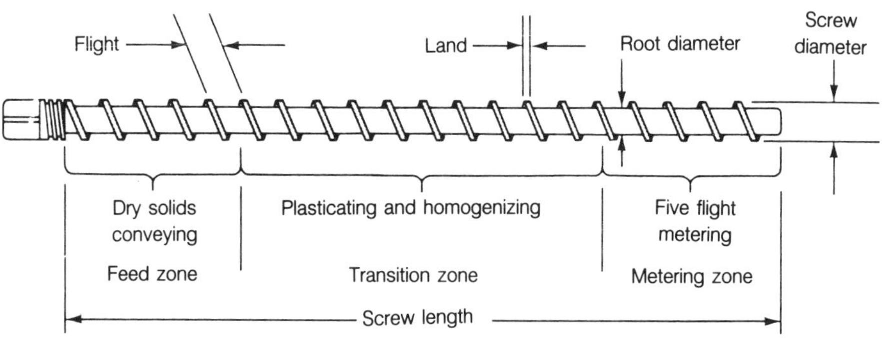

FIGURE 5-6 Three zones in a screw: feed section, transition section, and metering section.

TABLE 5-1. **The compression ratios of three screws.**

	Screw	*a*	*b* (mm)	*c*	*L/D*	Compression ratio
Medium	A	3.8	46	45	13	2
Shallow	B	2.2	46	45	13	2
Deep	C	6.5	46	45	9	2

screw usually having a compression ratio of 2.5 to 3.0 to 1. The compression ratios of three types of screws are given in Table 5-1. In defining screw performance, the compression ratio itself is less important than the flight depth in the metering section. In general, screws with high compression ratios (above 3:1) should be operated at lower screw speeds (10–50 rpm), whereas screws with compression ratios below 3:1 can be operated at higher screw speeds (50–150 rpm). The metering section controls the screw output in terms of its plasticating capacity. Its function is to convey softened material forward, at a controlled rate, without surges.

The screw length usually is expressed in terms of the ratio of the effective screw length to the major screw diameter—the $L:D$ ratio. On most reciprocating screw machines, depending on the length of the injection stroke, the effective $L:D$ ratio can be reduced by about two or three units. This reduction is less significant for high-$L:D$-ratio screws. The higher the $L:D$ ratio is, the less sensitive is the screw performance to variations in the properties of the material to be molded.

The channel depth is an important factor relating to the output rate and homogeneity. Because the feed section of the screw is basically a conveyor, the deeper the channel is, the larger the volume between the flights, and, therefore, the larger the output. However, other factors need to be considered in selecting the channel depth. In the metering section one such factor is the shear rate, defined as the surface velocity at the barrel wall divided by the channel depth:

$$\dot{\gamma} = \frac{DN}{h} \tag{5-2}$$

Where $\dot{\gamma}$ is the shear rate, D is the screw diameter, N is the rate of screw rotation, and h is the channel depth.

In general, current design practice is to use a shallow channel depth for less viscous materials and a deeper channel for mixes having a high melt viscosity. With material having a relatively low viscosity (100 Pa·sec–1000 Pa·sec at 100 reciprocal seconds and 120°C), experience with both deep

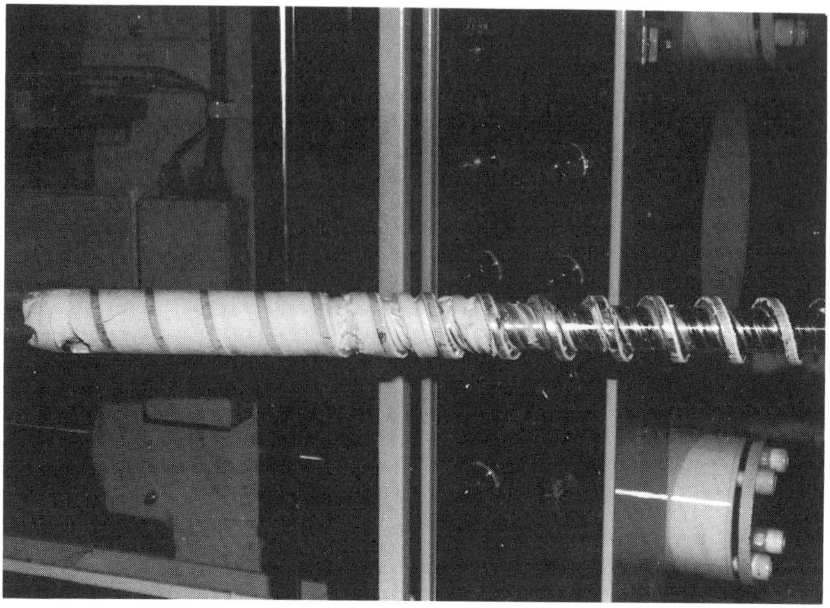

FIGURE 5-7a The melt length of a deep metering screw C. From Battelle Columbus, OH

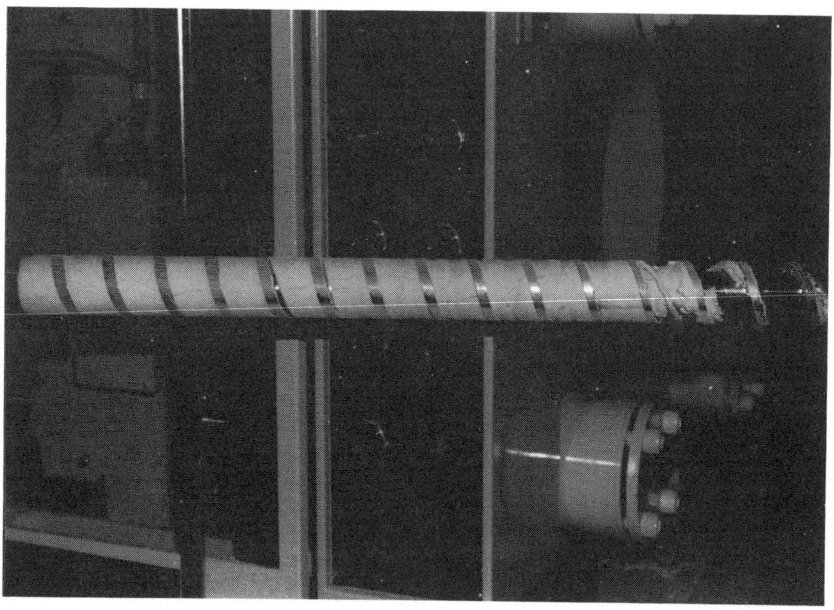

FIGURE 5-7b The melt length of a shallow metering screw B. From Battelle Columbus, OH

and shallow metering screws suggest that the total melt length of the deep metering screw (C) is shorter than that of the shallow metering screw (B) (see Figure 5-7a and b). Nevertheless, it is difficult to quantify the effects of the screw configuration on mixing and homogeneity. The results of an experiment to relate the weight of the molded parts to the mix homogeneity for three different screw configurations (A, B, and C), at injection rates of 8–40 cm³/sec, are shown in Figure 5-8. The consistently higher sample weights as a function of injection rate that were produced using a shallow metering screw (B) may be due to its better mixing action, which reduces any instability in the melt flow during molding.

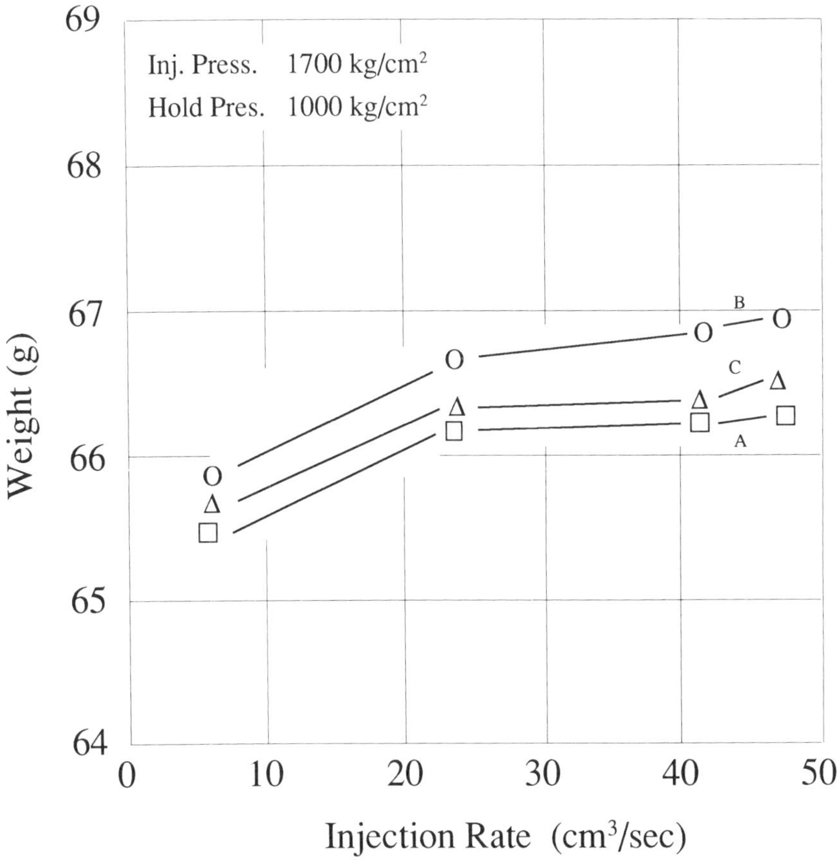

FIGURE 5-8 Effects of screw configuration on mixing and homogeneity. Screw configurations A, B, and C as in Figure 5-10. A = medium metering depth; B = shallow metering depth; C = deep metering depth.

Three different types of flow affect the output of a single-screw machine: drag flow, pressure back flow, and leak flow (Kruder and Nunn 1981). Drag flow occurs when softened material in the screw channel adheres both to the fixed cylinder and to the rotating screw. The material is being pushed or dragged toward the front of the machine by the action of the angled (i.e., helical) flights. Motion is generated in two directions: across the screw flight and along the screw flight. The component along the screw flight generates the drag flow along the screw.

As the drag flow is generating pressure in the material, there is a tendency for back flow through the open screw channel. This flow, which runs counter to the drag flow, reduces the net movement of the material along the cylinder. Figure 5-9 shows velocity profiles of the material generated individually by the drag flow and the pressure flow, and then by their combined flows.

The output of the screw metering section, which, in effect, is the output of the screw, Q, is related to the drag flow, Q_d, the pressure back flow, Q_p, and the leak flow, Q_l, as follows:

$$Q = Q_d - Q_p - Q_l \tag{5-3}$$

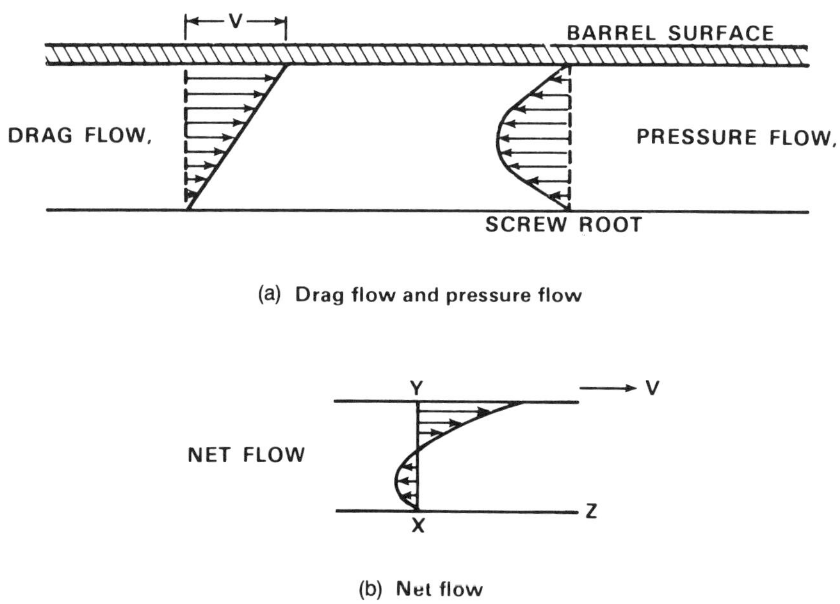

FIGURE 5-9 Velocity profile of the material: (a) generated by the drag flow and the pressure flow; (b) generated by their combined flows.

The three flow components can be expressed mathematically in terms of the drag flow (equation 5-4) and the pressure back flow (equation 5-5):

$$Q_d = \frac{\pi^2}{2} D^2 N h \sin\theta \cos\theta \tag{5-4}$$

where D is the diameter of the cylinder, N is the screw speed, h is the depth of the channel, and θ is the angle of the helix.

$$Q_p = \frac{\pi D h^3 \eta p \sin^2\theta}{12L} \tag{5-5}$$

where p is the increase in pressure, η is the viscosity, and L is the length of the metering section.

The leak flow usually is disregarded in these calculations. Its chief value is as an indicator of a screw's wear. As such, it provides a means of determining the point at which maintenance becomes economic.

The effect of different screw designs on the output can be seen by applying equations (5-4) and (5-5). The output, when operating at a constant speed with three different screws, is shown in Figure 5-10. For example, the output of screw C, which has a deep metering section, is sensitive to pressure because its deep channel constitutes a very large orifice through which pressure flow can take place. This accounts for the rapid decrease in output as the back pressure increases, as shown in Figure 5-10. Screw A, with a medium-deep metering section, shows a similar sensitivity to pressure. Screw B, with a shallow metering depth, is much less sensitive to changes in back pressure than the other two because its shallow channel is much more resistant to pressure flow.

The highest pressure occurs at the front end of the screw in the region of the screw tip; so it is essential to prevent back flow of the material into the rear flights. This is done during the injection and holding pressure stages by means of a closing element. The simplest design solution is to use a screwed-on tip that has a larger diameter than that of the screw root at the extreme front end of the screw. The narrow gap between the screw and the cylinder wall causes a rise in pressure that restrains the back flow.

A complete shutoff can be achieved only by using a nonreturn valve. As shown in Figure 5-11, this valve consists of three parts: the tip, which can be attached to the screw by means of a threaded shaft; the seat; and the sliding ring. The sliding ring is not firmly attached to the screw tip; should any binding occur between the ring and the cylinder screw, the screw is still free to turn, and probably will cause the sliding ring to free itself. Other types of screw tips in which the major diameter of the tip is rigidly attached to the

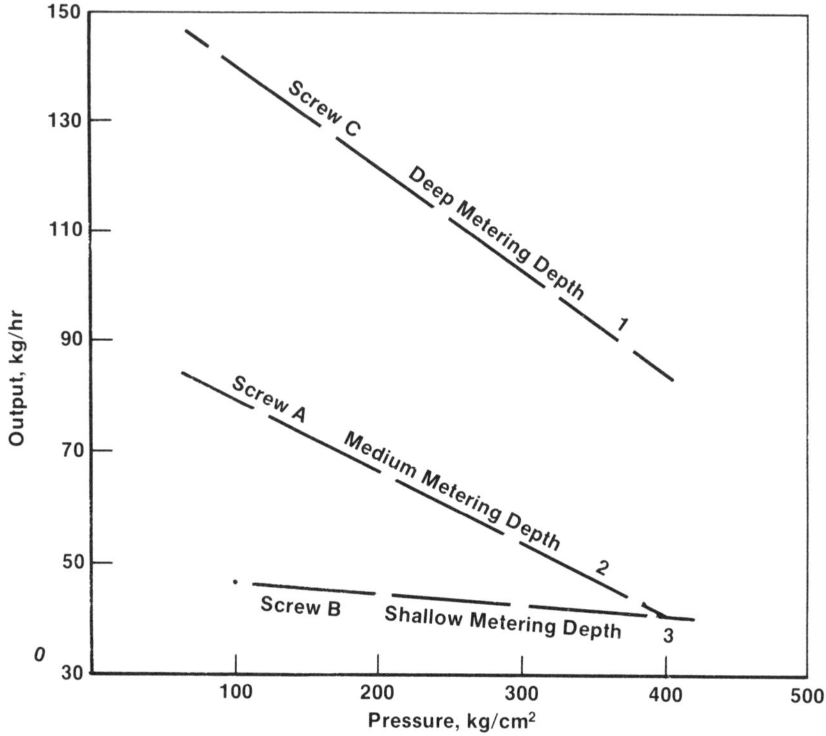

FIGURE 5-10 Output of three different screws operating at a constant speed.

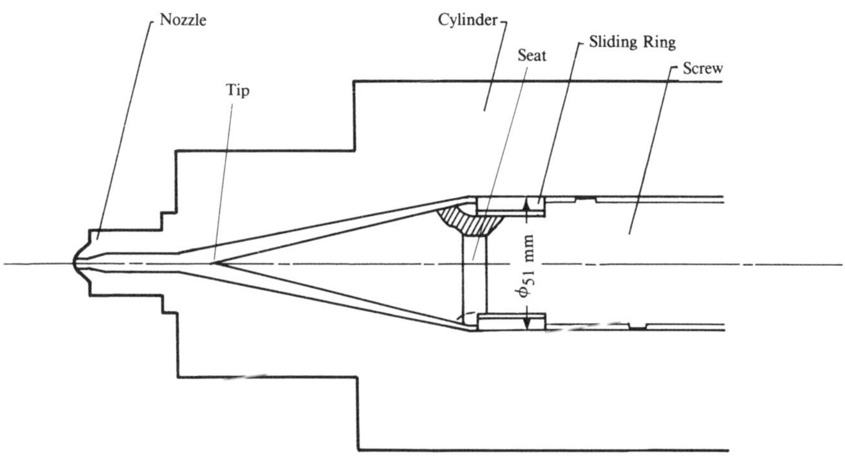

FIGURE 5-11 A nonreturn valve with the tip, the seat, and the sliding ring.

screw, such as a one-piece, ball-check screw tip, can cause problems. When binding between the cylinder and the tip occurs, the tip may break off, the tip and/or the cylinder may be scored, or the tip may become so tight that it is difficult to remove from the screw.

Although ball-check or floating-ring tips allow good control of the pressure, the problems they cause do not justify their use. However, the screw tips shown in Figure 5-12, which are used for molding rigid polyvinylchloride (PVC), have proved effective for molding ceramics.

Prior to injection, the nozzle, a component of the injection unit, is forced against the bushing of the mold to secure close locking. To accommodate the forces on the nozzle without undue wear, the tip must have a large spherical radius between 1.3 cm (0.5 in.) and 1.9 cm (0.75 in.). The nozzle has a channel that tapers toward the tip and connects the cylinder head to the mold's sprue bushing. An open nozzle is the most common type used because it can be the shortest length. Shutoff nozzles are used to prevent drooling and stringing of the material, or to make it possible to feed with the nozzle retracted.

The cylinder, which is the screw's outer casing, is made of steel, and it is heated externally to soften the ceramic mixture. The cylinder head plus the nozzle must be streamlined to prevent the material from being hung up in this section, which can necessitate the use of forced purging. In a plunger machine, a streamlined, torpedo-shaped spreader device is positioned inside the injection cylinder at the forward end just behind the nozzle. Its purpose is to reduce the cross section of the mass of the material by

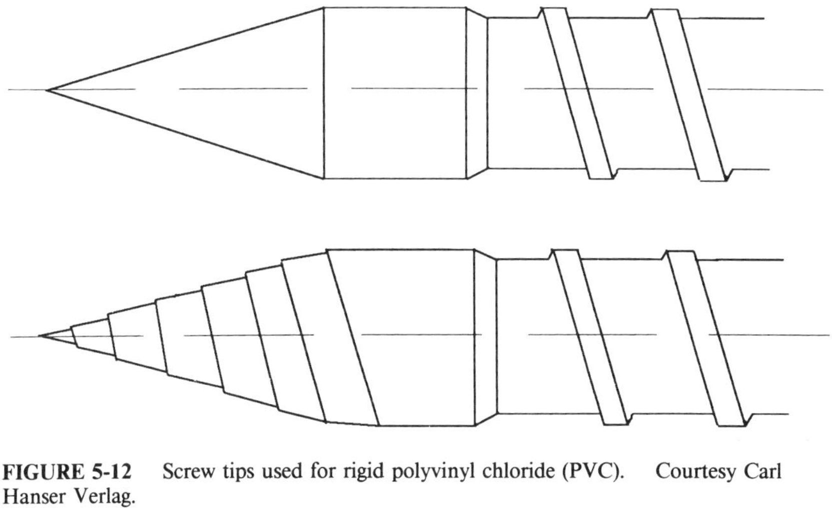

FIGURE 5-12 Screw tips used for rigid polyvinyl chloride (PVC). Courtesy Carl Hanser Verlag.

directing it through small slots parallel to the flow. The effect is to expose the material to more metal-heating surfaces, thereby speeding the plasticizing process. The spreader also creates a back pressure that causes any entrapped air to work its way back through the material, rather than moving forward to the mold.

The Clamping Unit

The clamping unit is the section of an injection molding machine in which the mold is mounted. It supplies the motion and the force to open and close the mold and to hold the mold closed forcefully during the injection cycle. There are three types of mold clamping mechanisms: straight hydraulic, toggle, and hydromechanical.

A common characteristic of all hydraulic clamping systems is the hydraulic clamping cylinder, which is generally positioned in the center, as shown in Figure 5-13a. There are no complicated mechanical linkages; a forward thrust clamps the mold together. To function well, the hydraulic cylinder must be large; and to achieve maximum efficiency and operating speed, the clamping action usually includes a fast traverse until just before the mold makes contact, followed by a slower but higher application of pressure that applies the full clamping force. Maintenance can be reduced by using smaller-capacity high-pressure pumps. The hydraulic clamping cylinder can withdraw the moving half of the mold rapidly. At the end of the stroke, the withdrawal speed slows to a low-velocity traverse, which reduces the possibility of the mold being damaged by a shock or an impact.

Using standard hydraulic equipment, motors, valving, and timers for the automatic control features of the hydraulic and clamping facilities guarantees a satisfactory pressure-generating system. In general, injection molding machines employ a dual pressure system: an intermediate- or lower-pressure hydraulic system that delivers large quantities of fluid, for example, for rapidly closing the mold shutting facilities; and a second high-pressure system that delivers a small volume of fluid for a tight squeeze, or final mold closing.

A toggle clamp, shown in Figure 5-13b, is a mechanical device used to amplify force. It consists of two bars joined together end to end with a pivot. The end of one bar is attached to a stationary platen, and the end of the second bar is attached to a movable platen. When the mold is open, the toggle is in a V shape; but when pressure is applied to the pivot, the two bars form a straight line. The linkages, which can be of equal or varying lengths and simple or complex design, are actuated by a screw or a hydraulic mechanism. Large clamping forces are developed only when the frame and the mold are very rigid. However, the clamping mechanism is reduced by

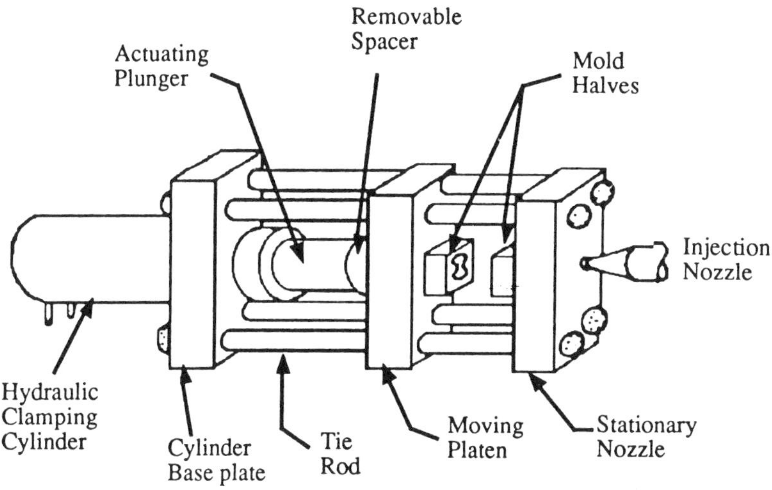

FIGURE 5-13a A hydraulic clamp unit. Courtesy Carl Hanser Verlag.

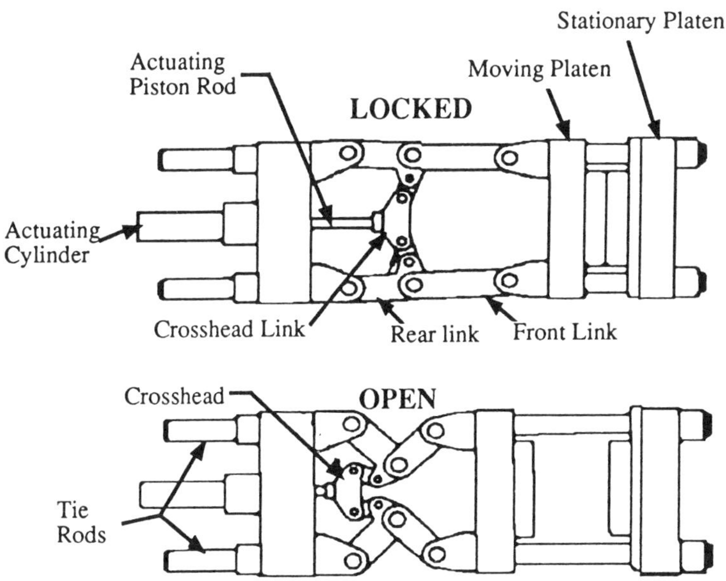

FIGURE 5-13b A toggle clamp unit. Courtesy Carl Hanser Verlag.

deflection of the various elements of the molding machine such as shear on the toggle pins and compression on the toggle arms. The smaller these deflections are, the greater the clamping force is. The frames of the molding machine that use toggle action are built very solidly. No force is experienced until resistance is encountered, and then it builds up very rapidly over a short distance, depending on the frame's rate of deflection.

Most molding machines employ a directed force parallel to the direction of the toggle clamping action, creating, in effect, a compound toggle action. Molding machines are actuated by a hydraulic piston that moves the toggle links forward. In order to benefit fully from the toggle action, the moving platen of the molding machine must be able to adjust for various die thicknesses. Clamping forces of 25 to 150 metric tons are common on laboratory-size machines with a shot capacity of 14 to 55 g. In further refinements of the toggle action, pressure pads on the ends of the toggle links can be used to relieve the toggle pins of shear forces. The locking features of the toggle mechanism should be noted because once the forward movement of the actuated link establishes full clamping pressure, there is no slippage backward.

The toggle system has two chief advantages: it is economical to run the small hydraulic cylinder, and because the toggle is self-locking, it is not necessary to maintain the hydraulic pressure throughout the molding cycle. The disadvantages of this system are that there is no indication of the clamping force, and the additional moving parts increase maintenance costs.

A preferred type of clamping action on injection molds combines hydraulic and mechanical linkage. Movement is initiated by a hydraulic cylinder that thrusts the moving platen and the mold mounted on it in a forward direction. When properly constructed with link pins of adequate size that are well lubricated, the clamping mechanism will give efficient, trouble-free service. Some machine designs introduce such refinements as an automatic lubricator that delivers a small amount of oil to moving parts during each cycle. In some designs, the final movement of the link pins brings the links into contact with one another so that the force resisting the spreading of the molds is resisted by the links themselves, rather than by the link pins or their bearings. This arrangement prolongs the life expectancy of the mechanism.

Process Control

The primary purpose for using a control unit is to eliminate or to control material variables. Many advances in the field of automatic controls for injection molding machines have been made in the last decade. The availability of powerful and affordable microprocessor-based controllers tied into a computer system has made these controls a standard feature in almost all injection molding machines.

There are two elements in process control: monitoring the performance of a machine and directing the machine functions. Usually both elements are combined in one microprocessor system. Monitoring helps relate the process to the part. During the process, the sensed molding parameters show the relationship between pressure, temperature, and position or movement. Monitoring can establish whether additional machine control is needed.

Monitoring

The Hydraulic Pressure Transducer

Monitoring the hydraulic pressure profile can help one to diagnose many machine problems such as hydraulic pressure changes during injection and the consistency of the screw return time. The hydraulic pressure transducer should be placed as close to the injection ram as possible. This location gives the most accurate pressure profile. It also is possible to sense the pressure of the material at the machine nozzle, but this can subject the transducer to high wear.

The Pressure Inside the Mold Cavity

Monitoring the buildup of pressure in the mold cavity provides information about functions such as filling time and peak pressure consistency. Transducers can be installed both in the mold's runner system and in the cavity. The type of transducer selected depends on the configuration of the part, mold construction, and the type of runner system. Transducers that measure the pressure of the material directly now are available, but it is important to select the optimal location to sense and monitor the material pressure. Monitoring at a point located halfway into the cavity is a good general rule. The maintenence of built-in transducers should be considered in designing a mold.

Barrel, Material, and Mold Temperatures

Barrel temperatures are sensed and controlled with thermocouples, with one thermocouple required for each zone. The nozzle usually has its own control. For accurate control of the temperature and the temperature set point, current-proportioning controllers are used. It is desirable to monitor and control the material's temperature at the nozzle. However, the highly abrasive nature of ceramic mixes limits the use of thermocouple-type sensors; so infrared and ultrasonic systems are now under development. A heat exchanger unit is used to maintain a constant mold temperature.

The relation of the melt temperature to the processing conditions can be expressed as (May 1972):

$$T_m^b = f(N, P_m, T_b, Y) \tag{5-6}$$

where T_m^b is the temperature of the mix/melt material in the barrel, N is the rotational speed of the screw, P_m is the pressure of the mix/melt material in the barrel, T_b is the temperature of the barrel, and Y represents the disturbance resulting from variations in the properties of the mix/melt and/or the screw geometry. This relation indicates that the temperature of the mix/melt is a function of the speed of the screw, the pressure exerted by the mix/melt, and the barrel temperature.

Ram Position

The ram position is monitored by a potentiometer mounted on the injection molding machine (both linear and rotary types). It provides information about the rate at which the material is being injected into the mold, the consistency of the ram profile during machine control, and the consistency of the screw return time.

Process Controllers

Sensors are central to process monitoring, and each one requires a power supply and an amplifier. A sensor is driven by an input voltage, a resultant output signal is generated as the sensor responds to the monitored parameter and an amplifier is used to boost the output signal's strength, as increased signal strength or amplitude is needed for recording.

Monitored molding parameters can be displayed as an analog signal or as a digital signal. Analog signals make it possible to see a continuous profile of the parameter being used. The profile is useful because it is time-related. The sensor's output signal in a digital monitoring device is conditioned to supply a discrete numerical readout, and data collectors are used to monitor multiple parameters digitally. Cathode ray tube oscilloscopes can be used to monitor repeating cycles. A selected starting point is used to trigger the scopes, and the storage scope display shows the excursion of a parameter over a period of time.

In a machine sequence control system, the input commands are set, and corrections are made to the machine output. The corrections can be made in real time—that is, a sensed deviation is corrected during the cycle—as rapidly as the machine's electrohydraulic valve and fluid system can respond. Alternatively, the corrections can be adaptive; that is, an adjustment is made for a sensed deviation on the next cycle. The system's ability to adjust depends on the sensitivity of the molding process and the capability of the controller to correct the deviation.

In closed loop control, a signal from the transducer that represents the parameter's value is compared with a desired preset value for the parameter. The difference is amplified before being sent to a control element for correction of the parameter.

Figure 5-14 (Rosato 1986b) illustrates the closed loop control of ram speed or pressure (force). A transducer (one is used for speed and one for pressure) measures the parameter being controlled. It creates a feedback voltage that corresponds to its transfer function, H, in volts per unit pressure or volts per unit velocity. A summing junction compares the feedback voltage to one commanded by the process controller. The difference is sent to the forward loop elements (amplifier, control valve, and ram position), which have a combined parameter transfer function, G, with units of speed per volt or pressure per volt. Using this combined parameter transfer function, the servoloop transfer function can be expressed as:

$$\frac{\text{Output}}{\text{Command}} = \frac{G}{1 + GH}\bigg|_{\lim G \to \infty} = \frac{1}{H} \tag{5-7}$$

It can be shown that as G gets very large, the servoloop transfer function becomes that of the transducer.

Figure 5-15 illustrates where the process controller fits into the overall machine function. The most efficient application of a process control strategy requires an understanding of the different aspects of variations in the process and their relationship to the quality of the part. Process control closes the loop between a process parameter and an appropriate machine control device to eliminate the impact of process perturbations.

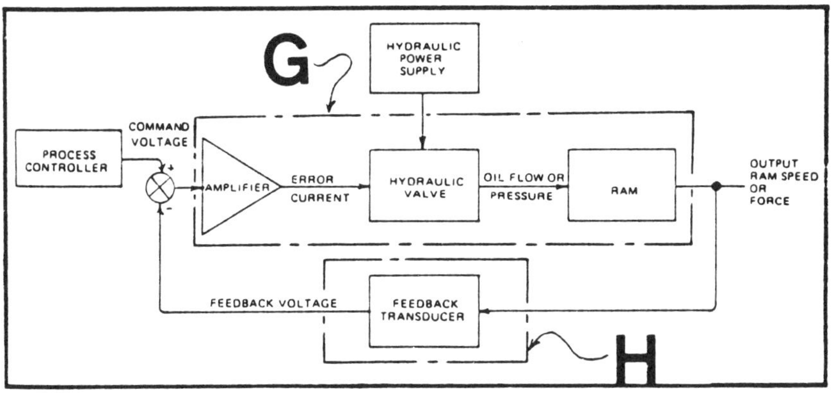

FIGURE 5-14 Closed loop control of the plunger speed. After Rosato 1986b.

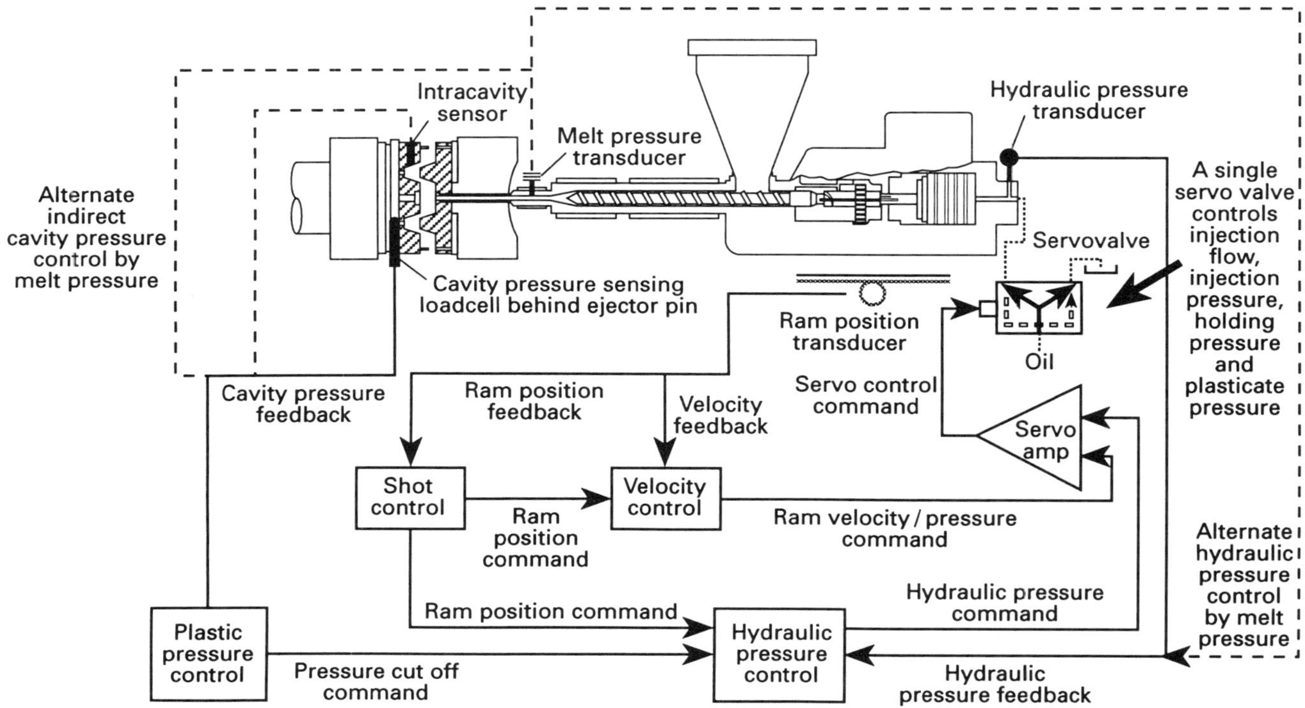

FIGURE 5-15 Process control closes the loop between process parameters and the machine function.

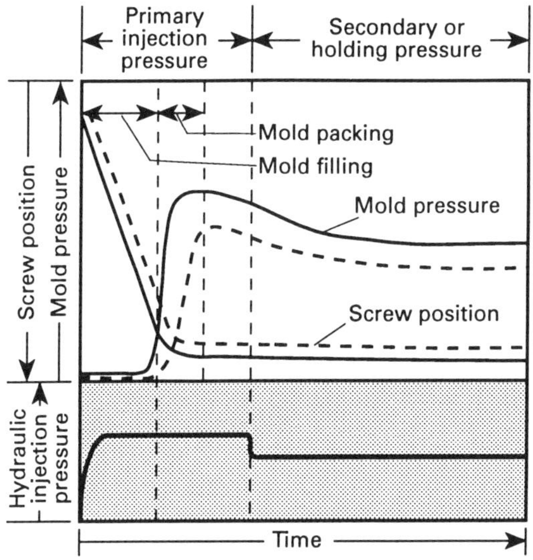

FIGURE 5-16 A change in the viscosity of the mix is reflected in a change in the mold pressure. The solid line indicates the mold pressure before a change in the viscosity of the mix, and the dashed line indicates the mold pressure after a change in the viscosity of the mix.

There are several levels of sophistication in process control, all employing different control parameters. One level measures the pressure in the cavity, which is the single most useful control parameter for injection molding. For example, a change in the viscosity of the mix is reflected as a change in the mix pressure, and can be detected by measuring the pressure in the cavity with respect to time, as shown in Figure 5-16.

Other variations in molding conditions such as hydraulic pressure, oil temperature, and mix temperature are manifested similarly, and can be detected by monitoring the pressure in the cavity and the position of the ram with respect to time.

MOLDING PARAMETERS

The Roles of Pressure, Temperature, and Time

In terms of the molding operation, the molding cycle can be broken down into the plunger/screw forward time, the mold closed time, and the mold open time. Although adequate for setting the controls on the machine, this sequence is not suitable for analytical purposes. A more useful presentation of the molding cycle for this purpose is shown in Figure 5-17 (Spencer and Gilmore 1950b). The sequence is as follows:

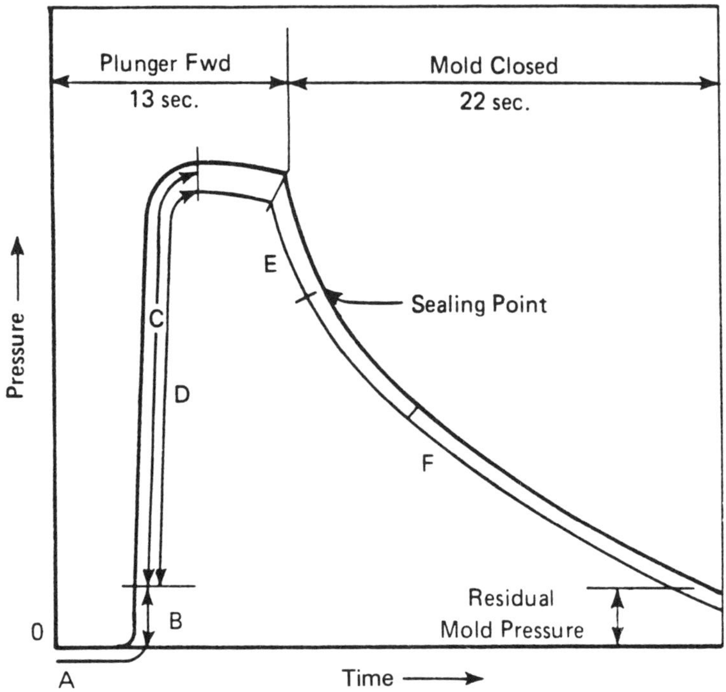

FIGURE 5-17 Molding cycle: A, dead time; B, mold-filling time; C, pressure buildup time; D, packing time; E, discharge; F, cooling with gate frozen.

- Period A, the dead period: the time (after the mold closes and the plunger/screw starts to move forward) before the material begins to flow into the sprue bushing.
- Period B, the mold-filling period: the time during which the material flows into the mold and fills the cavity.
- Period C, the pressure buildup period: the time (after the cavity is filled and the material begins to cool) during which pressure in the cavity builds up to the injection pressure.
- Period D, the packing period: the time (including period C) during which the material flows into the cavity at a relatively low rate. The injection cycle ends with this period, and the plunger/screw retracts.
- Period E, the discharge period: the time during which the material in the gate is still fluid, and, with zero pressure on the runner side and slightly less than hold pressure in the cavity, the material flow reverses.
- Period F, the cooling period: the time during which the material in the cavity cools. During this time there is no flow through the gate in either direction.

- Mold open, with residual pressure: after completion of the entire sequence during which the mold is closed (the mold closed time), the mold opens. There is still some pressure remaining in the cavity. This pressure is known as the residual mold pressure. The exact amount of the residual mold pressure depends on the part's shape and size, the mold design and cooling arrangement, and the thermal characteristics of the ceramic/binder composition.

Along with the time sequence, it is essential to have the correct temperature and pressure sequence. The pressure–temperature functions in molding can be related in the form of the ideal gas equation (Spencer and Gilmore 1950a):

$$(P + \pi)(V - W) = KT \tag{5-8}$$

where P is the hydrostatic pressure in MPa, V is the specific volume in cubic centimeters per gram (cm^3/g), T is the temperature in kelvin, and π, W, and K are material constants.

From equation (5-8) it can be seen that if the melt temperature at injection is constant in the cavity, the specific volume $(V - W)$ is inversely proportional to the effective applied pressure $(P + \pi)$. Once the gate is frozen, however, the effective specific volume $(V - W)$ remains constant, and the effective applied pressure $(P + \pi)$ drops in proportion to the drop in the temperature.

Mold Pressure and Temperature Defined

The molding process involves three different kinds of pressure: injection pressure, holding pressure, and cavity pressure. The injection pressure, one of the most important parameters, is the pressure exerted on the melt in front of the screw tip during the injection cycle, with the screw acting like a plunger. It affects both the speed of the advancing screw and the process of filling the mold cavity with material. The pressure rises within a short time from the lowest possible pressure in the system to a magnitude that corresponds to the flow resistance of the melt in the nozzle, sprue, runner, and cavity. The onset of the packing stage can readily be determined if there is little resistance in the nozzle and gating system. Hydraulic pressure is particularly important during the injection stage.

The holding pressure is the pressure exerted on the molded part during a secondary pressure stage. Both the magnitude and the duration of the holding pressure are critical for dimensional accuracy. As long as the sprue, gate, or any other narrow passage, is not yet frozen, changes in the magni-

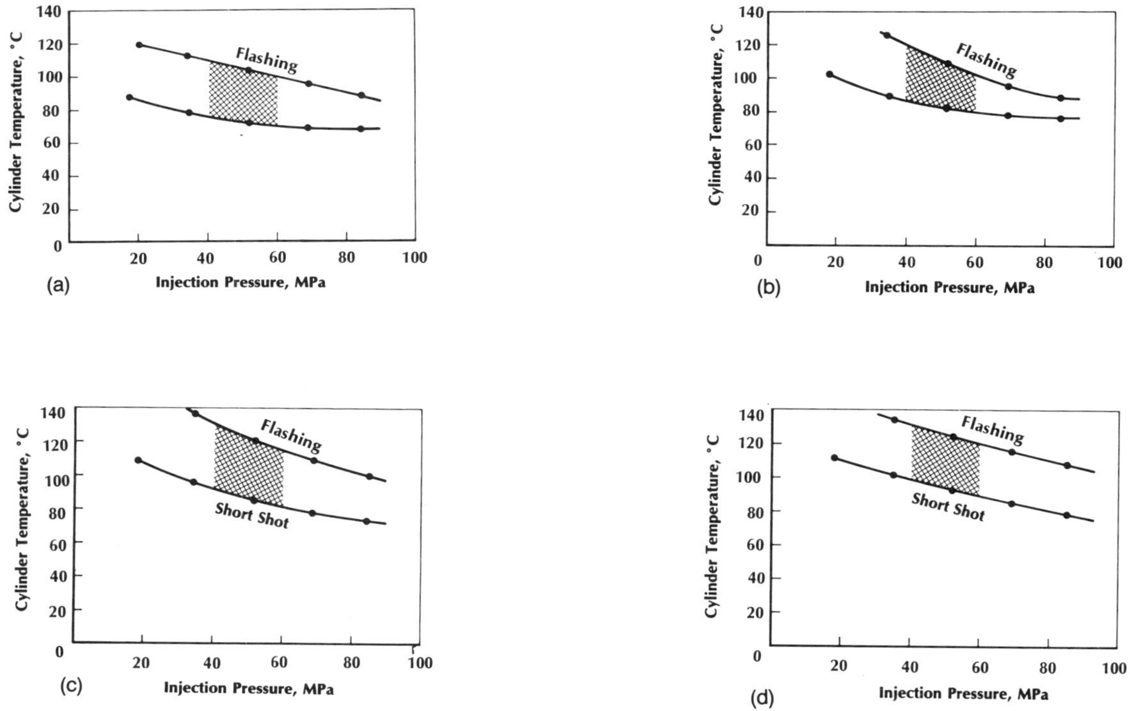

FIGURE 5-18 Molding range diagrams as a function of temperature and injection pressure for different mix formulations. (a) Zirconia; (b) Al6-SG alumina; (c) silicon; (d) silicon nitride.

tude and the duration of the holding pressure will have an effect on the cavity pressure. After the gate is frozen, no further influence can be exerted. The holding pressure in the mold will differ from point to point because pressure losses occur along the flow path.

It is not sufficient just to fill the cavity. Additional material must be pushed into the cavity to compensate for the volume change during cooling. This pushing is referred to as "cavity or packing" pressure. If too much material is forced in, it will be highly stressed, resulting in sticking or cracking, either on ejection or at a later stage. Therefore, the pressure should be uniform at the end of the packing stage.

The temperatures of the melt, the hydraulic fluid, and the mold also are extremely critical for maintaining reproducible properties and dimensions of the molded parts. All motions of the injection molding machine that are not pressure-compensated are influenced by the hydraulic oil temperature. For this reason, it is desirable for the oil to reach the operating temperature, or at least to be above the room temperature, before the molding starts.

The properties of softened material, such as viscosity and specific volume, change with temperature. Any change in viscosity will have a significant effect on the properties of the molded parts as well as on the molding cycle time. The temperature of the cavity wall is of major importance for the ultimate quality of the part, its dimensional stability and reproducibility, and the cost-effectiveness of the process. It is this temperature that, besides the thermal characteristics of the material, determines the cooling rate. The material starts to cool from the beginning of a molding cycle (Richardson 1983).

To apply the appropriate controls to the material cooling in the mold cavity, it is essential to obtain data on the specific heat, the thermal conductivity, and the thermal diffusivity of the ceramic mixes. Such data are unavailable at present because the research needed to determine the thermal properties of ceramic mixes has received very little attention to date. For this reason, specific data on heat transfer coefficients for cooling ceramic mixes in a mold cavity are not provided here. In practice, a constant mold temperature below the softening point of the material is achieved by circulating water through channels in the mold at a constant temperature. A temperature gradient across the mold face can cause differential cooling rates within the mix, resulting in thermal stresses in the molded parts (Hindle 1981; Williams 1981; Edirisinghe 1988).

Experimental Methods for Establishing Temperature and Pressure Limits

Two experimental methods illustrate how to establish temperature and pressure limits for molding a specific material in a specific mold. In the first experiment, a molding range diagram, as shown in Figure 5-18, is developed by plotting the cylinder temperature and injection pressure at which an A16-SG alumina mix (a mixture containing 86 wt% A16-SG alumina and 14 wt% organic binder) can be molded into a rectangular mold cavity using a plunger machine. The molding cycle (i.e., injection speed, mold open, and mold closed times) and mold temperature are held constant. The technique consists of gradually increasing the pressure at a constant temperature until the mold cavity is just full, then increasing the pressure further until the mix sticks or flashes (i.e., leaks). The short-shot point (partially filled mold) and the stick or flash (leakage) point are recorded. The cylinder temperature then is increased 20°C, and a new set of points is recorded. The tests are performed between 18 MPa and 85 MPa injection pressure and 80°C and 140°C cylinder temperature.

The mix can be molded successfully at any of the combinations of injection pressures and cylinder temperatures that fall between these curves. The shaded area is considered the optimal range (Rosato 1986a). A similar molding range is shown for an SN502 silicon nitride, a type C zirconia, and milled silicon mixes. (These ceramic powder designations are explained in Chapter 2.) This test can be performed using a new mix with the same ceramic filler or with a different filler. The mix formulations used in the test are shown in Table 5-2.

A spiral flow test provides the same type of information about cylinder temperature and injection pressure for a moldable ceramic mixture, and can be used as an alternative. In this test, a mold with a long channel in which flow lengths can be measured is substituted for the standard mold cavity. A convenient form of such a long channel is an elongated spiral with an increasing radius per revolution. Each section is a half round in diameter, as shown in Figure 5-19. The resulting channel length is 508 cm in a mold with overall dimensions of 35.56 cm by 27.43 cm. The mold temperature is controlled, and a full open gate is used.

One of the problems associated with this test is the feed setting. In normal injection molding, the feed is set to supply a sufficient quantity of material to fill the mold cavity under the conditions of temperature and pressure used; but with this spiral mold there is no fixed setting, because the object is not to fill the mold completely but rather to measure the extent of flow achieved.

Using fixed values of cylinder temperature, mold temperature, and injection pressure, the optimal feed setting can be established experimentally for each set of conditions by starting with a small quantity of feed and increasing it stepwise until the plunger just fails to complete its full forward stroke. This indicates the cessation of flow into the mold. If the feed then is held constant, succeeding shots yield spirals of about the same length, provided that all the other conditions remain unchanged. The procedure

TABLE 5-2.* **The composition of different mixes with indicated powders.**

	A16-SG Al_2O_3				Type C ZrO_2				SN502 Si_3N_4			
Composition[a]	K	Q	Y	Z	K	Q	Y	Z	K	Q	Y	Z
Powder	86.0	86.0	87.6	86.0	92.0	92.0	92.0	92.0	82.0	82.0	82.0	82.0
1865Q	9.0				4.25				11.25			
BASF-A		5.5	5.0			3.25	3.25			7.5	7.5	
Be[2] 170/175			5.0				3.25				7.5	
Shell-120	1.0				1.0				1.0			
Atactic polypropylene		5.5				3.25				7.5		
AC-400				9.1				5.2				12.8
AC-6	3.0				1.75				3.75			
Stearic acid	1.0				1.0				2.0	2.0	2.0	2.0
Butyl stearate				3.5				2.0				1.6
Castor oil		2.0	2.0	1.4		1.0	1.0	0.8				1.6
Corn oil		1.0	1.0			0.5	0.5			1.0	1.0	

a. Compositions in weight percent.
*From Battelle Columbus, OH.

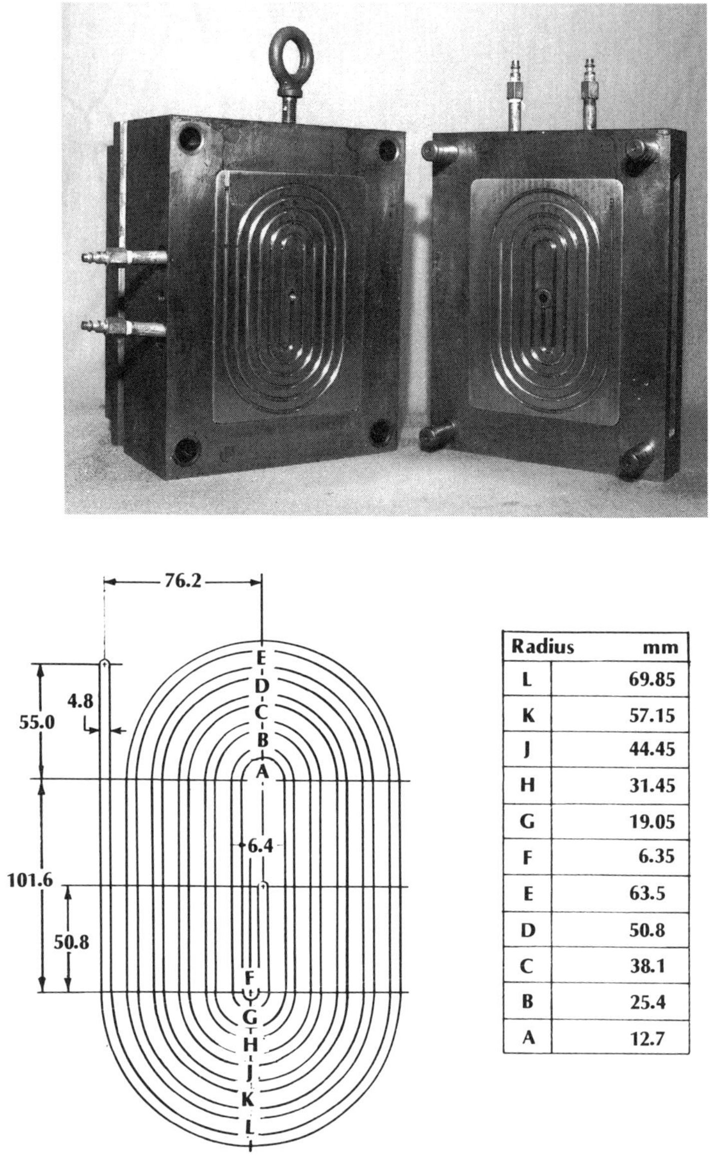

Radius	mm
L	69.85
K	57.15
J	44.45
H	31.45
G	19.05
F	6.35
E	63.5
D	50.8
C	38.1
B	25.4
A	12.7

FIGURE 5-19 Spiral flow mold for screening material/binder compositions. From Battelle Columbus, OH

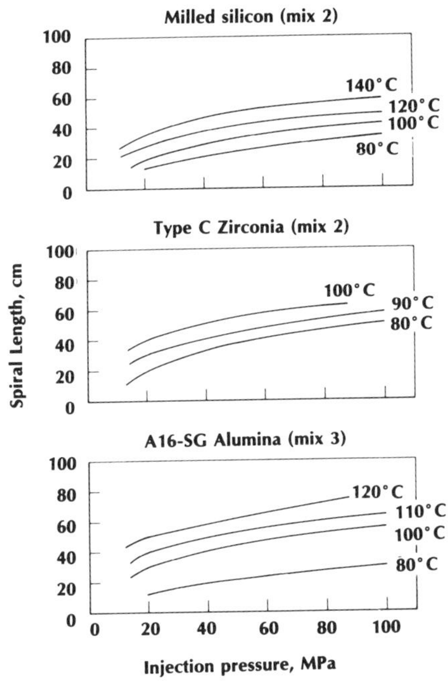

FIGURE 5-20 Spiral flow data for various molding formulations. From Battelle Columbus, OH

followed is that once equilibrium conditions are established, the average length of five successive spirals is taken as the flow value for that particular set of conditions.

The spiral flow data for three different ceramic powders and binders are shown in Figure 5-20. The ceramic powder and binder composition and the total concentration differ for the different ceramic powders. The data shown here are valid only for a specific powder with defined characteristics. Similarly, the data are applicable only for a specific binder formulation and concentration. However, as indicated in Figure 5-21, a correlation can be shown between fluidity (the reciprocal of capillary viscosity) and spiral flow data as a function of temperature (Mutsuddy and Kahn 1984).

To establish the maximum speed at which a material can be molded, multiple tests should be performed over a range of mold and cylinder temperature combinations. The shot size should not be greater than 75% of the machine capacity in order to avoid exceeding the cylinder's plasticating rate. At any one set of conditions, at least 50 to 100 shots are required to assure temperature equilibrium. The minimum molding temperature

TABLE 5-3.* **Molding conditions for various ceramic–binder mixes.**

Condition	Mix K Si	Mix K ZrO_2	Mix K Al_2O_3	Mix K Si_3N_4	Mix K Al_2O_3
Shot size (%)	74	55	68	66	53
Cushion (cm)	0.125	0.125	0.125	0.125	0.125
Injection time (sec)	3.5	3.0	3.5	3.5	3.0
Hold time (sec)	2.0	2.0	5.0	2.0	10.0
Cure time (sec)	10.0	15.0	10.0	15.0	15.0
Mold open time (sec)	2.0	2.5	2.5	2.5	2.5
Injection velocity (%)	80	80	80	80	80
Injection pressure (MPa)	34.5	34.5	34.5	69.0	34.5
Hold pressure (MPa)	6.9	6.9	6.9	6.9	6.9
Plasticating pressure (MPa)	3.8	3.8	3.8	3.8	3.8
Zone 1 temp. (°C)	125	80	105	105	125
Zone 2 temp. (°C)	125	80	105	105	125
Nozzle temp. (°C)	56	28	47	42	56
Mold temp. (°C)	55	31	40	45	40
Mold shape	Cup	Vane	Vane	Vane	Cup
Fully automated		Yes	Yes	Yes	
Mold releaser	None	None	None	Zinc stearate	Zinc stearate

*From Battelle Columbus, OH

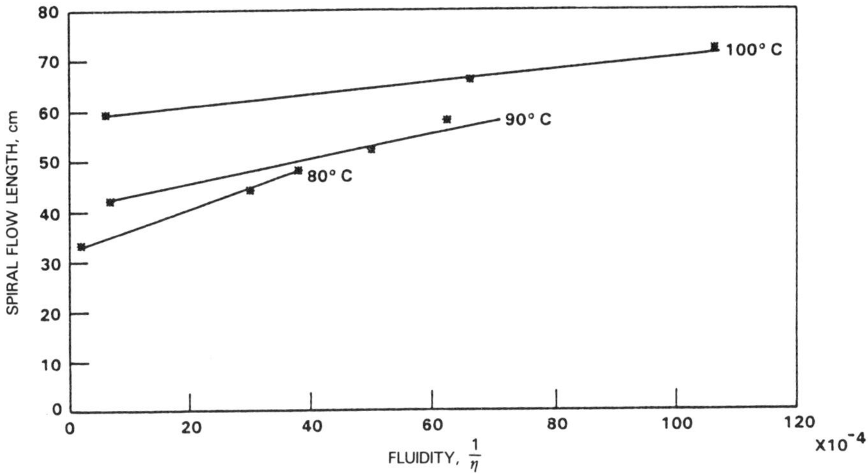

FIGURE 5-21 Fluidity as a function of spiral flow length at different loadings of A16-SG alumina.

ascertained from the molding range diagram (Figure 5-18) can be used as a basis for determining the minimum cycle time. The cycle time used for plotting the molding range diagram can be selected as the starting point, and the booster time should be added to facilitate filling.

The minimum screw/plunger forward time should be determined, while a constant cycle is maintained, by adding any time removed from the screw/plunger forward time to the mold closed time and by keeping the mold open time constant. After this time is established, the mold closed time should be decreased, thus reducing the overall cycle time. When satisfactory parts no longer are produced, the tests at these mold temperatures are completed. They should be repeated for several mold temperatures. The selection of cylinder and mold temperature combinations is based on using a lower cylinder temperature with a high mold temperature or a high cylinder temperature with a low mold temperature. Typical molding conditions for various ceramic/binder mixes are shown in Table 5-3.

In addition to the machine and mold variables, the properties of the ceramic–binder mix should be investigated thoroughly. This examination should include the mix's flow characteristics, its thermal properties, and its feed characteristics plus an evaluation of filling and packing behavior during actual molding (Chung and Ryan 1981; Agrawal et al. 1987). In this discussion of filling and packing, mix formulations, designated the K series, are used. Their compositions are given in Table 5-2.

Cavity filling characteristics of type C zirconia, A16-SG alumina, and milled SN502 silicon nitride compositions are illustrated with a rectangular-shaped cavity and an airfoil shape. A full-round gate with a cross-sectional area of 3.94 mm^2 is used, and the molding conditions are those given in Table 5-3. Partial shots are used to follow the sequence of flow. As shown in Figure 5-22a, the type C zirconia mix fills the cavity from its far end back toward the gate location. Because of their viscous nature, the A16-SG alumina and the milled SN502 compositions only partially fill the cavity until some resistance to the free movement of the material injected into the cavity is met. The fill is completed from the gate end of the cavity, as is shown in Figure 5-22b for the A16-SG alumina and in Figure 5-22c for the milled SN502 silicon nitride composition. Figure 5-23 shows the way in which the airfoil shape fills with both type C zirconia and A16-SG alumina mixes. The type C zirconia mix completely fills the platform end of the airfoil opposite the gate and then fills the airfoil, and, finally, the platform adjoining the gate. The A16-SG alumina mix only partially fills each of these three sections of the airfoil shape until the final stages of injection.

The results are indicative of the influence of the viscosity of a mix and of the complex flow patterns that occur even in a relatively simple mold. The

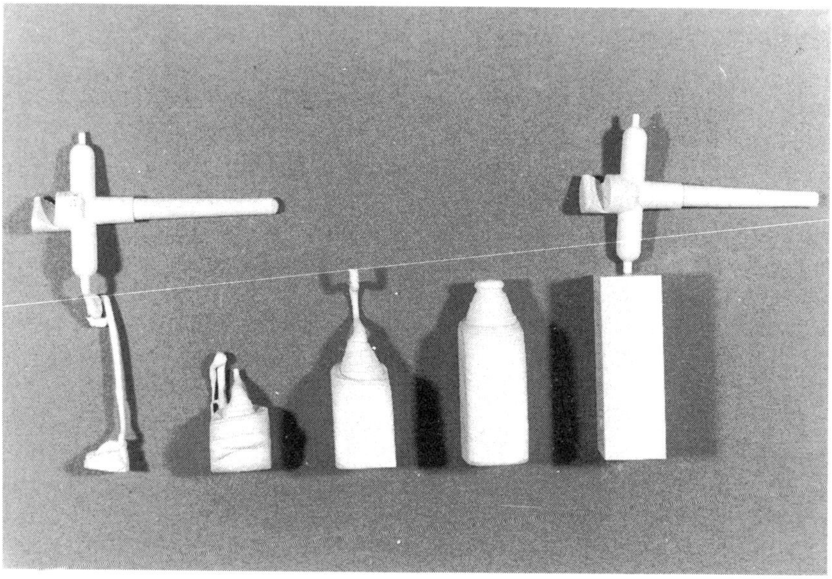

FIGURE 5-22a A type C zirconia mix fills the cavity from its far end back toward the gate. From Battelle Columbus, OH

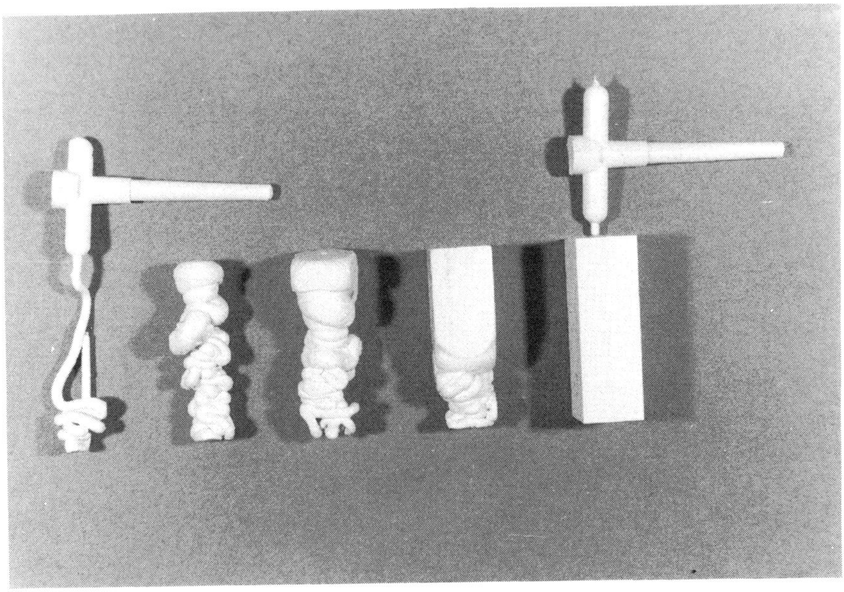

FIGURE 5-22b A viscous A16-SG alumina mix only partially fills the cavity until some resistance to the free movement of the material injected into the cavity is met. From Battelle Columbus, OH

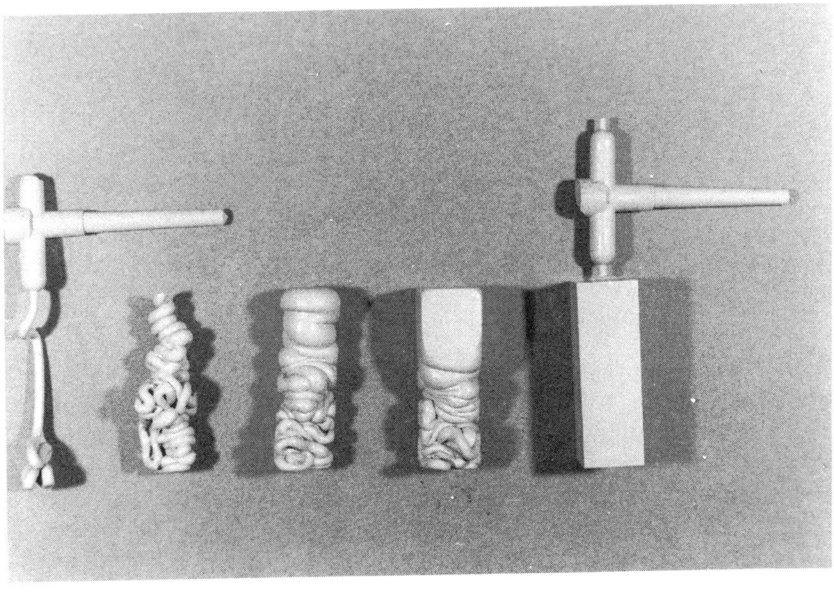

FIGURE 5-22c The behavior of a milled SN502 silicon nitride mix is similar to that of an A16-SG alumina mix. From Battelle Columbus, OH

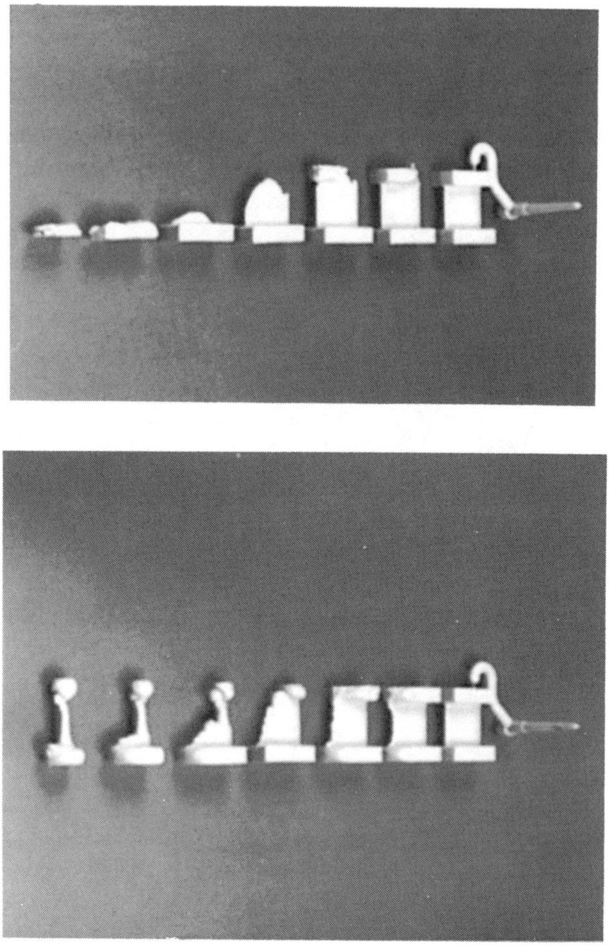

FIGURE 5-23 Airfoil shape fills with both type C zirconia and A16-SG alumina. From Battelle Columbus, OH

flow patterns illustrated in Figure 5-22 will cause internal flaws due to a folding effect and weld lines, but flaws can be minimized by adjusting the mix viscosity and by proper mold design. An optimal flow pattern is illustrated in Figure 5-24, which shows the cavity filling in the gate and progressing toward the opposite end as either a flat or a curved shape.

The packing or cavity pressure usually is recorded by placing one or more pressure transducers behind the mold cavity. Figure 5-25a follows the packing pressure in the airfoil mold cavity as a function of injection time. Although the injection time is the key influence on the packing stresses, any factor that allows excess material to enter the mold will have an effect.

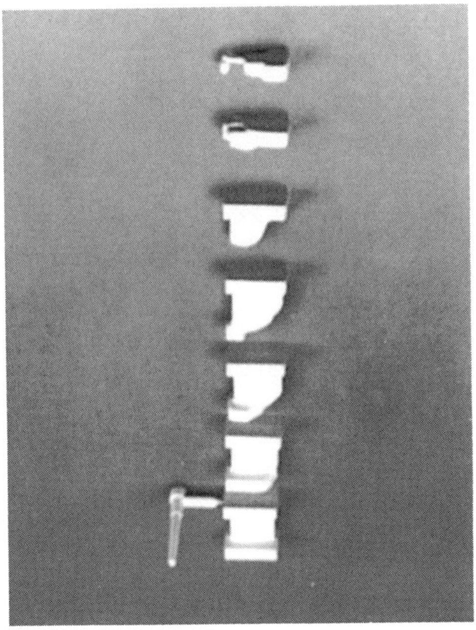

FIGURE 5-24 An optimal flow pattern shows the cavity filling in the gate and progressing toward the opposite end as either a flat or a curved shape. From Battelle Columbus, OH

As can be seen from the pressure transducer curves in Figure 5-25b, the pressure increases rapidly once the cavity is filled. As the injection time increases from 1 sec to 10 sec, the peak cavity pressure also increases. Figure 5-26 shows a recording of the pressure profile in a mold cavity as a function of time and location (Zhang et al. 1990). The pressure (P_1) near the runner and gate increases rapidly and is much higher than the pressure (P_2) at a location 420 mm away from the runner and gate. The increasing difference in the viscosity of the mix between the two locations creates a strong resistance to the transmitting pressure. The pressure losses in the mold cavity could be both dynamic and static.

Dynamic pressure loss is caused by overcoming the resistance of the flow of the melt, as shown in Figure 5-27. However, increasing the melt temperature reduces the dynamic pressure loss, and it becomes independent of the injection pressure.

Static pressure loss is due to frictional resistance when the melt stops flowing in the cavity. The distribution of pressure and residual stress in the cavity results from the static pressure loss shown in Figure 5-28. Here again

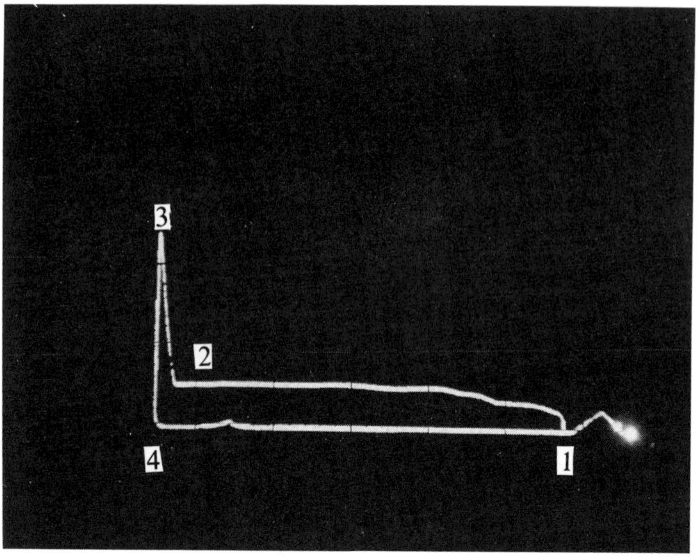

FIGURE 5-25a Packing pressure in the airfoil mold cavity as a function of injection time. The point designated 1 indicates the starting point of the molding cycle, 2 indicates the travel of the screw to fill the cavity, 3 indicates the maximum packing pressure, and 4 indicates the holding pressure cycle. From Battelle Columbus, OH

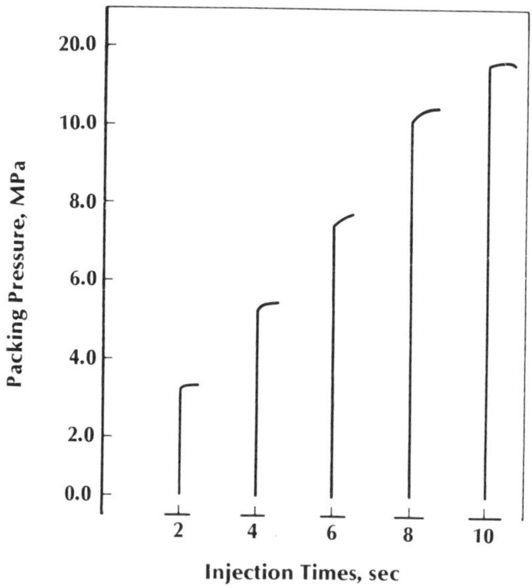

FIGURE 5-25b Recorded packing pressures in the mold cavity. From Battelle Columbus, OH

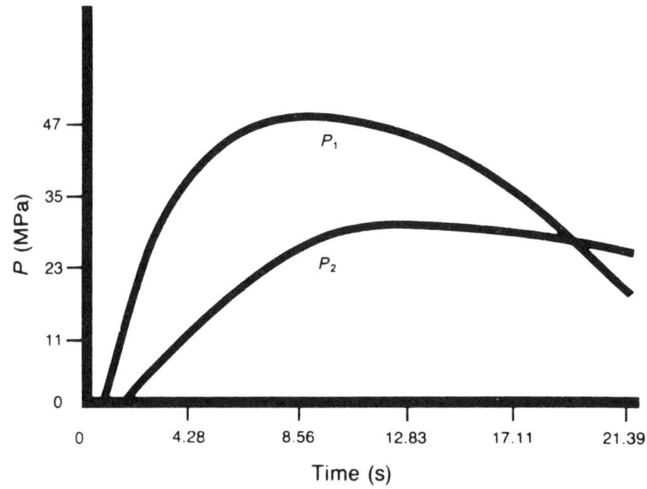

FIGURE 5-26 The difference in the mold cavity pressure with time between two locations in the mold cavity. After Zhang et al. (1990).

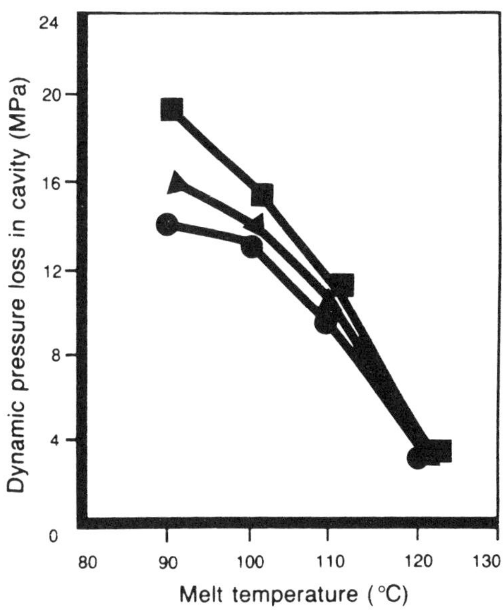

FIGURE 5-27 Dynamic pressure losses in the mold cavity with three injection pressures. After Zhang et al. (1990).

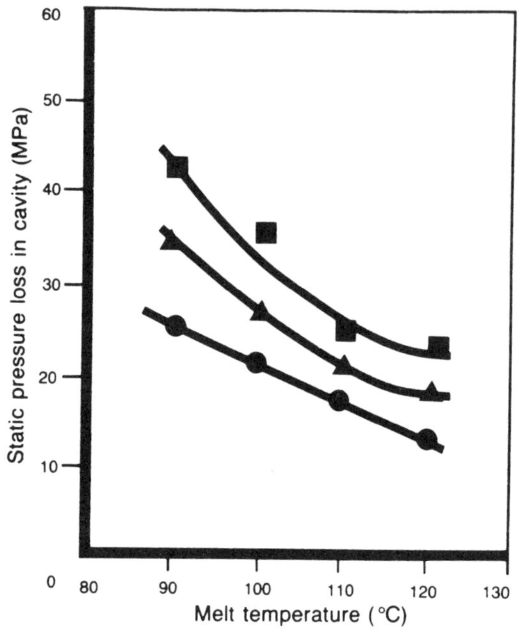

FIGURE 5-28 Static pressure losses in the mold cavity with three injection pressures. After Zhang et al. (1990).

the static pressure loss decreases with increasing temperature at all injection pressures. Furthermore, with an increase in the injection pressure, the friction resistance increases while the static pressure loss decreases. In general, the dynamic pressure loss is much smaller than the static pressure loss at high temperatures.

For ceramic injection molding, the cavity can be filled at a lower injection pressure because the flow length ratio required is usually smaller than that for unfilled polymers. If the molding is carried out at higher injection pressure, the pressure distribution in the cavity will be less uniform than at the lower pressure, and the residual stress will increase.

A calculation of the temperature distribution in a half-circle cavity using a finite difference equation method is shown in Figure 5-29 (Zhang et al. 1990). The boundary condition and the thermal diffusion coefficient were obtained from experimental data. The average thermal diffusion coefficient of the mix (silicon nitride plus an EVA wax binder) is 7 to 10 mm/sec,

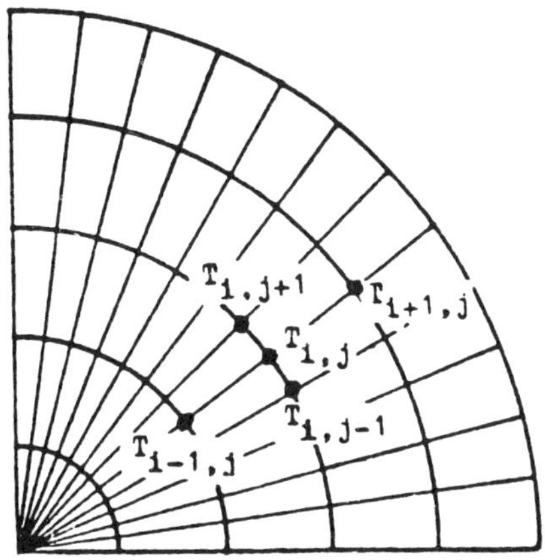

FIGURE 5-29 Temperature distribution in a half-circle cavity. After Zhang et al. (1990).

between 70°C and 120°C. The calculated results, in Figure 5-30, show the relation of time to temperature in the cavity.

As shown in Figure 5-31, maximum cavity pressure, as indicated by the ram position, varies widely from shot to shot. However, with proper control of temperature, flow, and machine sequence, reproducibility in the cavity pressure as a function of ram position can be greatly improved, as can be seen by comparing Figure 5-31a with the results shown in Figure 5-31b. Even tighter control of the maximum pressure and the flow profile can be obtained by using cushion control. Any variation in the cushion size from one shot to the next will cause the pressure and the flow profile to drift, resulting in nonuniformity in the molding.

These pressure curves only represent the pressure at the location in the cavity where the pressure transducer has been placed. Above a certain injection time, the cavity pressure reaches a level where parts will either stick to the cavity or show cracks at the gate area and along the parting line. Therefore, to understand the packing in the mold cavity, it is essential to record the pressure at several points: at the nozzle, at the end of the runner outside the gate, and at a minimum of two locations within the cavity. Monitoring the pressure at these points provides information on pressure differences within the cavity. Furthermore, during packing the material's temperature is dropping because of transient heat conduction to the relatively cold walls of the mold. Thus, these curves also represent the cooling of the material in the cavity.

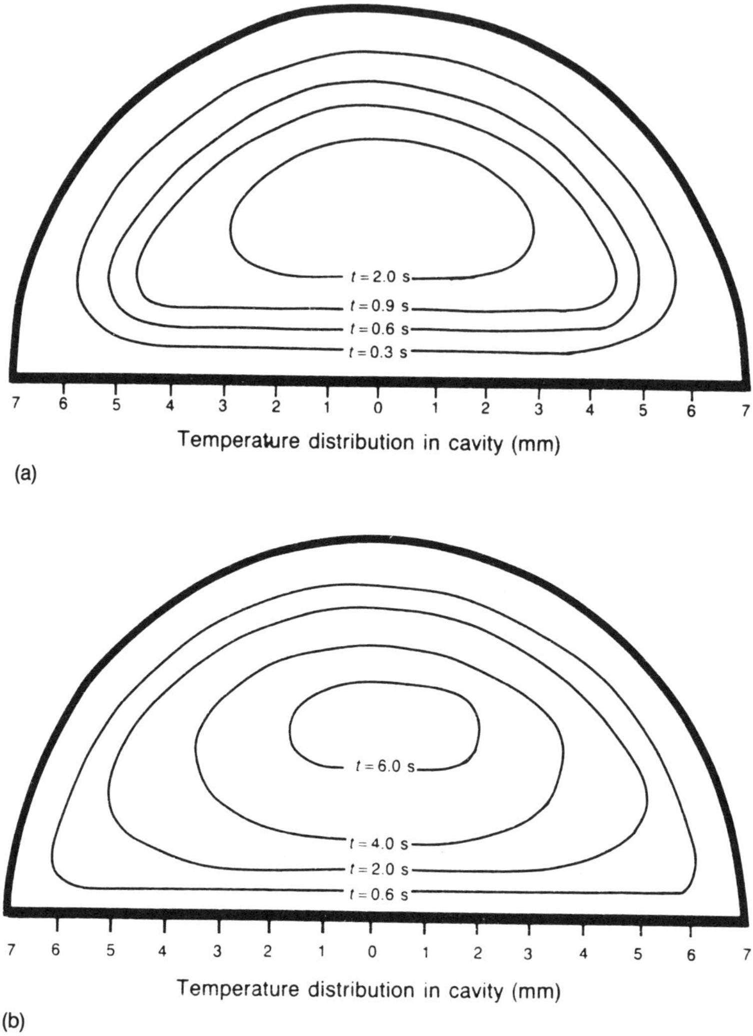

FIGURE 5-30 The relation between temperature and time in a mold cavity.

Flow Analysis

Because mold filling is a complex process, flow visualization studies are useful and necessary, both for the actual mold design and for producing defect-free parts (Wang et al. 1975, 1976; Stevenson 1978). There is very little information available about filling a mold with a polymeric system containing a large quantity of ceramic filler; however, a great deal of

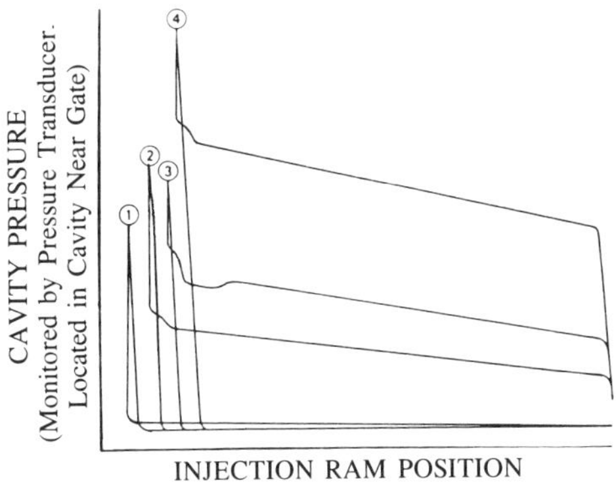

FIGURE 5-31a Variation in the maximum cavity pressure as a function of plunger position.

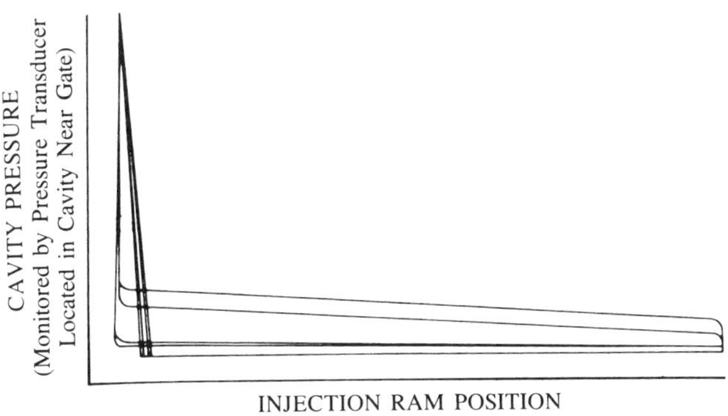

FIGURE 5-31b Improved reproducibility in the mold cavity pressure as a function of plunger position with proper sequence.

information and understanding can be obtained by modeling selected aspects of the filling problem in isolated flow regions, such as a gate area and around an insert with sharp angles (Kreuger and Tadmor 1980).

Each of the regions requires a unique approach and special mathematical tools. The major objective is to be able to predict the point where the cavity is full as a function of melt temperature, mold temperature, gate restriction, injection pressure, volumetric flow rate, and mix rheology.

Several numerical simulations of molding dynamics have been developed and shown to be in good agreement with experimental data (Richardson et al. 1980; Ryan and Chung 1980). The development of digital computers able to perform complex calculations in short periods of time was the breakthrough that made flow analysis a valuable tool that can be applied both to designing parts and molds and to the molding process.

The underlying principles of flow analysis involve complex mathematics. However, a useful modified finite element iterative technique has been developed that breaks down the part into a small number of triangular elements connected at points called nodes, and then forms a complete finite element grid by selecting one node as an injection molding point and specifing the fill time (Austin 1986). Because the volume is known, the flow rate into the cavity from the selected point or node is fixed.

The program runs a finite element analysis assuming a nodal flow rate at each node. This gives a pressure distribution throughout the cavity. In one sense, the computer program has "guessed" a filling pattern. The process can be repeated until the flow pattern becomes stabilized, and a total picture is formed. A temperature distribution profile is developed at the same time as the flow pattern. The final output is the pressure and the temperature at each nodal point. This technique now is being used by engineers working with injection molding.

Molding Faults and Probable Causes

Pressure on the material, temperature, and time during the molding process all affect the final part produced. Equilibrium conditions during molding are sensitive and highly interdependent, but measurements of these key parameters that must be controlled can only be made indirectly. The temperatures of the cylinder and the nozzle are measured instead of the material's temperature, which cannot be determined directly. Similarly, the hydraulic pressure of the oil behind the injection cylinder is measured instead of the pressure on the mix in the cylinder, nozzle, and mold.

Because the actual parameters needing control, such as the temperature of the material, cannot be directly measured, it is difficult to prevent the production of defective parts. Defects can appear at any point in the production process, from starting a new part through regular production operations. The first step in correcting a problem causing a defect is to pinpoint the problem area—the machine, the mold, the operating conditions, the material, or the part design. Table 5-4 lists possible molding faults. It also includes a matrix showing probable causes, which should be helpful in developing a systematic approach for identifying and solving a defect-causing problem.

MOLD DESIGN

The key contributors to the success of injection molding are the efficiency of the mold design and the quality of the mold construction (Sors et al.

TABLE 5-4. A matrix showing molding problems and probable causes.

Legend
A = Major
B = Secondary
C = Related

Causes			Problems								
			Jetting	Flowlines	Skin	Cracks	Shot Short	Sink Mark	Laminations	Sticking	Flashing
Process Adjustable		Cavity Pressure				A	A	C	B		B
		Injection Pressure	B	A		B			A		C
		Hold Pressure			C	A		C	B	A	B
		Back Pressure		C			C		C	C	
		Melt Condition	B	B	C	C			C		
		Cushion					A	C			
		Injection Profile	B	A	B	C		C			C
		Injection Speed	B	A	B	A	C	C			C
		Cool Time	C	C		C		B		A	
		Hold Time			C	B	C	C			
		Shot Size		B			A	C	B		
		Barrel/Nozzle Temperature	A	B		A		A	B		A
Mold		Mold Temperature		A	B	A	C	A	B	A	B
		Runner System	C	C			C	C	C		
		Gate System	A	A	B	C	C	C	C		
Material and Machine		Clamp Pressure									C
		Ejector System				A					
		Material Characteristics	A	A	A	B		A	B	B	

1971; Patton 1976; Du Bois and Pribble 1978). Before designing a mold, one should consider the design of the desired parts. The feasibility of fabricating a given part by injection molding should be based on factors such as its size and weight; its radius; the thickness of the section; anticipated shrinkage; required tolerances; its draft (or taper); and the presence, number, size, and location of threads and holes. After a thorough review of all information about the part that will be molded, the factors discussed in this chapter should be taken into account in designing the mold.

The Sprue

The sprue provides the path for the flow of the hot plasticized materials from the nozzle to the runner, which carries it to the cavity. The part of the mold that forms this path is called the sprue bushing. The top end of the sprue bushing contains a depression that fits the nozzle of the heated cylinder, as shown in Figure 5-32. The standard size for the small end of the sprue is 0.47 cm in diameter, and it is rarely larger than 0.64 cm in diameter. To ensure the ready removal of the sprue when the mold is opened and to permit withdrawing the plastic mix that forms in the orifice, the diameter of the small end of the sprue should be approximately 0.08 cm larger than the diameter of the nozzle. A nozzle with too small a diameter can increase the total cycle time, allow air to be entrapped, and introduce strains in the material being molded (Rosato 1986a).

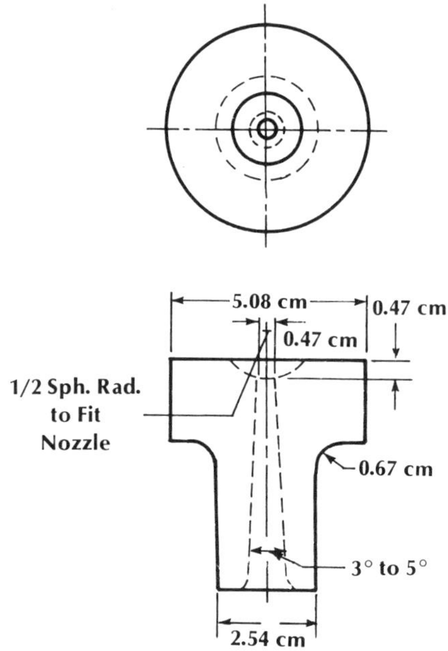

FIGURE 5-32 The design of a sprue bushing for injection molding.

Other dimensions of the sprue bushing in Figure 5-32 are related to its basic proportions and can be varied by the mold designer. However, the area allocated for the bushing must be large enough to allow the head of the sprue bushing to distribute the pressure rather than to transmit it directly to the opposite half of the mold. In general, an included angle of 5 degrees between the sprue and the sprue bushing allows a 0.16 cm taper along the length of the bushing. The sprue can be withdrawn readily when the mold is open with a taper of this size. If a smaller taper is used, any lines present in the bushing can cause the sprue to stick. With larger tapers, imperfections in the surface are less likely to present problems. Because the sprue bushing extends through the mold to the parting line, it should be hardened for wear resistance.

The Runner

The runner conveys the plastic mix from the large end of the sprue to just outside the gate, and it must be large enough to convey the material to the cavity quickly. The smaller the diameter of the runner is, the greater the pressure needed and the longer the time required to fill the mold. Smaller runners, less than 0.30 cm in diameter, produce less scrap and are easier to trim from the molded part than larger ones. Larger runners, greater than 1 cm in diameter, are more desirable, however, because they produce a smoother surface on the molded part than smaller ones do. Other advantages of a large runner are that the part shrinks less, and because the mold fills quickly, there are fewer flow or knit lines produced in the part. It takes longer for the material to solidify in a large runner; so the pressure must be maintained longer. This aids in minimizing the production of knit lines and voids by permitting additional material to flow into the cavity as the part cools. The optimal range for the diameter of a runner is between 0.32 cm and 0.95 cm.

To help retain the molded part on the ejector side of the mold, whenever possible the runner should be placed in the ejector half. It should be highly polished in the direction of flow and should not have sharp turns. In a multicavity mold, each mold cavity usually is fed by a separate runner. The total area of the main runner should equal the sum of all the branches stemming from it. Because the pressure drop in the runner is approximately proportional to the square of the increase in its length, the shortest runner that can be used is the optimal length for molding.

The arrangement of the mold cavities that results in the shortest runner is obtained when the runners are directed radially from the sprue. In addition, a runner should extend beyond its intersection with a secondary runner in order to provide a cold well for the flow front in the runner. A diagram of the design of a balanced runner for a multicavity mold is shown in Figure 5-33.

As can be seen in Figure 5-34, the optimal shape for a runner is round because this shape permits the maximum volume of material to pass while presenting the minimum surface area. A circular runner, however, requires an accurate fit between the mold halves.

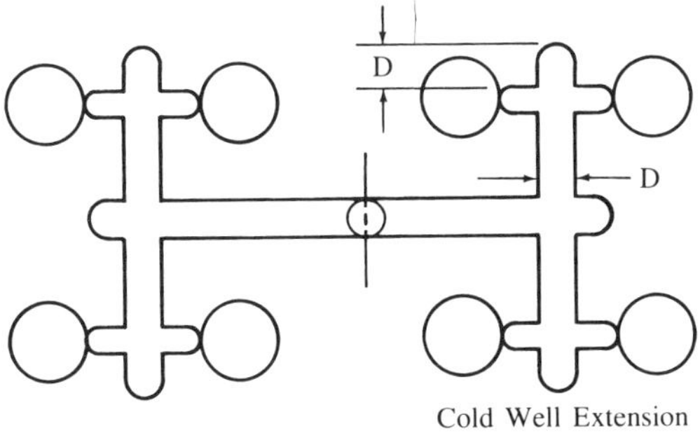

FIGURE 5-33 A balanced runner system for a multicavity mold.

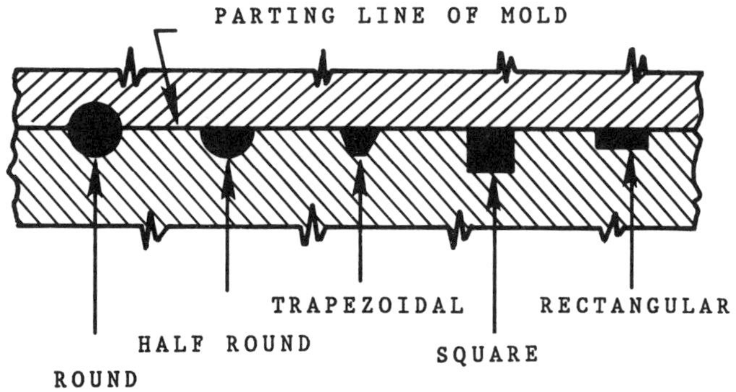

FIGURE 5-34 Various types of runners.

Figure 5-35 shows the cavity-filling characteristics of a 0.63-cm-diameter fully round runner for molding a tension bar. Figure 5-35a shows the way a tension specimen is filled from both ends using type C zirconia (see Table 5-2). As can be seen, the cavity fills the shape out to the ends as the shot size is increased. If the shape is filled from one end only, jetting occurs, and the cavity fills from the opposite end back toward the end with the runner, as shown in Figure 5-35b. If the material is introduced from the side, as shown in Figure 5-35c, filling begins at the location of the gate and progresses toward the opposite end with a steady melt front. It should be evident therefore, that besides the size and the shape of the runner, its location with respect to the geometry of the cavity also is important.

Runnerless molding is well established in plastic molding. The term "runnerless" refers to a runner system that holds the plastic mix in a fluid state. The material does not cool and solidify; so there is no material waste. Also runnerless molding allows a consistent flow of material into the mold cavity.

There are two design concepts in runnerless molding: the hot runner system and the insulated runner system. The hot runner system consists of a cartridge-heated manifold with interior flow passages (Rosato 1986a). The manifold is designed with various insulating features to separate it from the rest of the mold thereby preventing heat transfer.

In insulated runner molds oversized passages are formed in the mold plate. The passages are sufficiently large that, under operating conditions, the thermal characteristics of the mix combined with the heat applied with each shot maintain an open flow path (Rosato 1986a). None of these concepts is now being used in designing molds for ceramic injection molding, but the use of a hot runner could solve many of the flow problems encountered with ceramic injection molding.

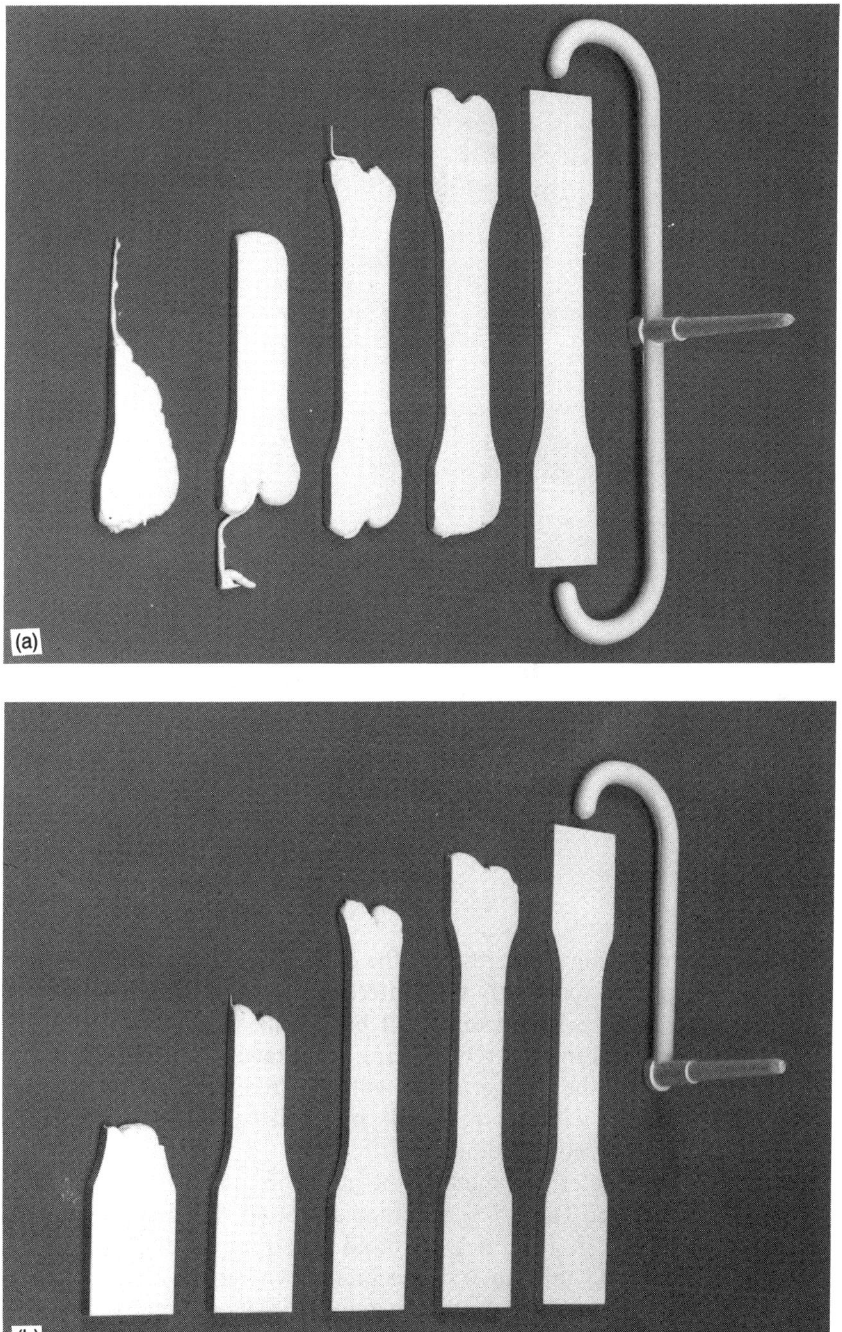

FIGURE 5-35. The effect of runner location in filling tension test specimens: (a) from both ends; (b) from one end; (c) from the side. From Battelle Columbus, OH

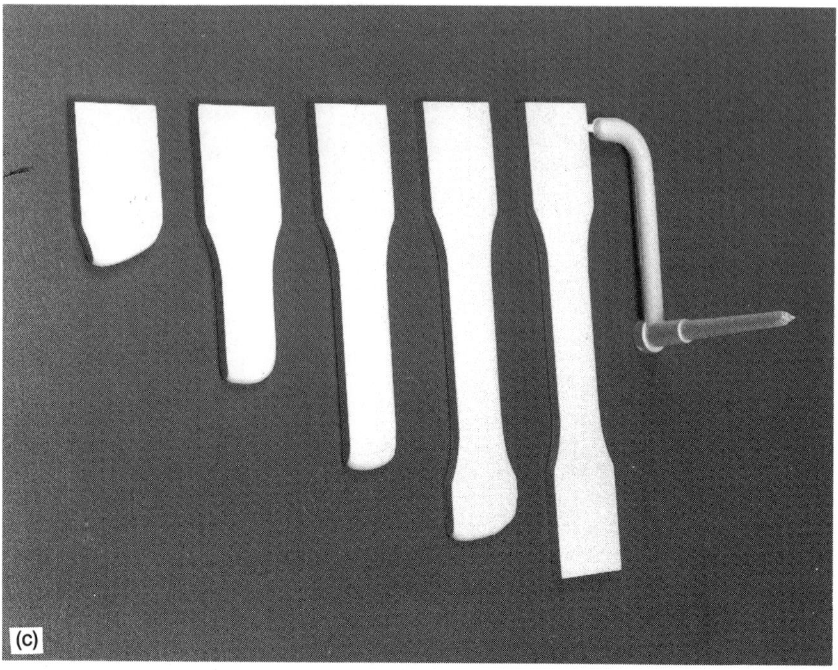

FIGURE 5-35 *contd*

The Gate

The gate is a small orifice that restricts the plastic mix flowing into the mold cavity. Its purpose is to solidify the material as soon as the cavity is filled; then the full mold under pressure will have time to solidify completely while the injection plunger/screw is being withdrawn. If the gate does not solidify the material, the plunger/screw will have to remain extended. When the plunger/screw is withdrawn, plastic mix will be discharged from the cavity if the mix has not solidified.

Several kinds of gates and methods of gating are possible, four of which are discussed here (see Table 5-5): rectangular, round, fan, and fin gates. All have been evaluated in a multicavity mold. The four cavities in the mold consist of a standard tension test specimen, two rectangular cavities of different thicknesses, and a complex shape that includes an airfoil section. A drawing of the mold's layout is shown in Figure 5-36. The tension test specimen and the complex airfoil shape are injected individually as in a single-cavity mold. The two rectangular shapes are injected simultaneously.

TABLE 5-5.* **Gate shapes, dimensions, and areas.**

Gate Shape	Gate Dimensions	Area
Rectangular 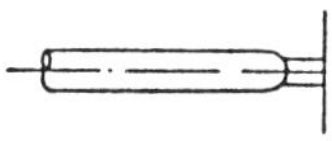	2.95 × 4.5 mm 0.116 × 0.177 in. (Gate for large cavity)	13.28 mm^2 0.0205 in.^{2}
	1.93 × 3.73 mm 0.076 × 0.147 (Gate for small cavity)	7.20 mm^2 0.112 in.^{2}
Round 	2.24 mm diameter 0.088 in. diameter (Gate for large cavity)	3.94 mm^2 0.0061 in^2
	1.70 mm diameter 0.067 in. diameter (Gate for small cavity)	2.27 mm^2 0.0035 in.^{2}
Fan 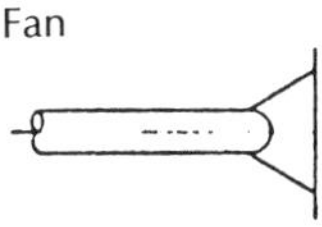 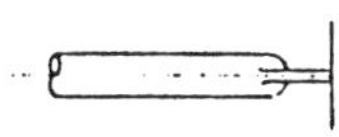	0.89 × 13.3 mm 0.035 × 0.525 in. (Gate size for both cavities)	11.84 mm^2 0.0184 in.^{2}
Fin 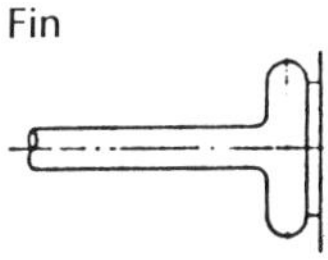 	0.41 × 17.6 mm 0.016 × 0.693 in. (Gate size for both cavities)	7.22 mm^2 0.0111 in.^{2}

*From Battelle Columbus, OH

A photograph of this mold separated at the parting line is shown in Figure 5-37.

The configuration and the dimensions for each of the four cavities are shown in Figure 5-38. A round gate has been used for both the tension test specimen and the airfoil shape, and an optional side gate is provided for the tension test specimen. Four different gates can be used with the two rectangular shapes; the gate location is at the parting line along the narrow dimension. As shown in Figure 5-39, the airfoil shape can be injected from four gate locations.

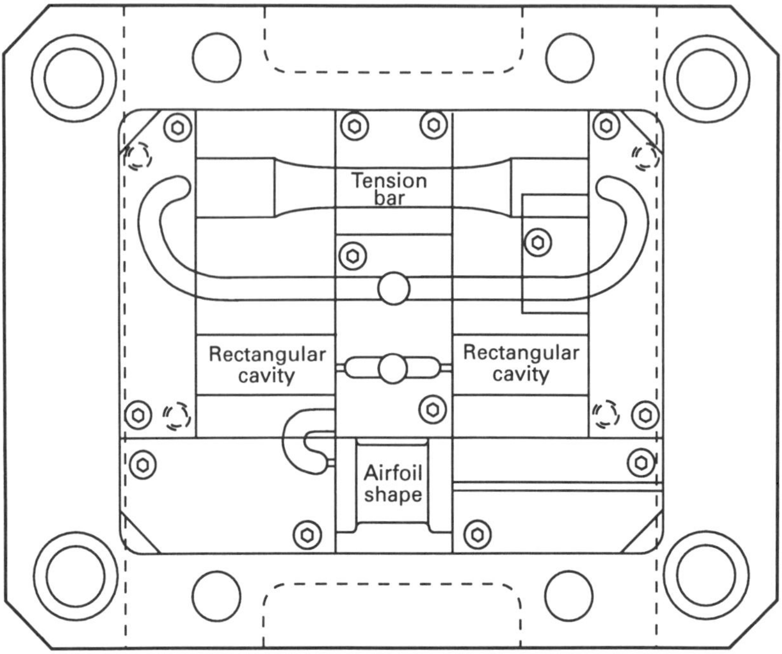

FIGURE 5-36 The layout for a four-cavity mold.

FIGURE 5-37 A four-cavity mold separated at the parting line.

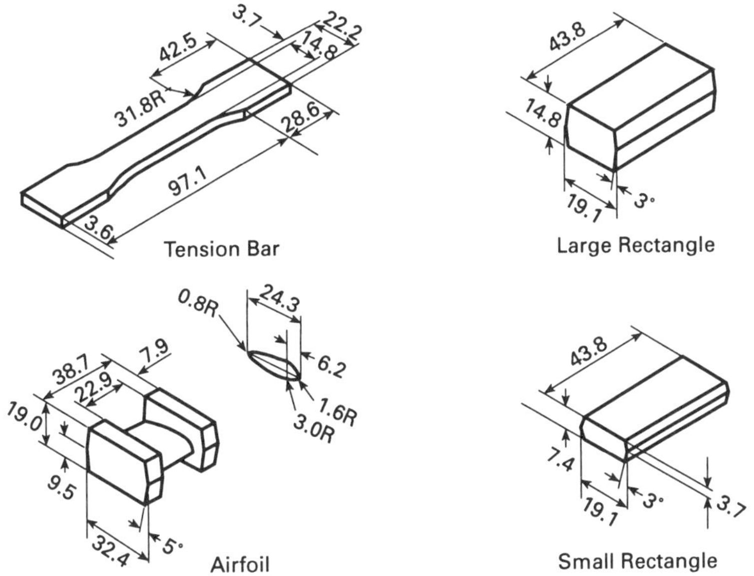

FIGURE 5-38 The configuration and the dimensions for four cavities (clockwise): a tension bar, a large rectangle, a small rectangle, and an airfoil.

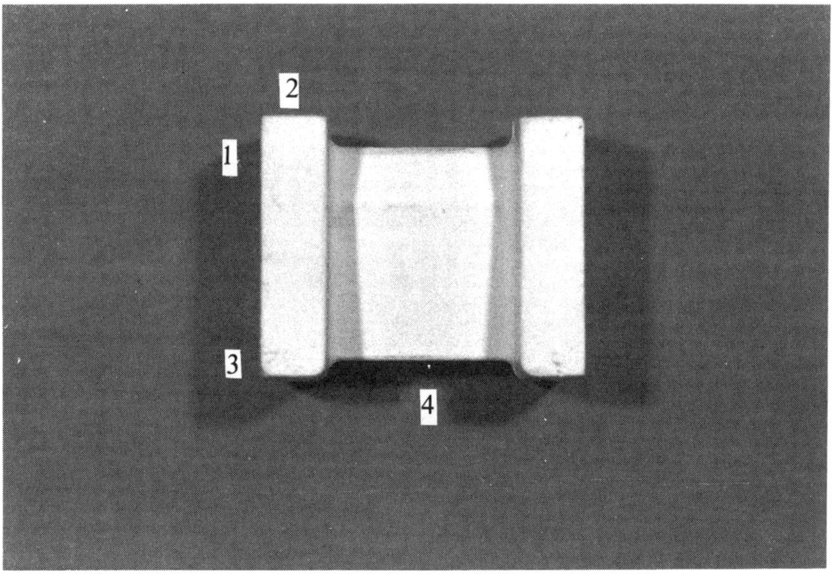

FIGURE 5-39 The location of the gates for an airfoil shape: (1) top platform, (2) side platform, (3) bottom platform, and (4) fan gate. From Battelle Columbus, OH

The Effect of the Gate Size and Shape

The effect of gate size and shape is demonstrated with the four gate shapes described in Table 5-5. Mix compositions of type C zirconia, A16-SG alumina, and milled SN502 silicon nitride (see Table 5-2) are used to illustrate differences in cavity filling. In a full-round gate with a cross-sectional area of 3.94 mm, the type C zirconia mix fills the cavity from its far end back toward the gate location, as shown in Figure 5-22a. The other two mixes, A16-SG alumina and milled silicon nitride, partially fill the cavity until some resistance is met to the free movement of the material injected into the cavity. The fill is completed from the gate end of the cavity, as shown in Figure 5-22b and c.

When changes in both the gate's shape and its size are made in a rectangular gate with a cross-sectional area of 13.28 mm (this is 3.37 times the area of the round gate), the differences in the filling characteristics of the three materials are less apparent. Rather than filling from the end opposite the gate back to the gate (as is the case with the round gate), after partial filling takes place and some resistance to flow within the cavity is met filling occurs from the gate end to the opposite end of the cavity, as shown in Figure 5-40. Type C zirconia, A16-SG alumina, and milled SN502 silicon

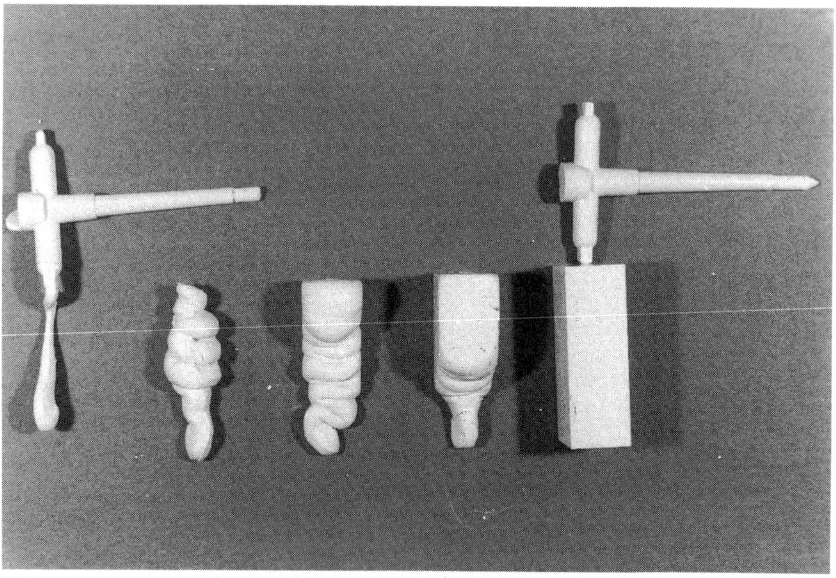

(a)

FIGURE 5-40 Filling the rectangular cavity from a rectangular gate: (a) with a zirconia K mix; (b) with an alumina K mix; (c) with a silicon nitride K mix. From Battelle Columbus, OH

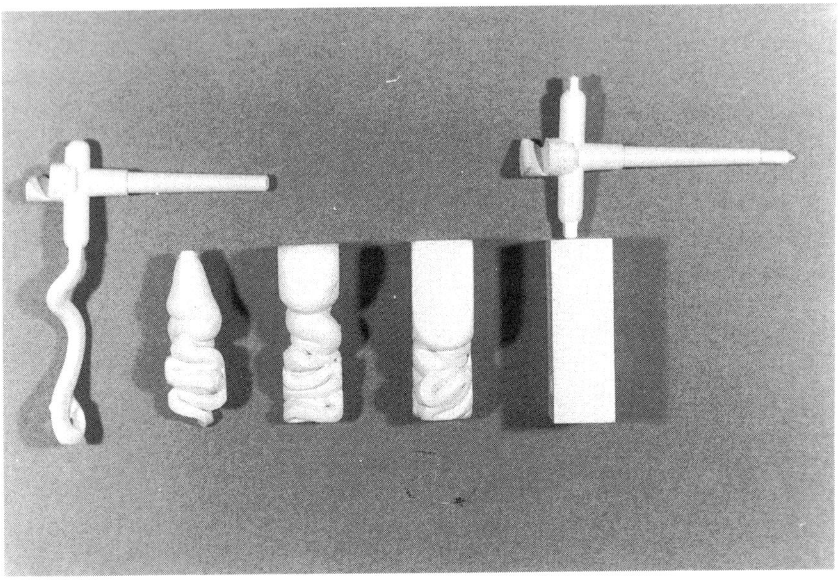

(b)

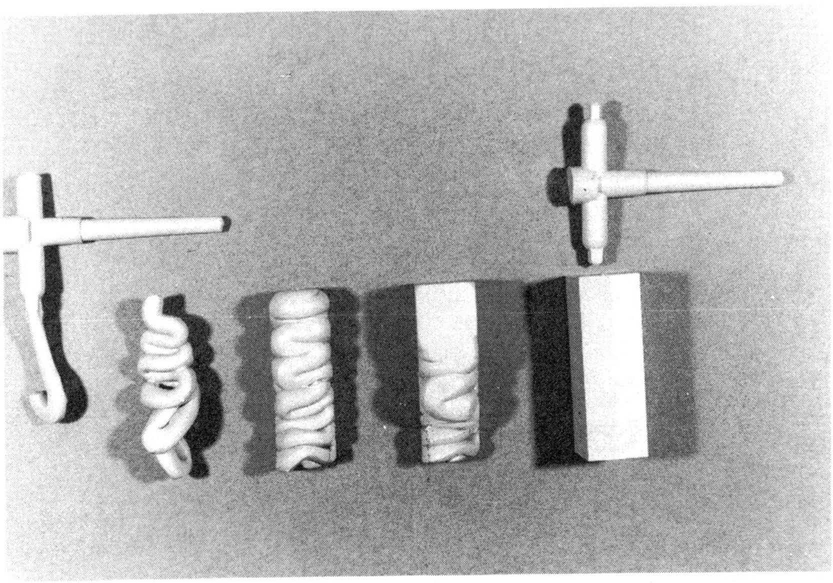

(c)

FIGURE 5-40 *contd*

nitride mixes all act in very much the same manner. As the material fills the cavity, the knit lines in the length of the material previously injected into the cavity are not eliminated until the cavity is full. With type C zirconia, these knit lines are not evident in the material that fills the entire cross-section of the cavity as it moves through the length of the cavity.

For the tension bar cavity with the end gate, the A16-SG alumina mix first loosely fills most of the cavity, and then the remaining areas of the cavity are filled. This type of filling, shown in Figure 5-41a, is not consid-

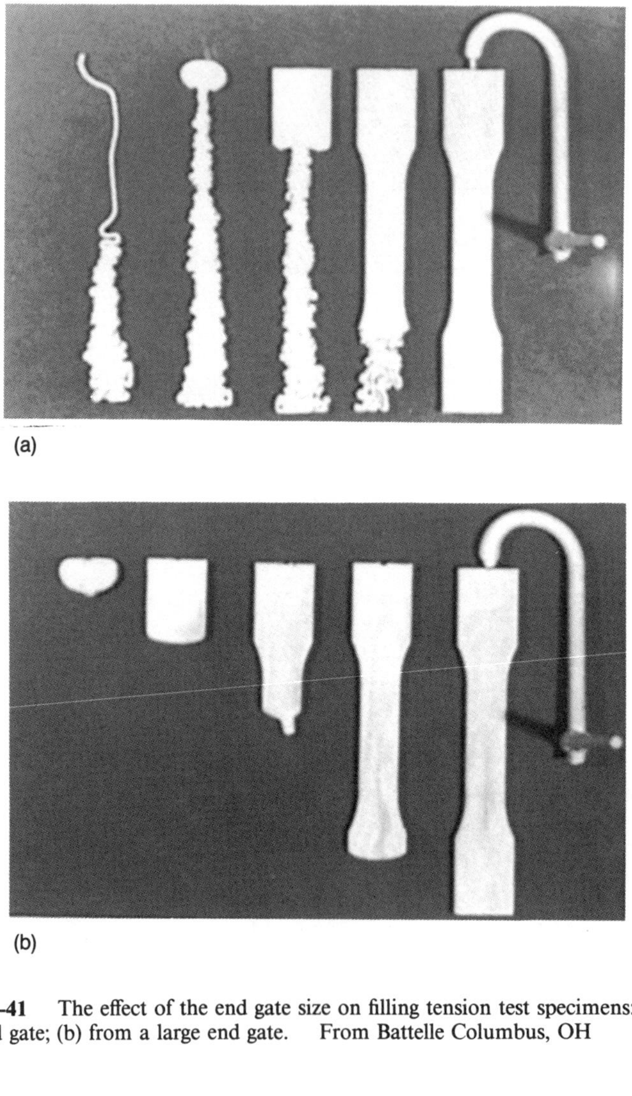

FIGURE 5-41 The effect of the end gate size on filling tension test specimens: (a) from a small end gate; (b) from a large end gate. From Battelle Columbus, OH

ered desirable because the material that is injected first into the cavity can cool before the remaining cavity volume is filled. To reduce the tendency for jetting and to obtain a more controlled filling, the end gate area is increased from a round gate with a 2 mm diameter (6.28 mm^2 in area) to a rectangular area of 3.2 mm $\times$ 4 mm (12.8 mm^2). Increasing the area of the gate decreases the velocity of the material entering the cavity (Boldizar et al. 1990); so when material is injected for molding the tension test bar, which is relatively thin, the material is more likely to contact the sides of the cavity adjacent to the gate and flow into the cavity in a controlled manner. The change in filling behavior with the larger gate is shown in Figure 5-41b. Another advantage of using the increased gate size is that although the velocity is decreased, the volume of material per unit time is increased; thus the chance that the material will cool before the cavity becomes completely filled is minimized. Therefore, when filling is done from one side, a large gate is advantageous if the mold geometry requires injection from the end of the cavity.

The fan and the fin gates are very similar. Both types fill the cavity with ribbon-shaped material. This configuration has a large surface-to-volume ratio compared with the material from the rectangular and the round gates. With all the mixes, the material fills the cavity loosely until sufficient resistance is encountered, which causes the cavity cross-section to become completely filled. The filling then proceeds to the end of the cavity opposite the gate. The ribbon-shaped material that loosely fills the far end of the cavity then is fully compacted.

Knit lines are evident between the ribbon sections of the material first entering the cavity. This is shown in Figure 5-42a for the milled SN502 silicon nitride shape made by using the fan gate, and in Figure 5-42b for the A16-SG alumina shape made by using the fin gate. Also, because of the smaller cross-sectional area of the fin gate, the material tends to jet and fill from the end opposite the gate. When the area of the fin gate is increased to that of the fan gate, the filling behavior is similar for both materials.

Gate Location

For the best molding results, gates should be placed at a relatively thin section of the molded part in order that the flow can be directed toward the thicker section. The effect of gate location is illustrated with the airfoil shape, and the locations of these gates are shown in Figure 5-39. Three of the gates are round, with cross-sectional areas of 2.0 mm^2. The original rectangular gate, 0.8 mm $\times$ 2.3 mm with a cross-sectional area of 1.84 mm^2, was subsequently changed to a fan gate (the fourth gate), 0.8 mm $\times$ 12.5 mm with an area of 10.0 mm^2.

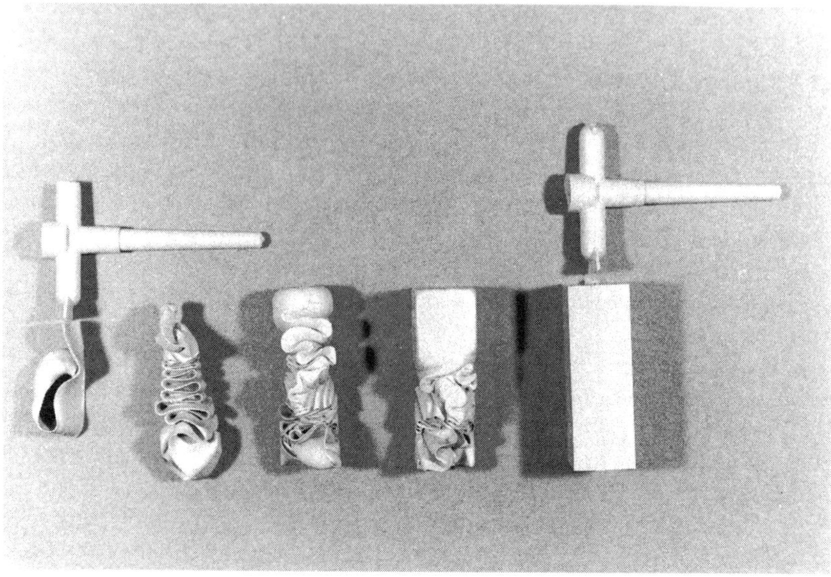

(a)

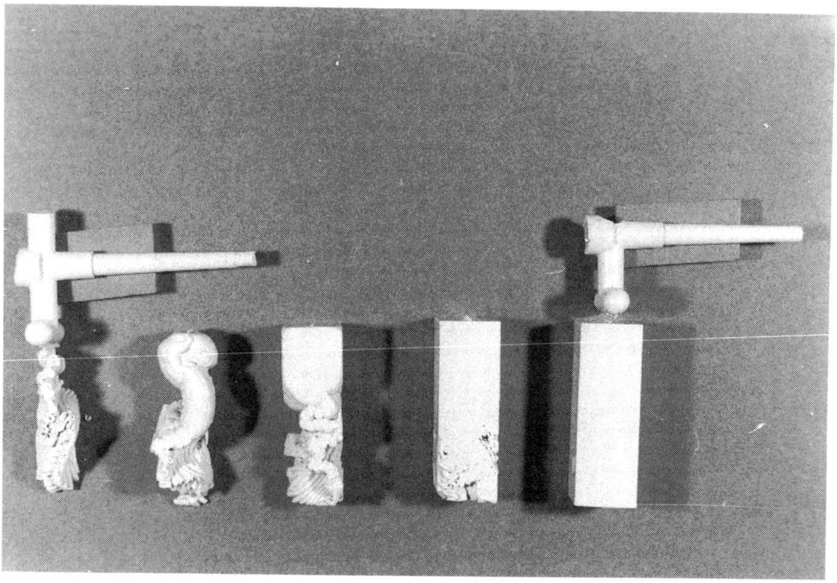

(b)

FIGURE 5-42 Knit lines can be seen when a rectangular cavity is filled from (a) fan and (b) fin gates. From Battelle Columbus, OH

Figure 5-43a shows the airfoil shape filling from the platform where the gate is located (the side gate location). The filling then progresses through the remainder of the cavity. When the gate is moved to the top of the part, as shown in Figure 5-43b, the platform opposite the gate location is filled first, then the airfoil section, and finally the platform adjacent to the gate location. When an A16-SG alumina mix is used with this gate location, the way in which the cavity fills is quite different because of the high viscosity of the mix. As shown in Figure 5-44, this mix only partially fills each of the three sections of the airfoil shape until the final stages of injection. The less viscous type C zirconia mix, however, completely fills the platform end opposite the gate, then fills the airfoil, and, finally, the platform adjoining the gate, as shown in Figure 5-43b. Numerous knit lines are formed in this shape, particularly with the A16-SG alumina.

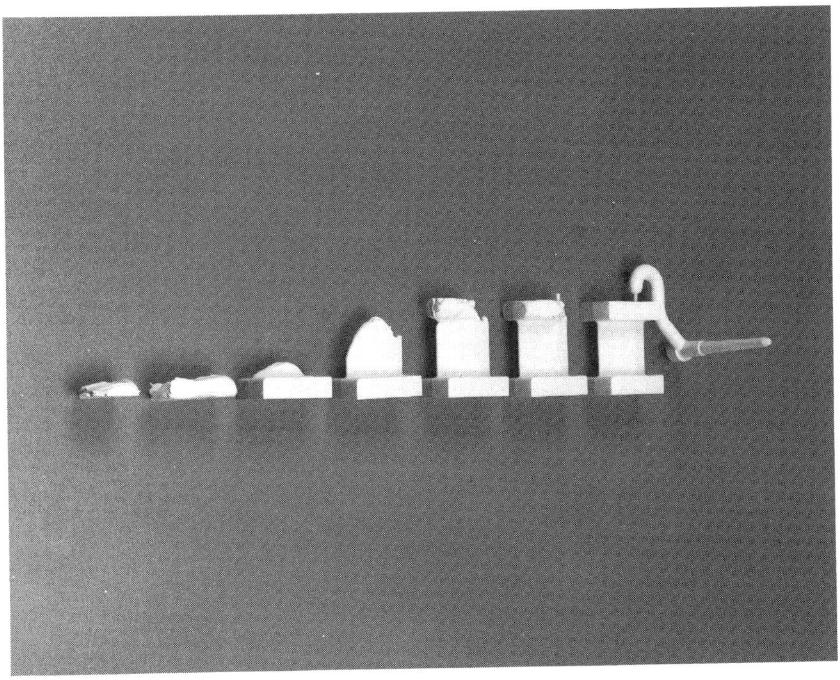

(a)

FIGURE 5-43 The effect of the gate location on filling an airfoil shape with a zirconia K mix: (a) filling from a side gate; (b) filling from a top gate. From Battelle Columbus, OH

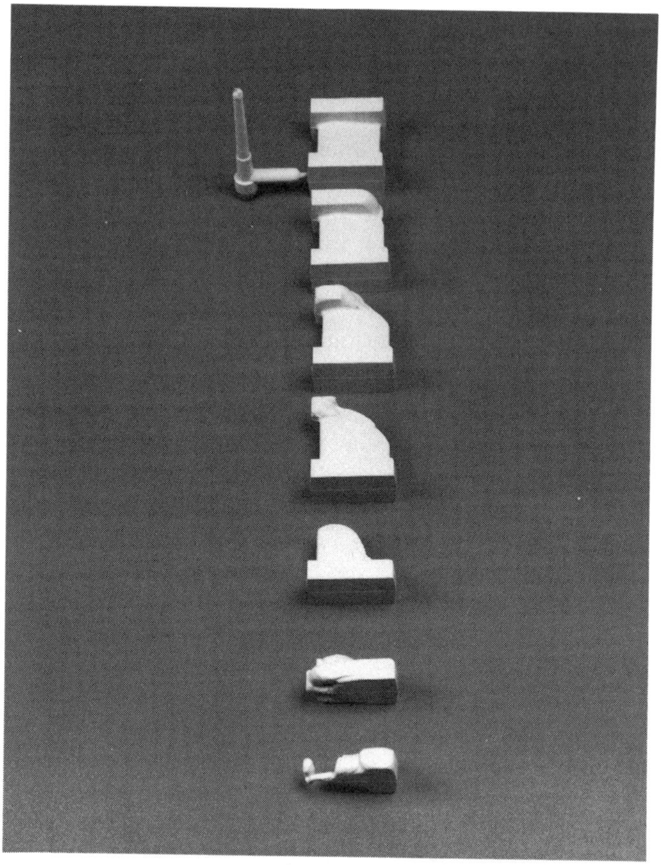

(b)

FIGURE 5-43 *contd*

To reduce the formation of knit lines with viscous mixes, the area of the gate is increased to about 9.5 mm^2, and the gate is repositioned so that the material will impinge on a surface close to the gate location. Figure 5-45a shows the results of this change in the side gate location when an A16-SG alumina mix is used. Although the platform at the gate location fills rather loosely with material before it becomes completely filled, the large gate size reduces the amount of surface area that needs to be joined together. It also results in more rapid filling. Increasing the gate size reduces the number of knit lines and other defects experienced with the smaller diameter round gate. The result of increasing the diameter of the top gate is shown in Figure

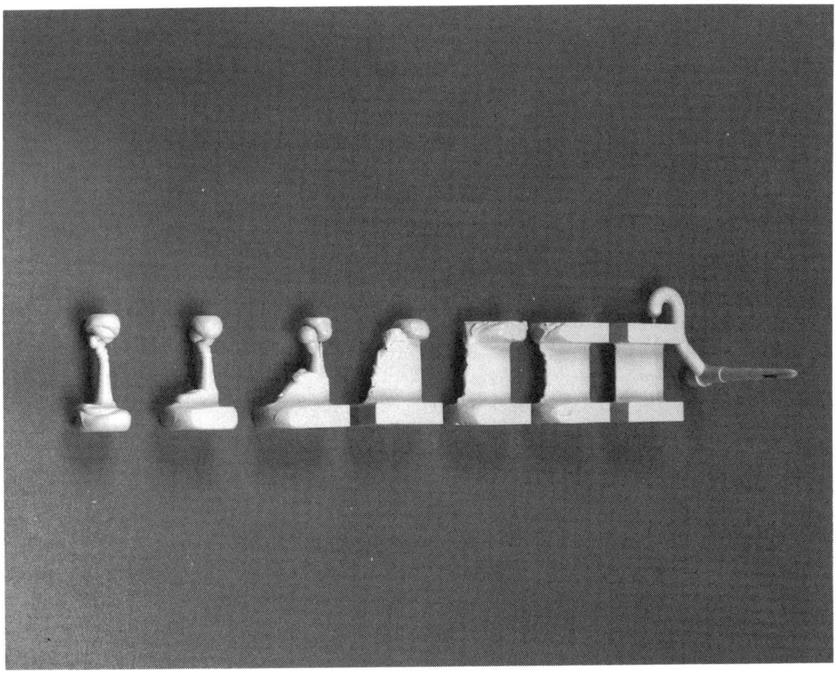

FIGURE 5-44 Filling the airfoil shape with an alumina K mix (compare with Figure 5-43b). From Battelle Columbus, OH

5-45b. As indicated earlier, the direction of the flow of material into the mold also is changed so that the material will impinge on a surface close to the gate rather than be directed toward the platform opposite the gate location as shown in Figure 5-43b. Greater control is possible in filling the cavity with the modified gate than with the smaller-diameter gate, and the number of knit lines is reduced considerably.

Because of the airfoil mold's design, the third gate location cannot be placed in the optimum location, but the results indicate that if the material entering the cavity can come into contact with a surface, cavity filling will be better controlled. The way that the A16-SG alumina mix fills the cavity is shown in Figure 5-45c.

The fourth gate is centrally located along the trail edge of the airfoil section. This was originally a small rectangular gate, 0.8 mm $\times$ 2.3 mm, with a cross-sectional area of 1.84 mm^2. However, as noted earlier, the rectangular gate configuration was subsequently replaced with a fan gate.

(a)

FIGURE 5-45 The effect on the flow pattern of using various gate locations: (a) a larger side gate; (b) a larger top gate; (c) a bottom platform gate; and (d) a fan gate. From Battelle Columbus, OH

The size of this gate was increased incrementally, improving each time until it reached its final size, 0.8 mm x 12.5 mm (10.0 mm^2 in area). The way that this shape fills when an A16-SG alumina mix is used is shown in Figure 5-45d. Because at this gate location the mold section is relatively thin, and also because the material contacts the surfaces of the cavity immediately after leaving the gate, a fan gate is the appropriate one to use here.

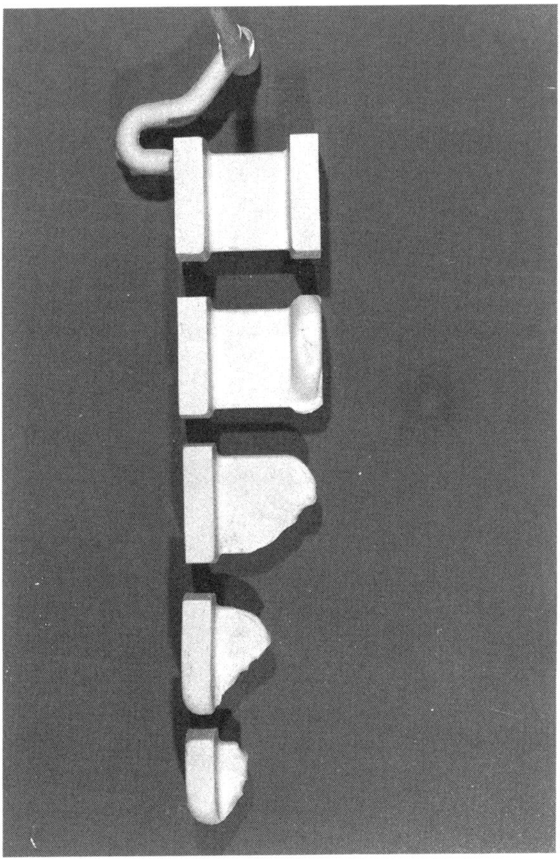

(b)

FIGURE 5-45 *contd*

Effects of Number and Size of Cavities on Gate Design

When multiple cavities are filled during a single injection cycle, two effects can be observed. If the gate size is the same for each cavity, as is the case with the fan and the fin gates, the volume of material injected into each cavity per unit of time is the same. Therefore, the smaller cavity fills first.

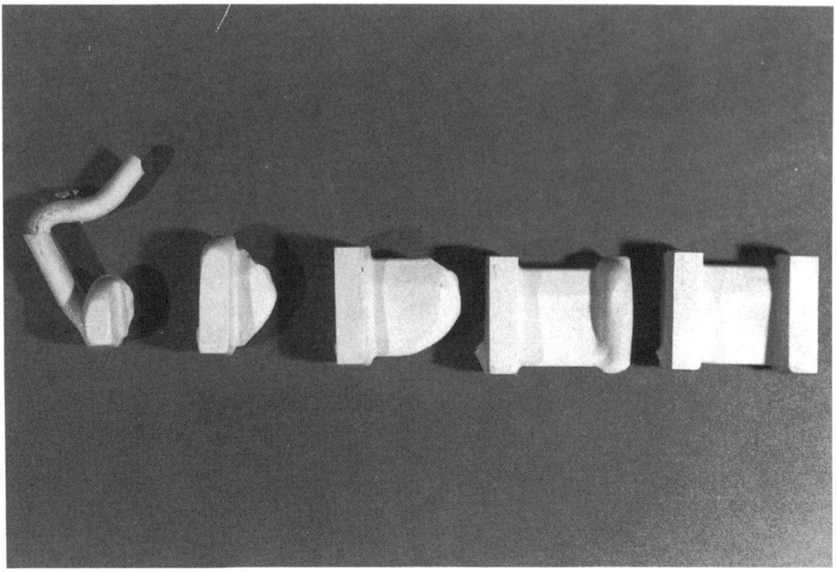

(c)

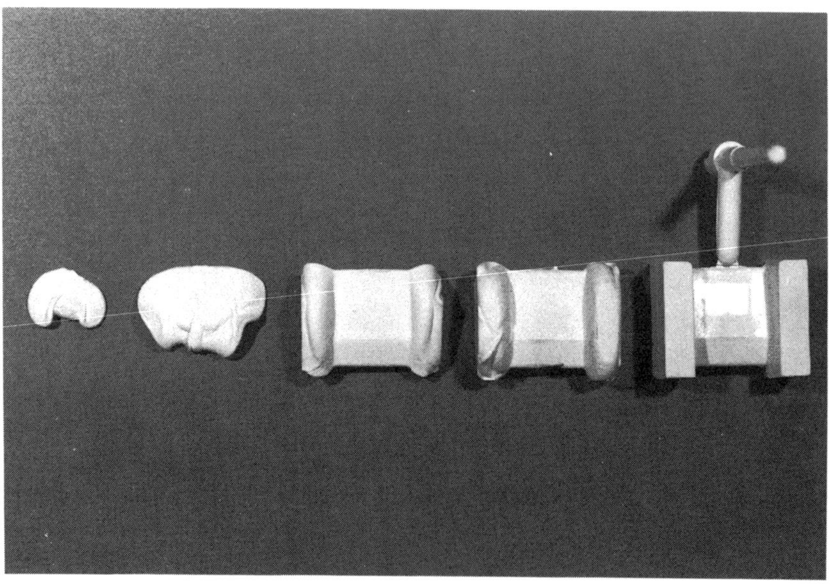

(d)

FIGURE 5-45 *contd*

However, no pressure is exerted within the cavity until the larger cavity is filled. If freezing begins in the smaller cavity before pressure is applied, this will create surface and internal defects. In addition, if the gate size of the smaller cavity is reduced, then in order to balance the filling time for the two cavities, the material will take the path of least resistance by first filling the larger cavity and then the smaller cavity. This will also create surface and internal defects in the molded part.

Venting the Mold

Mold venting should be addressed in association with gating, as experience shows that a vent usually is required opposite every gate. The purpose of a vent is to allow the cavities to fill as quickly as possible without trapping air. Vents should be located at junction points of the streams of material or at parting lines, as shown in Figure 5-46. Loose-fitting knockout pins can be used, and in some cases even scratch marks on fitted surfaces are adequate vents. Experience with the airfoil cavity suggests that a vent can be as large as 0.07 mm deep and 0.7 mm wide. Various schemes for applying a vacuum to mold cavities also have been tried to prevent trapping air. It is highly likely that in the future vacuum molds will be used to prevent venting problems.

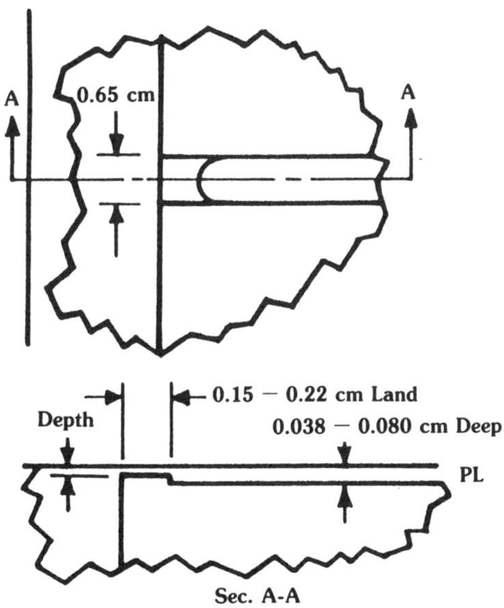

FIGURE 5-46 Schematic of a parting line (PL).

EJECTING THE MOLDED PART

Most molds use ejector pins to push the molded part out of the cavity. Although the part can be removed without the aid of such pins, they should be used to economically produce parts in volume. The multicavity mold described earlier is fitted with four ejector pins. Two are located in the airfoil shape, with one each in the rectangular cavities. A pressure transducer located in the ejector plate is in contact with one of the ejector pins. The signal generated can be used to control its function or to display the pressure in the cavity on an oscilloscope, during operation.

In deciding how many knockout pins to use, the designer should bear in mind that because the material may still be soft when it is ejected from the mold, a number of knockout pins may be required. The pins should be fitted loosely in the knockout-pin plate to prevent binding as they move to eject the molded part. For parts with thin walls, it may be necessary to use a knockout plate because knockout pins might pierce the thin walls.

A sprue puller, which is a device used to pull or draw a molded sprue out of the sprue bushing, is related to an ejector. It is usually a straight round pin with the end machined in the form of an undercut, and it is attached to the knockout or ejector mechanism. It runs through the movable part of the mold in a direct line with the mold entrance (Beck 1980).

Computer Assisted Design and Manufacturing (CAD/CAM) of Molds

During the past decade, CAD/CAM software has been applied extensively to mold design and manufacturing with improved precision resulting in faster turnaround time and overall cost savings.

A key to the implementation of a CAD/CAM program is the effective use of a data base. One of the largest elements in the information structure hierarchy, a data base can be organized in order of increasing content as:

- *Data item*: single unit or specific piece of information.
- *Data record*: collection of data items in a logical sequence.
- *Data file*: collection of data records, generally organized around some common attribute.
- *Data base*: collection of files, organized around some common attribute or organized for a specific purpose.

To construct an effective data base for designing molds, the data must integrate and consolidate all of the information requirements of the engineering and manufacturing operations subsequent to the initial mold design.

The methodology of creating mold designs via CAD/CAM technology differs from drawing the design in that the model is generally a three-dimensional representation of the real part, which completely describes the part without requiring additional views or supplementary information. Because the product model data base is always to scale, information can be extracted directly from the data base without the introduction of scaling errors.

Figure 5-47 illustrates the CAD/CAM mold-making sequence. A preliminary analysis of the product requirements should be initiated before the actual mold design and fabrication process is begun. Specific software programs are available to perform this type of analysis. Once the product requirements have been met satisfactorily, a systematic analysis of various mold components should be undertaken, starting with the location, size, and type of gating necessary for the part. A flow analysis program is used for this purpose.

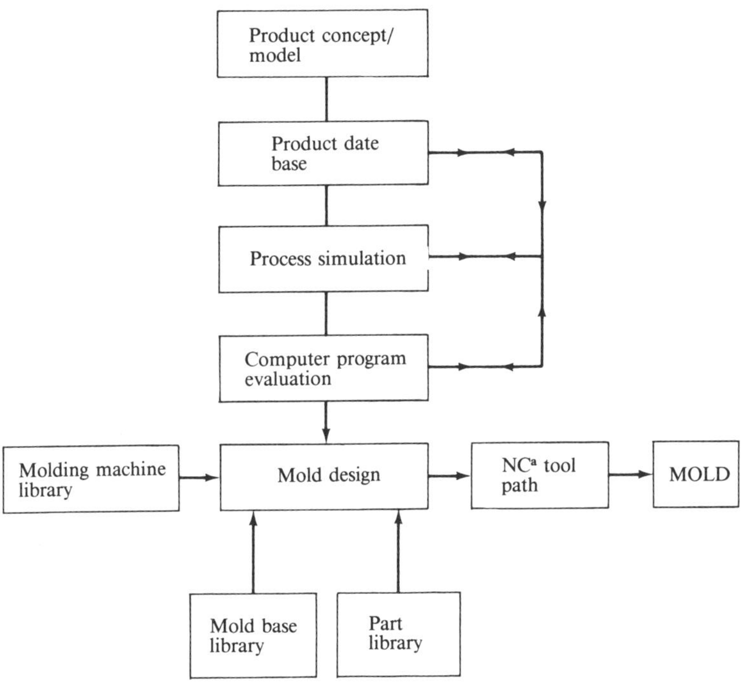

FIGURE 5-47 A CAD/CAM mold-making sequence. a. Numerically controlled. After J. Gosztyla (1986).

Once the gate location and the number of cavities are determined, the computer software can be used to aid in establishing the minimum size of mold base in order to select the appropriate mold base (process simulation sequence). Then, specific mold base information will be generated automatically by the CAD/CAM system from the mold base library. Once the part geometry has been merged into the mold, the cavity blocks can be defined. Attention then should be focused on the design of the runner system. Again, the flow analysis packages come into play.

The next step in the mold design process should be the design of the heating/cooling system. From the description of the waterline geometries and locations, the computer will be able to organize the flow circuits for the optimal combination of pressure drops and constant mold temperature.

The remaining components, such as ejector pins, support pillars, and parting line locks, then can be incorporated into the mold design. Libraries of mold components are available either to create components or to retrieve components on file in the data base. Both types of library methods help to promote the standardization of mold design.

The final steps of the process include creating the necessary detailed drawings and generating numerically controlled (NC) toolpaths to machine the actual mold.

Machine Wear—the Cylinder, the Screw, and the Screw Tip

Relatively little is known about wear on a screw-type machine. The information that does exist, however, suggests that new design concepts should be investigated, and a better, more refractory, grade of metal should be used to manufacture these machines.

Although the following information on machine wear is based on only a limited number of test results, it is included to exemplify the concern about machine wear. After the 100th, 500th, and 1000th shots, the inside diameter of the cylinder, which is made of a nickel-base alloy, has enlarged by 0.014 mm, 0.018 mm, and 0.022 mm, respectively. After about 9000 shots, this diameter has further enlarged about 0.125 mm. The wear is not uniform along the bore but is concentrated at a distance of about 40 mm from the end of the cylinder. This point is about midway in the distance that the check ring is displaced. The screw does not show as much wear as the cylinder; only about 0.0025 to 0.038 mm is worn from the screw's outside diameter. The wear on the sliding-ring screw tip, which is made of a sintered hard alloy of a multiple boride (for example, tungsten and titanium boride) containing iron, after the 100th, 500th, and 1000th shots, has been found to be 0.007 mm, 0.010 mm, and 0.015 mm, respectively. In this case, the wear is fairly uniform along the axis of the screw tip.

Mold Wear

The mold design study indicates that the gate should be located so that the material entering the cavity contacts a surface as soon as possible. The material then will flow into the cavity in a controlled, uniform manner instead of jetting or flowing nonuniformly into the cavity and thus exposing a large surface area of material that subsequently must be compacted when the cavity is filled. A problem could result from the rapid impingement of the material on the surface of the cavity, as cavity wear could occur close to the gate due to the high velocity of the injected material. However, such wear was not observed in this study, perhaps because of the limited number of shots. To prevent such wear, hard metal coatings or hard metal inserts, such as tungsten carbide, are suggested for those mold surfaces that first would contact the injected material.

ALTERNATIVES TO INJECTION MOLDING

Two-stage injection molding was introduced in the plastics industry as early as 1950, both to increase productivity and to minimize degradation of the melt. Recently, a similar two-stage screw-plunger machine has been used to fabricate silicon nitride parts with high reliability (Pasto and Natansohn 1990). In this type of machine, a plasticizer unit is located above an injection unit, as shown in Figure 5-48. After mixing and plasticizing the ceramic–binder mixture, the plasticizer unit feeds it into the injection unit, where the injection ram forces the plasticized mass forward into the mold

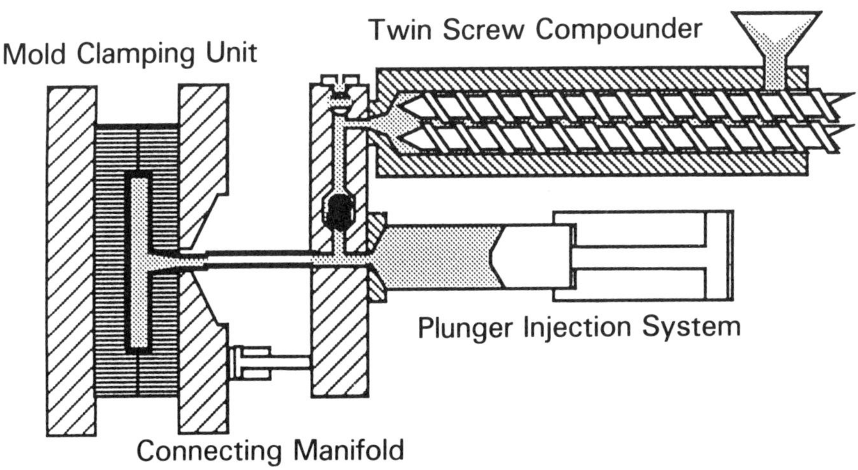

FIGURE 5-48 Two-stage screw-plunger machine. After Pasto and Natansohn (1990).

cavity. The plasticizer unit contains a screw extrusion device, and the extrusion unit operates as a shooting plunger. The ceramic–binder mixture passes over the full length of the screw and through an admission valve into the injection unit, which serves as a combination holding, measuring, and injection chamber.

Two other plastic forming techniques are also being adopted for shaping ceramics—compression molding and transfer molding.

Compression Molding

This process is best suited for ceramic mixes containing a thermosetting binder. There are five steps in the compression molding process:

- Powder, pellets, or preformed discs of the ceramic polymer mix are heated to raise the temperature closer to the curing range.
- The mix is loaded directly into the mold cavity, which is usually maintained between 165°C and 225°C, depending on the material.
- The mold is partially closed so that the heat plus the pressure can soften the mix and cause it to start flowing into the recesses of the mold.
- The mold is fully closed, causing the mix to complete its flow and cure.
- After the cure is completed, the mold is opened, and the part is ejected.

The relatively simple compression molding equipment consists of a matched mold, a means of heating the mix and the mold, and some method of exerting force on the mold halves to cause them to close. The molds are made from polished, hardened steel because they must withstand very high pressures as well as the abrasive action of the mix as it softens and flows.

The mold halves usually are cored or hollowed, and electrical heating elements are inserted. Any compression device can be used to close the mold halves. The press has one stationary platen (a flat metal platform) and another platen that is moved by a hydraulic or a pneumatic cylinder. As the cylinder moves the platen, it produces a squeezing action. The capacity of a press is rated in terms of the total tons of force it can exert on the mold. Compression molding is best suited for parts that have a large area and a deep draw.

Parts that are compression-molded usually have a parting line, which tends to be located at the point of their maximum dimension. Compression molding was used during the early days of the development of ceramic molding.

In a variation of this process, freeze compression molding, a ceramic slurry containing 40 to 50 vol% water is extruded to form solid blanks, which are sliced into discs. One or two of these discs are placed in the mold cavity, which is held at −15°C to −20°C under pressure until the material in the mold cavity is completely frozen. The frozen molded part then is ejected, dried, and sintered (Nakagawa et al. 1987). This process eliminates the lengthy time ordinarily required for binder removal.

Transfer Molding

In compression molding the mix is quite viscous as it is being softened and forced around the mold cavity. When producing parts with delicate sections or where inserts are to be molded in place, compression molding can distort the sections or move or break the inserts. An adaptation of compression molding called transfer molding eliminates these shortcomings.

There are four steps to transfer molding. First, the molding mix is placed in a transfer chamber, which usually is located above and connected to a closed mold by a sprue. The mix is softened in the transfer chamber by the application of heat. Second, the softened mix is forced from the transfer chamber to the heated closed mold through the sprue. Third, the heat of the closed mold causes the mix to cure. Fourth, after the cure is complete, the mold is opened, and the part is ejected.

Transfer molding equipment essentially consists of a heated matched mold with a gate for the transferred mix, a heated transfer chamber where the mix is softened, and a mechanical or a hydraulic device both to hold the mold closed during molding and to provide a force to push the mix from the transfer chamber into the mold.

The molds used in transfer molding are quite different from those used in compression molding because they are completely closed during molding. The sprue from the closed mold allows the softened mix to be forced from the chamber into the mold. The mold is simply a hollow block that has been split in half. Each mold cavity is exactly the size and the shape of the section of the part to be molded. The transfer chamber is a cavity, usually cylindrical, that contains a heating element and a ram to push the mix from the transfer chamber into the mold.

Transfer molding offers certain advantages over compression molding: The mix is less viscous as it enters the mold and therefore is less likely to distort molded-in inserts. It also is much easier to push the mix into thin sections with fine detail. Because the mold is fully closed during molding, there is no flash line to be removed although there is a sprue that must be removed in a subsequent process step. Transfer molding is used extensively in ceramic core manufacture for investment casting with both thermoset and thermoplastic resins.

COMPETING PROCESSES

During the past decade a number of forming processes have emerged, among them pressure slip casting, the Quickset™ process, slip injection molding, and gelcasting. All of these processes are based on the use of suspensions or slurries, and in certain aspects they resemble traditional slip casting. Both aqueous and nonaqueous fluids have been used as the carrier in these processes.

Slip casting involves pouring a slip (or slurry) into a plaster of Paris mold of the desired shape. The capillary action of the micropores in the mold removes the water from the cast, causing the slip to solidify so that a layer of the ceramic part forms along the inner wall of the plaster mold.

In drain casting, hollowware is formed by pouring off the excess slip when the solid layer has reached a specified thickness. In solid casting, the entire mold space is filled with the slip, which has to be poured into the mold until complete solidification is reached. Because the cast acts as a barrier for water penetration, the casting rate decreases with increasing wall thickness of the cast. The rate at which the part builds up has to be kept in equilibrium with the ability of the plaster mold to absorb water by capillary forces. The part is removed from the mold after it shrinks due to some initial drying. The viscosity of the slip, which determines the pouring characteristic, is adjusted by the addition of deflocculants. The mechanism of deflocculation is based on ion exchange.

Some of the disadvantages of slip casting are that the process is slow and density gradients can develop in the green part, the dried part is not strong enough to be machined prior to firing, and porous molds must be dried before reuse. Furthermore, slip casting is labor intensive and time-consuming.

Pressure Slip Casting

Early investigation demonstrated that there is no advantage to using pressure above 4.2 MPa (600 psi) with a heated clay slip that has little or no electrolyte addition and is in a heated mold. Temperature and the degree of deflocculation of the slip were found to be the important factors in pressure casting (Adcock and McDowell 1957; Gainer and Carter 1964; Nelson and Andrews 1969; and Hogg 1985). In subsequent work, high pressure slip casting was successfully used for silicon carbide, and an increase casting rate and green density were observed (Rogers 1970; Freedman and Millard 1986).

Models have been proposed for the consolidation kinetics and pressure filtration based on Darcy's law (Scheidegger 1974; Aksay and Schilling 1984; Lange and Miller 1987):

$$J = -\frac{\eta}{A}\frac{\mathrm{d}p}{\mathrm{d}x} \tag{5-9}$$

where J is the linear flux of a fluid through a porous medium, that is, the filtration rate; η is the viscosity of the filtrate; A is the specific porous medium resistance; and dp/dx is the hydraulic pressure gradient.

The specific porous medium resistance A can be expressed as (Carman 1956):

$$A = B\,(1 - E)\,\frac{S}{E} \tag{5-10}$$

where B is the ratio of the shape factor to the tortuosity constant; S is the specific surface area, which is usually taken as the ratio of the total solids surface area to the apparent volume of the consolidated system; and E is the void ratio.

The pressure filtration process is expected to follow a parabolic rate behavior while the hydraulic resistances are at a minimum. Besides producing increased understanding of the basics of pressure casting, the technique is gaining rapid commercial acceptance (Numberger 1988; Segerer 1988; Forster and Weiser 1989). The sanitary ware industry, which developed the process and originated the first commercial application, is expected to drive its commercialization.

A typical pressure casting machine demonstrates two essential components of this process: the mold, which has a special microstructure and is composed of a polymer or hard plaster or a mixture of both; and the machine, which is equipped with a precise, electronically controlled hydraulic system. In the most recent development, positioning the casting molds in batteries is combined with pressure casting.

The Quickset Process

In this process, introduced by Ceramic Process Systems Corporation, Milford, Massachusetts, U.S.A. (Novich et al. 1989), a ceramic suspension is injected under low pressure (less than 0.35 MPa, 50 psi) into a closed mold cavity. The part solidifies in the mold by "pore fluid freezing." The pore fluid acts as the binder and provides the green strength necessary for removal of the part from the mold.

The combination of additives, the tool design, and the freezing process results in a "cryptocrystalline pore fluid microstructure" in which there is a minimum change in volume during solidification. The pore fluid is removed as a vapor directly from the molded part without forming liquid, thereby reducing the capillary stresses and particle rearrangements that contribute to warpage and cracking in the part. Because of the high gas permeability, the drying by sublimation enhances rapid binder removal.

Tight dimensional tolerance can be achieved by closely controlling the green microstructure throughout the Quickset process. Because the slurry preparation for the Quickset process is similar to pressure slip casting, a slip casting slurry can be used in this process.

Slip Injection Molding (SIMSE)

In principle, slip injection molding is the same as low pressure injection molding, mentioned in Chapter 1 (Peltsman 1986). In this process, ceramic powder is mixed with the binder in a closed tank mixer under controlled temperature to form a slip. During mixing, a vacuum pump is used to de-air the slip, which is next pumped from the tank, by low pressure nitrogen or air (0.2 – 1 MPa), through a heated transfer tube to a mold. In molding by SIMSE, shown in Figure 5-49, the slip flows up gently from the bottom to the top end of the mold without creating flow marks. No jetting occurs despite sudden, discontinuous increases in a section or an area from narrow gates to a wide cavity (Nishio et al. 1989). Efficiency is in no way sacrificed because it takes only 8 to 10 sec to complete the molding cycle, even for complex parts. As indicated by the graph of the log of the viscosity as a function of shear rate in Figure 5-50, a high production rate can be achieved with low injection pressure and low viscosity by using a number of cavities.

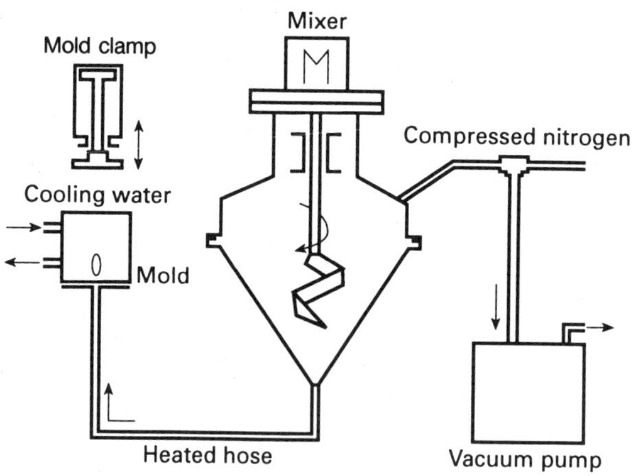

FIGURE 5-49 Schematic of a slip injection molding machine. After Nishio et al. (1989).

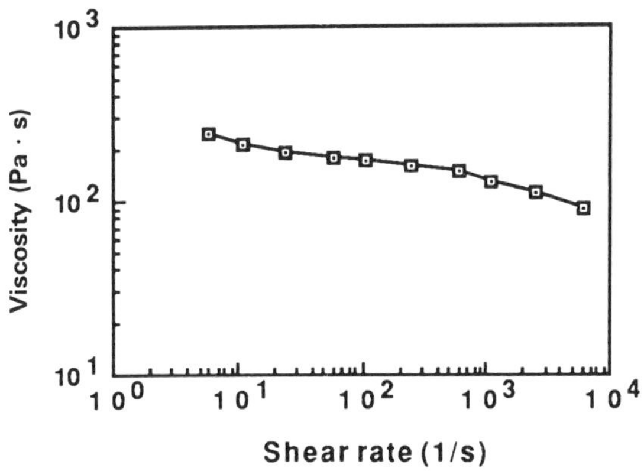

FIGURE 5-50 The viscosity of a silicon nitride mix used in the SIMSE process as a function of shear rate.

Gelcasting

This process, which was developed at the Oak Ridge National Laboratory, Oak Ridge, Tennessee, U.S.A. (Omatete et al. 1991), is based on the use of a monomer solution (trimethylolpropane triacrylate or 1,6 hexanediol diacrylate), which can be polymerized to form a strong, cross-linked polymer–solvent gel. The monomer solution provides a low-viscosity vehicle for carrying the ceramic powder, and cross-linking to form a polymer–solvent gel provides a mechanism for permanently immobilizing the ceramic slurry in the desired shape after it has been poured into a mold.

Gelcasting has a number of advantages: it is easy to remove the solvent from the gelled part during drying because the cross-linked polymer–solvent gelled vehicle contains only 10 to 20 wt% polymer; the polymer cannot migrate with the solvent during drying; molds can be made from metals, glass, plastic, or wax; the green strength is high enough for machining; and the polymer binder is removed during sintering.

References

Adcock, D. S. and I. C. McDowell. 1957. Mechanism of Filter Pressing and Slip Casting, *J. Am. Ceram. Soc.*, 40(10), 355–360.

Agrawal, A. R., I. O Pandelidis, and M. Pecht. 1987. Injection Molding Process Control—Review, *Poly. Eng. & Sci.*, Mid-Oct., 27(18), 1345–1357.

Aksay, I. A. and C. H. Schilling. 1984. Mechanics of Colloidal Filtration, in *Advances in Ceramics*, Vol. 9, *Forming of Ceramics*, ed. Mangels and Messing, Am. Ceram. Soc., 85–93.

Austin, C. 1986. Flow of Plastic into a Mold, in *Injection Molding Handbook*, ed. D. V. Rosato and Donald V. Rosato, Van Nostrand Reinhold, New York, 349.

Beck, R. D. 1980. *Plastic Product Design*, Van Nostrand Reinhold, New York.

Boldizar, A., J. Kubat, and M. Rigdahl. 1990. Influence of Mold Filling Rate and Gate Geometry on the Modulus of High Pressure Injection Molded Polyethylene, *J. Appl. Poly. Sci.*, 39, 63–71.

Carman, P. C. 1956. *Flow of Gases through Porous Media*, Butterworths, London.

Chung, T. S and M. E. Ryan. 1981. Analysis of the Packing Stage in Injection Molding, *Poly. Eng. Sci.*, 21, 271.

DuBois, J. H and W. I. Pribble. 1978. *Plastics Mold Engineering Handbook*, Van Nostrand Reinhold, New York.

Edirisinghe, M. J. 1988. Computer Simulation Solidification of Ceramic Injection Molding Formulations, *J. Matl. Sci. & Letters*, 7, 509–510.

Fischer, P. 1976. Auslegung von Einschneckenextrudern auf der Grundlarge Verfauren Stenchnischer Enndaten (Modelltheorie).

Forster, H. J and K. Weiser. 1989. Formenwerkstoff für das Mitteldruckgiebverfahren, *Keram. Z.*, 41(10), 742–744.

Freedman, M. R and M. L. Millard. 1986. Improved Consolidation of Silicon Carbide, *Ceram. Eng. Sci. Proc.*, Jul–Aug. 884–892.

Gainer, T. M and J. A. Carter. 1964. Pressure Casting of Sanitary Ware, *Am. Ceram. Soc. Bull.*, 43(1) 9–12.

Gosztyla, J. 1986. The CAD/CAM for Plastic Part and Mold Design, *Injection Molding Handbook*, ed D. Rosato and D. V. Rosato, Van Nostrand Reinhold, New York.

Hindle, C. S. 1981. Characterization of Injection Moldings: Residual Stresses, Orientation and Distortion, *SPE-ANTEC Papers.*

Hogg, C. S. 1985. A Comparision of Plaster Casting and Pressure Casting of Sanitary Ware with Particular Reference to Clay Properties, *Ceram. Internat.*, 11(1) 32–36.

Kreuger, W. L and Z. Tadmor. 1980. Injection Molding into a Rectangular Cavity with Inserts, *Poly. Eng. Sci.*, 20, 426–431.

Kruder, G. A and R. E. Nunn. 1981. Extruder Efficiency, How You Measure It—How You Get It, *Plast. Eng.*, June, 20.

Lange, F. F and K. T. Miller. 1987. Pressure Filter: Consolidation Kinetics and Mechanics, *J. Am. Ceram. Soc.*, 66(10), 1498–1504.

May, C. Y. W. 1972. *SPE ANTEC. Tech. Papers*, 18, 851.

McKelvey, J. M. 1962. *Polymer Processing*, John Wiley & Sons, New York.

Mutsuddy, B. C. and L. R. Kahn. 1984. A Practical Approach in Relating Capillary Viscosity and Spiral-Length Data For Molding Ceramic Mixes, in *Advances in Ceramics*, Vol. 9, *Forming of Ceramics*, ed. Mangels and Messing, Am. Ceram. Soc., 251–258.

Nakagawa, T. et al. 1987. Freeze Injection and Compression Molding of Fine Ceramic Powder by Using Water Binder, *J. Japan Soc. Powder & Powder Metall.*, 34, 383.

Nelson, J. A and A. I. Andrews. 1959. Rheological Properties of Clay–Water Systems under Pressure, *Am. Ceram. Soc. Bull.*, 38(9), 447–455.

Nishio, H., A. Miyamoto, and T. Miyashita. 1989. Slip Injection Molding and Supercritical Extraction—SIMSE, *J. Aust. Ceram. Soc.*, 26(2), 127-132.

Novich, B. E, R. R. Lee, G. V. Franks, and D. Ouellette. 1989. Quickset Injection Molding of High Temperature Gas Turbine Engine Components, in Proc. 27th Automotive Technology Development Contractors' Coordination Meeting, SAE Inc., 400 Commonwealth Drive, Warrendale, PA, 311-318.

Numberger, G. 1988. Entwichlung und Einfuhrung des Schlicker-Druckgiebverfahrens in der Sanitakeramik, *Keram. Z.*, 40(4), 227-232.

Omatete, O. O., M. A. Janney, and R. A. Strehlow. 1991. Gelcasting—A New Ceramic Forming Process, *Am. Ceram. Soc. Bull.*, 70(10), 1641-1649.

Pasto, A. E. and S. Natansohn. 1990. Improved Processing Methods for High Reliability Silicon Nitride, in Proc. Ann. Automotive Technology Development Contractors' Coordination Meeting, Dearborn, MI, 143-160.

Patton, W. J. 1976. *Plastics Technology: Theory, Design and Manufacture*, Reston Publ. Co.

Pearson, J. R. A. 1985. *Mechanics of Polymer Processing*, Elsevier Applied Science Publishers, New York.

Peltsman, M. 1986. Low Pressure Injection Molding and Mold Design, *Metal Powder Report*, 41, 367-369.

Potente, H. 1981. *Auslegen von Schneckenmaschinen Baureihen, Modellgesetze und ihre Anwendung*. Carl Hanser Verlag, Munich.

Richardson, S. M. 1983. Injection Molding of Thermoplastics, Freezing during Mold Filling, *Rheol. Acta*, 22, 223.

Richardson, S. M., H. J. Pearson, and J. R. A. Pearson. 1980. Simulation of Injection Molding, *Plastics & Rubber Processing*, 5, 55.

Rogers, M. G. 1970. High-Pressure Slip-Casting of Silicon Carbide, in *Special Ceramics*, 5, 87-97, ed. Popper.

Rosato, D. V. 1986a. Injection Mold Design, in *Injection Molding Handbook*, ed. D. V. Rosato and Donald V. Rosato, Van Nostrand Reinhold, New York, 160-234.

Rosato, D. V. 1986b. Process Control Technology, in *Injection Molding Handbook*, ed. D. V. Rosato and Donald V. Rosato, Van Nostrand Reinhold, New York, 277.

Rubin, I. I. 1973. *Injection Molding: Theory and Practice*, SPE Monographs, John Wiley & Sons, New York.

Ryan, M. E. and T. S. Chung. 1980. Conformal Mapping Analysis of Injection Mold Filling, *Poly. Eng. Sci.*, 20, 642.

Scheidegger, A. E. 1974. Darcy's Law, in *The Physics of Flow through Porous Media*, 3rd edn. University of Toronto Press, Toronto, Canada, 73-98.

Segerer, H. 1988. Das Schlicker-Hochdruckgieben in der Keramischen Industrie, *Keram. Z.*, 40, 235-238.

Sors L., L. Bardocz, and I. Radnoti. 1981. *Plastic Molds and Dies*, Van Nostrand Reinhold, New York.

Spencer, R. S and G. D. Gilmore. 1950a. Equation of State for High Polymers, *J. Appl. Phys.*, 21, 523.

Spencer, R. S. and G. D. Gilmore. 1950b. Role of Pressure, Temperature and Time in the Injection Molding Process, *Modern Plastics*, 27, 143.

Stevenson, J. F. 1978. A Simplified Method for Analyzing Mold Filling Dynamics, Part 1: Theory, *Poly. Eng. Sci.*, 18, 577.

Wang, K. K., S. F. Shen, J. F. Stevenson. 1975, 1976. Computer-aided Injection Molding Systems, 2nd and 3rd Progress Reports, Cornell University, Ithaca, New York.

Whelan, A. and J. H. Craft. 1981. *Development in Injection Molding*, Vol. 2, Applied Science Publishers, London, England.

Williams, J. G, 1981. On the Prediction of Residual Stresses in Polymers, *Plastics & Rubber Processing*, 359.

Zhang T., Z. Jiang, J. Wu, and Z. Chen. 1990. Influence of Rheological Behavior of Ceramic Mixes on Injection Molding of Ceramic Compacts, *J. Am. Ceram. Soc.*, 73(7) 2171–2175.

6

Binder removal or dewaxing

Binder removal is a key step for successful ceramic injection molding. The binder must be removed from the molded part prior to its densification by sintering. Binder systems can consist of several organic ingredients, each with a markedly different melt viscosity as well as with different decomposition characteristics. These ingredients must be removed slowly, in a process that requires the gradual formation of passages within the part with increasing temperature and time. These passages allow the major binder ingredient to escape at higher temperatures without causing the part to fail.

Thermal degradation involves three basic stages (Barone et al. 1989; Barone and Ulicny 1990). During the early stages of heating, thermal expansion of the liquid binder induces hydraulic pressure in the fully saturated part. As the temperature rises, binder removal due to evaporation from the surface increases. When the saturation level of the binder is sufficiently reduced, the liquid remaining in the mix is driven to the surface by capillary pressure, where it can evaporate. As binder removal by liquid transfer continues, gas pockets begin to coalesce, forming a network of interconnected pores. Eventually, this internal structure creates passages through which gas can flow, thereby allowing diffusion to play an important role in binder removal. If the debinding occurs in air, chemical reactions also can take place.

With respect to the distribution of the binder in the mix during debinding, two disparate conditions are possible (Calvert and Cima 1988). In one, debinding reactions take place at the surface of a shrinking core of the binder phase. In the other, the binder burns out homogeneously throughout the material. If the first mechanism dominates, there is a buildup of pressure at the surface of the binder phase caused by the presence of gaseous reaction products. This can result in delamination and the formation of internal cracks. If the other mechanism dominates, there is a possibility that small particles will be transported to the surface of the part by capillary pressure. This can result in the formation of a hard outer shell, which has been suggested as the cause of the skin peeling effect that can occur during sintering (Bandyopadhyay et al. 1986; Mutsuddy 1987).

One way to absorb low-viscosity liquid components of the binder during debinding is to surround the molded parts with an inorganic powder (Schreiber et al. 1980; Ikejima and Imai 1984; Kato 1985; and Tanikawa and Kurata 1985). Another suggestion is to use a debinding atmosphere with a partial oxygen pressure higher than that of the ambient atmospheric pressure. The thinking here is that because the surface region of a molded part is generally more dense than the bulk, in a high-oxygen partial pressure it will burn open, thereby reducing the buildup of internal pressure (Gilissen et al. 1985). Other recommended ways to minimize problems during debinding include programming the total furnace pressure so that the rate (by volume) of the production of gaseous binder components is constant (Sakai et al. 1985), or equalizing the composition of the furnace atmosphere to match that of the gaseous burnout products (Ando 1986).

Proposals also have been made to directly improve the process control for a particular binder formulation and associated debinding conditions via feedback control of the furnace temperature by monitoring weight change with sensors. The monitored rate of weight loss (dw/dt) can be related to a rate of temperature increase at which dw/dt remains constant (Sakai et al. 1985). Weight loss monitoring has been used to stop the increase in temperature when a specified rate of weight loss (dw/dt) is exceeded (Johnsson et al. 1983; Verwerji and Bruggink 1988).

This chapter describes processes and methods for binder removal. It presents tests to determine removal rates and products, gives examples of typical schedules of binder removal or dewaxing, discusses mechanisms of binder removal, suggests alternatives to thermal methods, and reviews commonly encountered defects.

PROCESSES AND METHODS

Thermal degradation is the principal method for removing the binder. Solvent extraction is an alternative process. Four different approaches are used for thermal degradation:

- Oxidation and evaporation.
- Vacuum extraction.
- Draining through a porous bed.
- Pyrolysis in an inert atmosphere.
- High-pressure extraction.

During binder removal, injection molded ceramic parts undergo chemical and physical changes as the organic binders start to break down. A number of analytical tools are available that provide information about decomposition rates, decomposition orders, and products.

ANALYZING THERMAL DECOMPOSITION OF THE BINDER

Differential thermal analysis (DTA), differential scanning calorimetry (DSC), and thermal gravimetric analysis (TGA), are the techniques used most often to establish the glass transition temperature, the thermal characteristics, the chemical nature, and the rate of decomposition of binder compositions with respect to time and temperature.

Differential thermal analysis (DTA) measures the difference in temperature between a test material and a reference material while both are subjected to a program in which the temperature is changed under controlled conditions.

In differential scanning calorimetry (DSC) the difference in the energy consumed by a test material and a reference material is measured as a function of temperature while the test material and the reference material are subjected to a program in which the temperature is changed under controlled conditions.

In thermogravimetric analysis (TGA), the mass of a test material is measured as a function of the temperature while the material is subjected to a programmed time rate of change.

The type of information generated from these tests is illustrated by several mix formulations shown in Table 5-2 for A16-SG alumina, type C zirconia, and milled SN502 silicon nitride compositions.

Mass spectrometric analysis also is discussed in this section.

Differential Thermal Analysis (DTA)

The DTA history of an A16-SG alumina mix with increasing temperature is presented in Figure 6-1, which shows a number of peaks. The size of each peak is proportional to the amount of heat evolved or absorbed by the mix. The mix analyzed in this figure undergoes a number of transformations, including loss of moisture (indicated by the first peak), loss of milling aid (second peak), softening (glass transition) of the binder (third peak), and decomposition of the binder components (fourth peak).

The figure qualitatively illustrates the approximate order of magnitude of the reactions that can be anticipated to occur with each mix formulation. Note that after each of the first two peaks the baseline shifts. This occurs

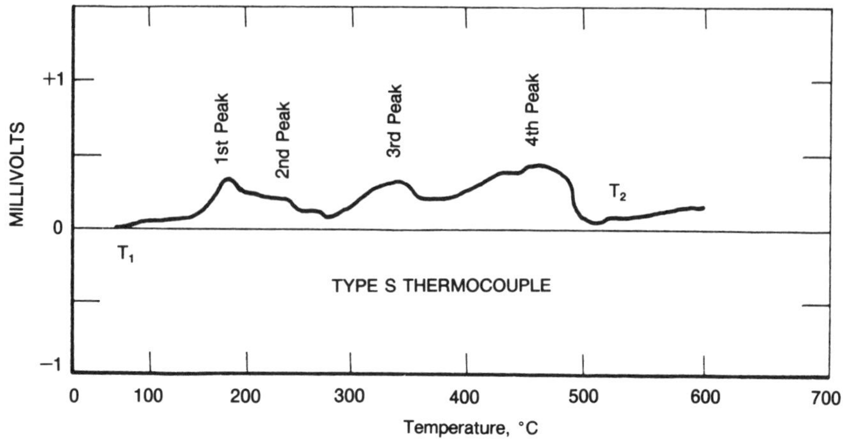

FIGURE 6-1 The differential thermal analysis (decomposition) history of an A16-SG alumina mix with increasing temperature. From Battelle Columbus, OH.

because of changes in the density of the sample as a consequence of weight loss due to the loss of both water and volatilized organics as the temperature is increased. The resulting change in the thermal path within the sample changes the rate of heat flow into the sample, slightly offsetting the temperature increase.

Figure 6-1 also presents a qualitative picture of the thermal decomposition process in a sample over a range of temperatures for a given time period. The significance of DTA can be understood quantitatively by calculating the heats of enthalpy within a specified temperature range for each mix formulation, which indicates the amount of heat energy involved during the thermal treatment.

The heats of enthalpy for a mix formulation can be calculated by using the following equation:

$$\frac{m(\Delta H)}{gK} = \int_{t_1}^{t_2} \Delta T \, \mathrm{d}t \tag{6-1}$$

where m is the mass of the reacting material, ΔH is the heat of reaction, g is the geometric shape factor for the apparatus, K is the thermal conductivity of the sample, ΔT is the differential temperature in °C, and t_1 and t_2 are the integration limits of the differential curve.

Equation (6-1) is only an approximate relationship. It relates the heat of reaction to the area under the peaks in the curve (see Figure 6-1) through

TABLE 6-1. Heats of enthalpy calculated from DTA test results showing exothermic trend.

Base material	Mix formulation	Heats of enthalpy (cal/g)	Beginning exothermic (°C)	Ending exothermic (°C)
A16-SG				
Al_2O_3	Q	4.10	150	455
	Y	3.95	180	455
	Z	3.04	180	490
Type C				
ZrO_2	Q	0.315	155	440
	Y	0.192	160	448
	Z	0.144	140	593
SN502				
Si_3N_4	Q	3.82	140	445
	Y	4.64	155	508
	Z	4.48	160	485

the constants g and K. It neglects differential terms and the temperature gradients in the sample, and considers the area under the peaks to be independent of the sample's specific heat.

The results of the calculations of the heats of enthalpy are given in Table 6-1. They indicate an exothermic tendency in all the mix formulations that is independent of the type of ceramic filler. The differences in maximum exothermic temperatures between the various mixes for each powder are evidence of an oxidative breakdown of the cross-linked and cyclic structures, which could have occurred during the earlier stages of degradation in air.

What the DTA results show is that thermal decomposition is an exothermic process for all the ceramic binder formulations. Therefore, the decomposition rate depends primarily on restraining the heat input rather than on inducing it more rapidly. Furthermore, it suggests that the ability to control the overall dewaxing cycle is limited by the extent of the exothermic reaction.

Differential Scanning Calorimetry

In power-compensated differential scanning calorimetry (DSC), the sample and the reference material are maintained isothermally by individual heaters, whereas in DTA a difference is detected in the temperatures of the sample and the reference material. The DSC curve resembles the DTA curve in all respects with the exception of the units of the ordinate axis, as

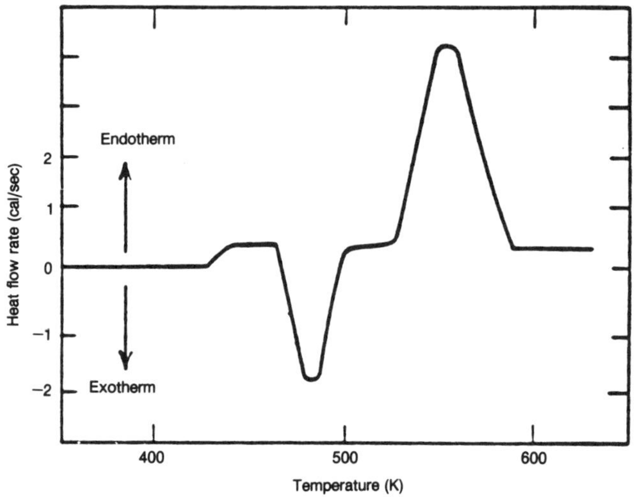

FIGURE 6-2 A typical differential scanning calorimetry curve.

shown in Figure 6-2. As with the DTA curve, the area enclosed by the peak of the DSC curve is directly proportional to the enthalpy change:

$$A = \frac{\Delta H m}{C} \tag{6-2}$$

where A is the area under the peaks in the curve, ΔH is the heat of reaction, m is the mass of the reacting material, and C is the calibration coefficient. The calibration coefficient is related to the geometry and the thermal conductivity of the sample holder, and usually is determined by calibrating the system with compounds that have known heats of reaction. However, it does not change with temperature.

Thermal Gravimetric Analysis

Thermal gravimetric analysis (TGA) is the most frequently used method for measuring the rates of thermal decomposition of organic binders. The percent weight loss of polyethylene, a commonly used binder material, as a function of time is shown in Figure 6-3. To obtain the activation energies of decomposition, the logarithms of the rate constants are plotted as a func-

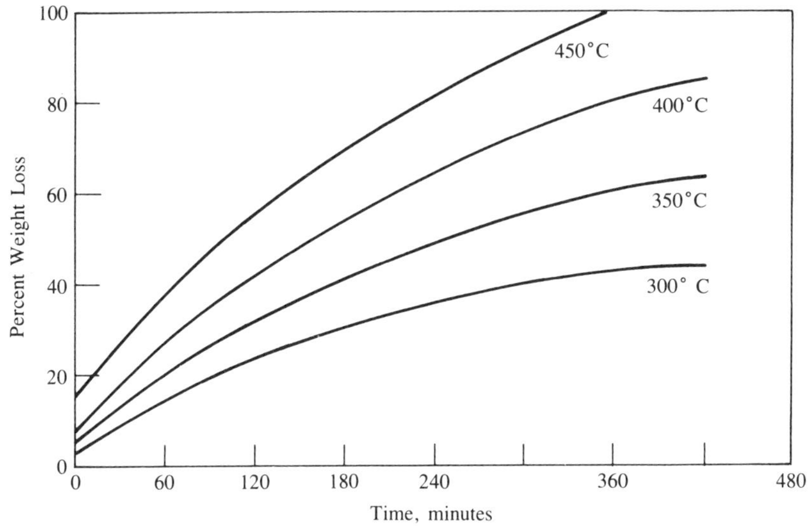

FIGURE 6-3 Percent weight loss of polyethylene as a function of time.

tion of the inverse of the absolute temperatures according to the Arrhenius equation. This equation can be written simply as:

$$-E = 2.303R\,\frac{\mathrm{d}\log K}{\mathrm{d}(1/T)} \tag{6-3}$$

where K is the rate constant, E is the activation energy, T is the absolute temperature, and R is the universal gas constant. In the Arrhenius plot shown in Figure 6-4, the slope of the straight line corresponds to an activation energy of 70 kcal/mole.

Figure 6-5 illustrates the results of thermogravimetric analysis of different mix formulations of A16-SG alumina and milled SN502 silicon nitride. The plot of weight loss versus time is shown as solid lines. The dotted lines represent the temperature gradient. The temperature scale is along the right Y-axis. Only three lines need to be drawn to determine the weight loss of a particular mix at a particular temperature. First, a horizontal line is drawn from the temperature axis to the point where it intersects the dotted line representing the temperature gradient. Next, a vertical line is drawn from the intersection point to the curve of the desired mix. Finally, a second horizontal line is drawn from the intersection point on the mix line to the weight loss scale on the Y-axis to read the weight loss.

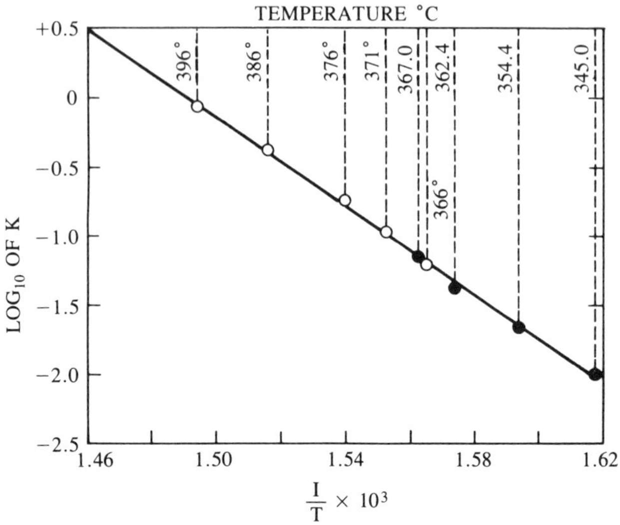

FIGURE 6-4 Arrhenius plot for the thermal degradation of polyethylene.

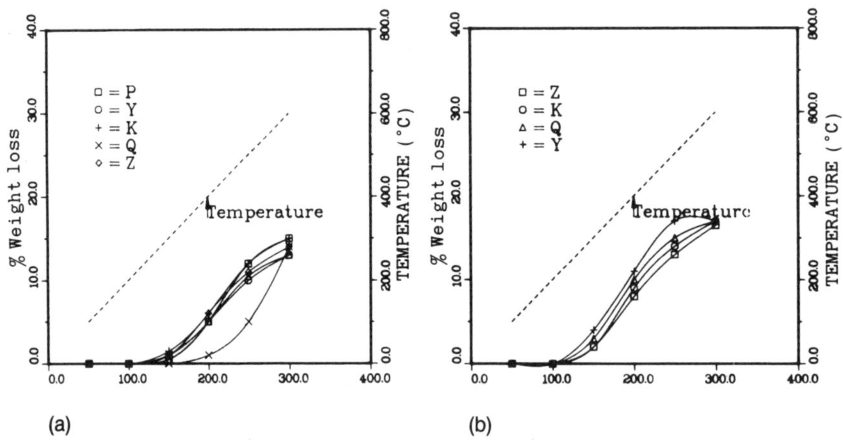

FIGURE 6-5 Thermogravimetric analysis of different mix formulations of (a) A16-SG alumina (mixes P, Y, K, Q, and Z) and (b) milled SN502 silicon nitride (mixes Z, K, Q, Y). From Battelle Columbus, OH.

From these data, it can be shown that the critical temperature range for dewaxing these mixtures lies between 350°C and 420°C, and that the variation within this temperature range is influenced by the nature of the ceramic powder filler. The nature of the metal–oxygen bond in the ceramic is a key factor that determines the degree of interaction between the ceramic and the polymeric binder during thermal degradation. For example, in a material such as aluminum oxide (Al_2O_3) in which the metal–oxygen bond is more ionic than covalent, a surface reaction occurs (Sun et al. 1988).

Mass Spectrometric Analysis

In addition to determining the rate of volatilization it is important to characterize the volatile products. The volatile components of several different mixes detected after heating for 30 minutes to 150, 300, and 600°C, are given in Tables 6-2 through 6-4. A mullite tube and a zirconia boat are used for the blank run, and pure A16-SG aluminum oxide or silicon nitride is used as the reference points.

As can be seen in Table 6-2, at 150°C the composition of the volatilized gas is very close to that of air. Evidently there is very little oxidation of the binder components when these mixes are heated at 150°C.

TABLE 6-2. Volatile oxidation products detected at 150°C in air. Results expressed as volume percent.

Oxidation products	Blank run, mullite tube	Al_2O_3	Mix K	Mix Q	Mix Z
Oxygen	20.9	21.0	20.8	20.9	20.9
Nitrogen	78.0	77.9	78.0	78.1	78.0
Argon	0.9	0.96	0.94	0.95	0.95
Carbon dioxide	0.1	0.2	0.1	0.1	0.1
Water	<0.1	0.1	0.1	<0.1	<0.1

Oxidation products	Si_3N_4 wet milled	Mix K	Mix Y	Mix Z
Oxygen	20.8	21.0	20.7	20.8
Nitrogen	78.0	77.9	78.3	78.1
Argon	0.95	0.96	0.94	0.94
Carbon dioxide	0.1	0.1	0.1	0.1
Water	0.1	—	<0.1	<0.1

TABLE 6-3. Volatile products detected at 300°C in air. Results expressed as volume percent.

Products	Blank run, mullite tube	Al_2O_3 A16-SG	Mix K	Mix Q	Mix Z
Oxygen	20.9	21.2	9.6	14.6	11.8
Nitrogen	77.9	77.6	82.7	82.4	82.2
Argon	0.96	0.98	1.0	0.99	0.99
Carbon dioxide	0.2	0.2	2.8	1.7	2.4
Water	<0.1	<0.1	0.2	<0.1	<0.1
Carbon monoxide	<0.1	<0.1	3.2	0.1	2.3
Methane	—	—	0.02	—	0.03
Ethane	—	—	0.13	0.13	—
Ethylene	—	—	—	—	0.04
Propylene	—	—	0.03	0.03	0.04
Methanol and ethanol	—	—	0.07	0.03	0.06
Various hydrocarbons	—	—	0.2	—	0.1
Total organics	—	—	0.45	0.19	0.27

Products	Silicon nitride, wet milled	Mix K	Mix Y	Mix Z
Oxygen	20.8	8.6	11.3	7.9
Nitrogen	78.0	84.1	82.8	84.1
Argon	0.94	1.01	1.00	1.0
Carbon dioxide	0.1	2.7	2.1	3.0
Water	0.1	0.2	0.1	0.2
Carbon monoxide	—	2.9	2.3	3.1
Methane	—	<0.1	—	0.01
Ethane	—	—	0.1	0.13
Ethylene	—	0.06	—	0.01
Propylene	—	0.06	0.04	0.08
Methanol and ethanol	—	0.09	0.08	0.2
Various hydrocarbons	—	0.2	0.1	0.2
Total organics	—	0.44	0.32	0.63

At 300°C, the blanks and the alumina and silicon nitride reference samples are unchanged, but all the mixes are oxidized. Their principal decomposition products, carbon dioxide (CO_2) and carbon monoxide (CO), result either from the direct oxidation of the binder components or from the oxidation of the products of their pyrolysis. Table 6-3 gives the volatile products of the various mix formulations and their volume percents are given in Table 6-5.

TABLE 6-4. Volatile products detected at 600°C in air. Results expressed as volume percent.

Products	Blank run, mullite tube	Al_2O_3 A16-SG, as-received	Mix K	Mix Q	Mix Z
Oxygen	17.4	19.7	< 0.1	0.2	0.2
Nitrogen	80.4	77.7	25.0	13.5	7.9
Argon	1.03	0.97	0.3	0.3	0.2
Carbon dioxide	1.1	0.4	2.3	2.2	1.8
Water	—	1.2	57.1	71.0	74.0
Carbon monoxide	—	—	5.4	6.2	6.0
Methane	—	—	2.3	2.0	1.0
Ethane	—	—	2.0	0.8	1.0
Ethylene	—	—	1.5	0.5	0.6
Propylene	—	—	1.0	0.6	0.5
Methanol and ethanol	—	—	0.5	0.2	0.3
Toluene	—	—	0.3	0.8	0.6
Benzene	—	—	0.5	0.7	0.6
Propionic and acetic acids	—	—	0.4	—	1.1
Various hydrocarbons	—	—	0.3	0.3	0.3
Hydrogen	—	—	0.8	0.7	3.8
Propane	—	—	0.3	—	0.1
Total organics			9.1	5.9	6.2
Total gas (cm^3)[a]	118	132	161	144	144

a. Cubic centimeters at standard pressure and ambient temperature.

Products	Si_3N_4, wet milled	Mix K	Mix Y	Mix Z
Oxygen	18.6	< .1	< 0.1	0.5
Nitrogen	78.3	24.8	20.3	22.5
Argon	0.96	0.4	0.3	0.3
Carbon dioxide	0.8	2.1	1.7	2.3
Water	1.3	58.0	65.0	61.0
Carbon monoxide	—	3.9	3.8	5.2
Methane	—	2.9	2.1	2.1
Ethane	—	1.2	1.7	1.1
Ethylene	—	1.9	0.8	1.5
Propylene	—	0.9	1.2	1.0
Methanol and ethanol	—	0.2	0.3	0.5
Toluene	—	0.7	0.7	0.2
Benzene	—	0.9	0.6	0.4
Propionic and acetic acids	—	0.8	0.2	0.6
Various hydrocarbons	—	1.1	0.3	0.1
Hydrogen	—	0.8	0.8	0.4
Propane	—	0.8	0.2	0.3
Total organics	—	—	8.1	7.8
Total gas (cm^3)	124	157	161	152

TABLE 6-5. Oxidation of mix formulations at 300°C, shown as volume percent.

Mix formulation	O_2	CO_2 plus CO	Total organics
Al_2O_3/K	9.6	6.0	0.45
Al_2O_3/Q	14.6	1.8	0.19
Al_2O_3/Z	11.8	4.7	0.27
Si_3N_4/K	8.6	5.6	0.44
Si_3N_4/Y	11.3	4.4	0.32
Si_3N_4/Z	7.9	6.1	0.63
Si_3N_4 SN502	20.8	0.1	<0.1
Al_2O_3 A16-SG	21.2	0.2	<0.1

At 600°C most of the oxidation is completed. Table 6-4 shows the volatile products and the total volume of gas produced after heating at 600°C for 30 minutes. Subtracting the volume of gas produced by the reference samples of alumina or silicon nitride from the total volume of gas produced by the respective mixes, and correlating this difference with the total amount of organics formed, provides an indication of the relative extent of decomposition of these mixes. Data taken from Table 6-4 are reproduced in Table 6-6 to emphasize the differences in the extent of decomposition among these mixes. The total quantity of water evolved from each mix also reflects these differences.

The production of light hydrocarbons is a result of the pyrolytic degradation of the large polymer molecules and also of radical-induced hydrogen transfer between the molecular species present. This accounts for the production of methane (CH_4), ethane (C_2H_6), and propane (C_3H_8), the dehydrogenated compounds ethylene (C_2H_4) and propylene (C_3H_6), and hydrogen.

TABLE 6-6. The difference in the extent of decomposition among mixes at 600°C in air.

	Gas volume (cm^3)	Percent organics	Percent water
Al_2O_3/K	29	9.1	57.1
Al_2O_3/Q	12	5.9	71.0
Al_2O_3/Z	12	6.1	74.0
Si_3N_4/K	33	9.8	54.4
Si_3N_4/Y	17	8.1	65.0
Si_3N_4/Z	28	7.8	61.0

Benzene and toluene also can be produced, as they are in petroleum refining, by the cyclization and dehydrogenation of the aliphatic hydrocarbons hexane and heptane, which contain six and seven carbon atoms, respectively. Organic acids (propionic and acetic acids) can be produced from the partial oxidation of small hydrocarbon molecules or the fragmentation of larger ones by oxidation. However, if the supply of oxygen is kept sufficiently high, most of the pyrolytic fractions probably will be converted to CO_2 and water. The results of the mass spectrometric analysis of silicon nitride mixes in a nitrogen atmosphere at 600°C are given in Table 6-7.

Table 6-7 also indicates that pyrolytic fractions with various compositions of silicon nitride mix formulations Q, Y, and Z, produce the same amount of total gas, but silicon nitride M produces far less total gas. Table 6-8 ranks the five different mix formulations (M, K, Z, Y, and Q) with respect to their degree of pyrolysis. Total organics, methane, and hydrogen are used for this ranking because they are the most likely pyrolysis products. Although the mix formulations Q, Y, and Z produce the same volume of gas, the amounts of hydrogen and methane in mix formulation Q are noticeably different from amounts in the other two. Furthermore, the organic content and the amounts of hydrogen and methane in formulation M are relatively lower than in the other formulations. This finding reflects a difference in pyrolytic behavior.

The main purpose of DTA, TGA, DSC, and mass spectrometric analysis is to establish some initial guidelines on the nature of the decomposition reaction, the threshold dewaxing temperature, and the rate and products of volatilization. All of these tests are extremely useful in providing a full understanding of the mechanisms of binder decomposition.

EVAPORATION AND OXIDATION AS A PRACTICAL DEWAXING APPROACH

The analyses discussed above usually are performed on a small sample of the mix formulation. Therefore, the knowledge gained from these experiments should be cautiously applied in developing a dewaxing procedure for injection molded parts. Experiments should be performed with actual parts in order to establish a heating rate from the heats of enthalpy data determined by DTA, from TGA temperature profiles, and from mass spectrometric data on the products of decompositon. Two linear heating rates of 5°C/hr and 10°C/hr illustrate the binder removal process with real parts.

Figure 6-6a shows a gradual change in weight loss for an A16-SG alumina mix containing the plasticizer AC-1702 as the temperature increases at a rate of 5°C/hr. At a rate of 10°C/hr, the weight loss is fairly rapid between 300°C and 400°C, and then it slows as the temperature approaches

TABLE 6-7. Chemical components detected at 600°C (nitrogen cover gas).

	Volume percent					
Component	Blank on N_2 and mullite Tube	Silicon nitride M	Silicon nitride Q	Silicon nitride Y	Silicon nitride Z	Silicon nitride K
Nitrogen	99.9	99.4	67.7	68.1	69.5	72.6
Oxygen	<0.01	<0.01	<0.01	<0.01	<0.01	<0.01
Carbon dioxide	0.02	0.08	0.5	0.6	1.2	0.3
Hydrogen	0.2	0.3	2.7	2.0	2.0	1.6
Water	<0.1	<0.01	0.2	0.5	0.5	<0.1
Argon	0.04	0.05	<0.02	0.2	<0.02	0.02
Misc. hydrocarbons				0.3	0.4	0.3
Methane		0.03	12.3	9.8	9.7	8.4
Ethylene		0.03	5.3	6.4	5.8	6.3
Ethane			4.1	3.8	3.5	3.3
Propylene		0.1	2.7	3.0	2.7	3.0
Butyne			0.1	0.1	0.1	0.1
Butene			1.0	1.0	0.2	0.9
Butane			0.1	0.2	0.3	0.1
Pentyne			0.2	0.3	0.2	0.3
Pentene			0.1	0.3	0.1	0.2
Benzene			1.7	2.1	2.4	1.7
Toluene			1.1	1.5	1.4	0.9
Total hydrocarbons (%)		0.16	28.7	28.8	26.8	25.5
Total gas (cm^3)[a]	95	95	133	133	133	123
Less blank			95	95	95	95
Net cm^3 gas			38	38	38	28

a. Cubic centimeters, standard pressure, and ambient temperature.

TABLE 6-8. **Ranking of different mix formulations in relation to degree of pyrolysis.**

Mix Formulation	Total cm^3 Gas	Net cm^3 Gas	Total Organics %	Total Organics cm^3	Methane %	Methane cm^3	Hydrogen %	Hydrogen cm^3
Silicon nitride M	95	0	0.2	(0.2)	0.03	(0.03)	0.3	(0.3)
Silicon nitride K	123	28	25.5	(31)	8.4	(10)	1.6	(2)
Silicon nitride Z	133	38	26.8	(36)	9.7	(13)	2.0	(3)
Silicon nitride Y	133	38	28.8	(38)	9.8	(13)	2.0	(3)
Silicon nitride Q	133	38	28.7	(38)	12.3	(16)	2.7	(4)

500°C. In the case of the A16-SG alumina mix containing the processing aid menhaden oil and the plasticizer butyl stearate, shown in Figure 6-6b, a more gradual but relatively rapid weight loss occurs at 10°C/hr than at 5°C/hr. At the higher heating rate, a large amount of volatiles is generated during a relatively short period of time. In air, this can substantially lower the local partial pressure of oxygen during the time period when most of the weight loss occurs because there is not enough time for volatiles to escape and for oxygen from the surrounding atmosphere to reach the sample's interior.

The change in partial pressure of the oxygen in the vicinity of the sample changes the degradation behavior. If the degradation were independent of the heating rate, the entire weight loss curve would shift to a higher temperature range as the heating rate increased. In Figure 6-6a and b, it can be seen that there are intermediate temperatures at which more weight loss occurs at 10°C/hr than at 5°C/hr. This can happen only when the degradation behavior changes with the heating rate. These results indicate that varying the heating rate with temperature is the most efficient way to dewax in air.

The results with two SN502 silicon nitride compositions are shown in Figure 6-6c and d. The two mixes contain the same binder ingredients and total binder content, but they differ in the concentration of each ingredient. As these figures show, however, their overall binder dewaxing profiles are similar at the two heating rates.

The slow dewaxing rate as well as the TGA and mass spectroscopy results show that below 200°C there is relatively little thermal breakdown. Consequently it should be possible to markedly reduce the heating time from ambient temperature to 200°C.

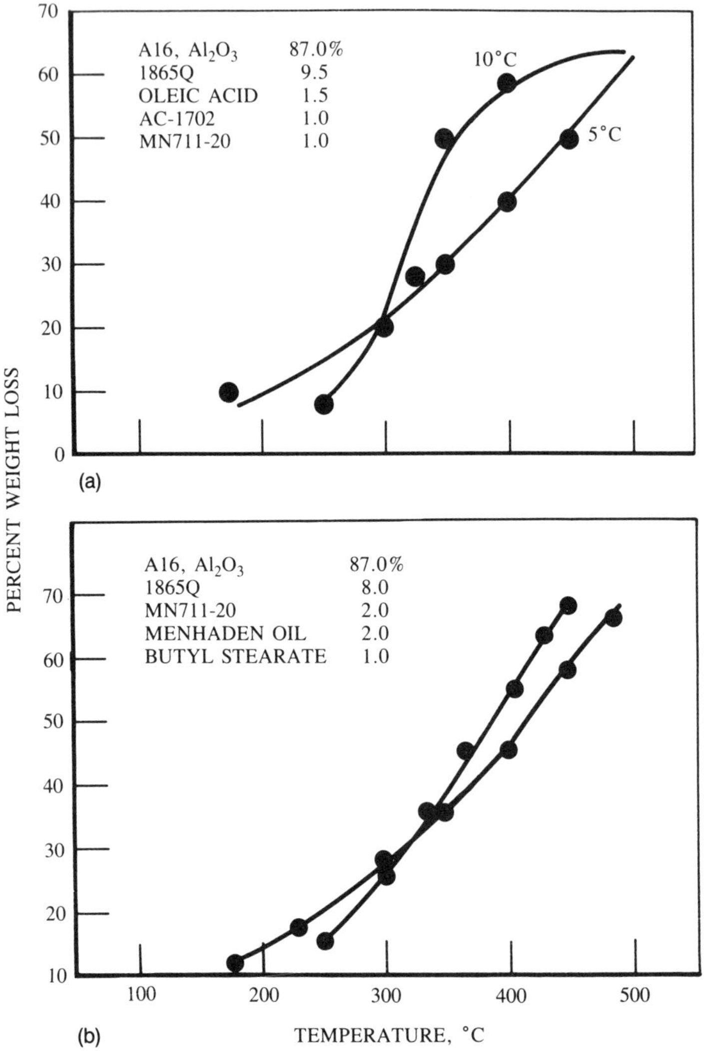

FIGURE 6-6 Weight loss as a function of temperature for different mix formulations: (a) an A16-SG alumina mix containing the plasticizer AC-1702; (b) an A16-SG alumina mix containing the processing aid menhaden oil and the plasticizer butyl stearate; (c,d) mixes of SN502 silicon nitride containing different concentrations of the same binder ingredients. From Battelle Columbus, OH.

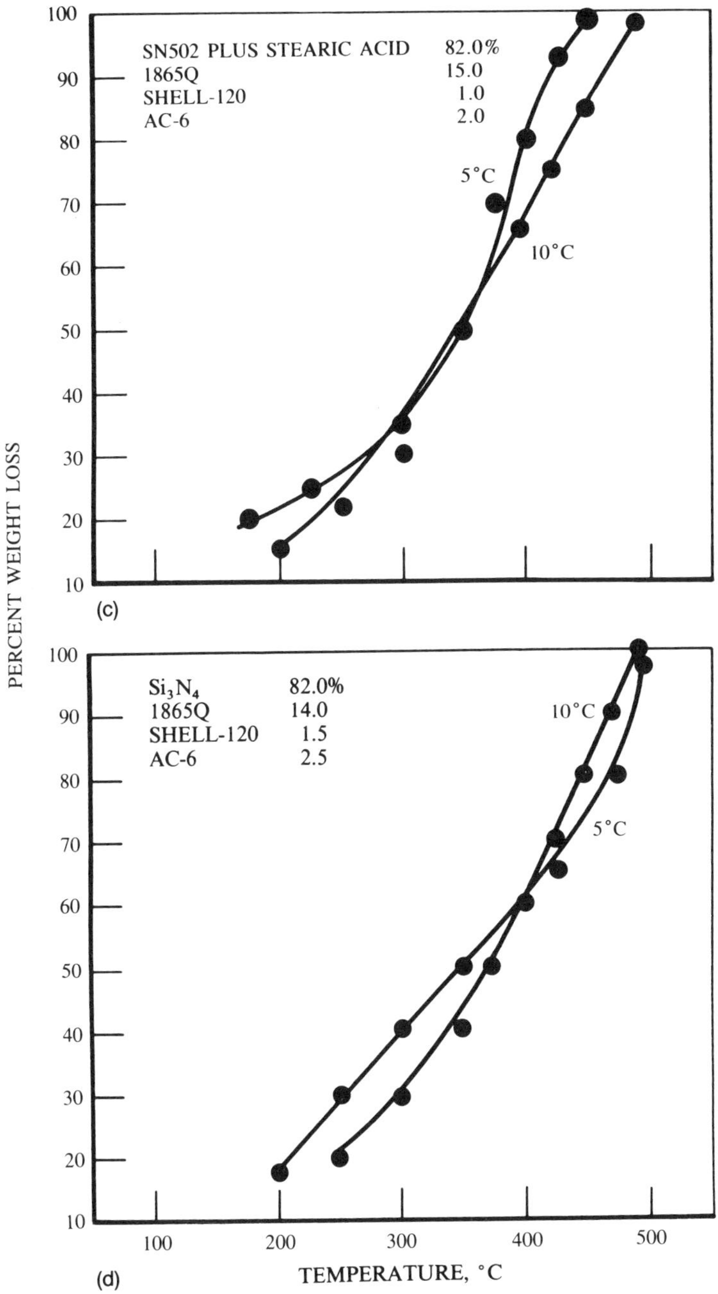

FIGURE 6-6 *contd*

At some temperature before the thermal breakdown is initiated, the binder will undergo a transition from a highly viscous to a fairly fluid state. If the heating rate is very rapid (about 25°C/hr) close to this transition temperature, deleterious effects such as bloating and distortion can occur. By testing heating rates of 25, 50, and 100°C/hr from ambient temperature to 200°C it has been found that there is a threshold heating rate (25–50°C/hr) that, when used during the initial part of the dewaxing cycle, maintains the integrity of the parts. From 200°C to 500°C the safest heating rate is either 5°C/hr or 10°C/hr, depending on the size and the shape of the parts.

This is illustrated with an SN502 silicon nitride and A16-SG alumina mix in Figure 6-7. The weight loss with the A16-SG alumina mix at a heating rate of 5°C/hr is gradual from about 200°C up to 375°C, becoming fairly rapid up to 500°C. Similarly, with SN502 silicon nitride, the weight loss becomes rapid above 400°C. With both these mixes, at a heating rate of 10°C/hr the weight loss is gradual from about 200°C up to the peak temperature. The data suggest that the dewaxing cycle can be shortened from 100 hours to between 30 and 60 hours by rapid heating at 25 to 50°C/hr up to about 200°C, followed by slow heating at 5 to 10°C/hr to a peak temperature of 500°C.

In designing a dewaxing cycle, both the shape and the size of the parts must be taken into consideration. A larger part degrades more slowly than a smaller one because of difficulty both in removing volatiles from the part's interior and in getting oxygen into it. (The local partial pressure of oxygen within the sample is lower during the degradation of the larger sample.)

As shown in Figure 6-8, there is a rapid decrease in the time it takes for air to diffuse when the rate of binder extraction exceeds 25 vol% of the binder, and the parts become porous. From experimental data it has been determined that the average diameter of the pores formed after binder extraction, D, is proportional to the average diameter of the ceramic powder particles, R (Wada and Oyama 1989):

$$D = (0.3 - 0.4)R \text{ for silicon nitride} \tag{6-4}$$

$$D = (0.2 - 0.3)R \text{ for silicon carbide} \tag{6-5}$$

The numerical value of the constant depends on the ratio of the volume of the ceramic powder to the volume of the binder and the particle size distribution of the ceramic powder.

The heating rate for parts with large surface-to-volume ratios can be higher than for parts with small surface-to-volume ratios without creating internal flaws, as indicated in Figure 6-9. Flaws in parts with small surface-to-volume ratios are created by the high gas pressure of the decomposed and

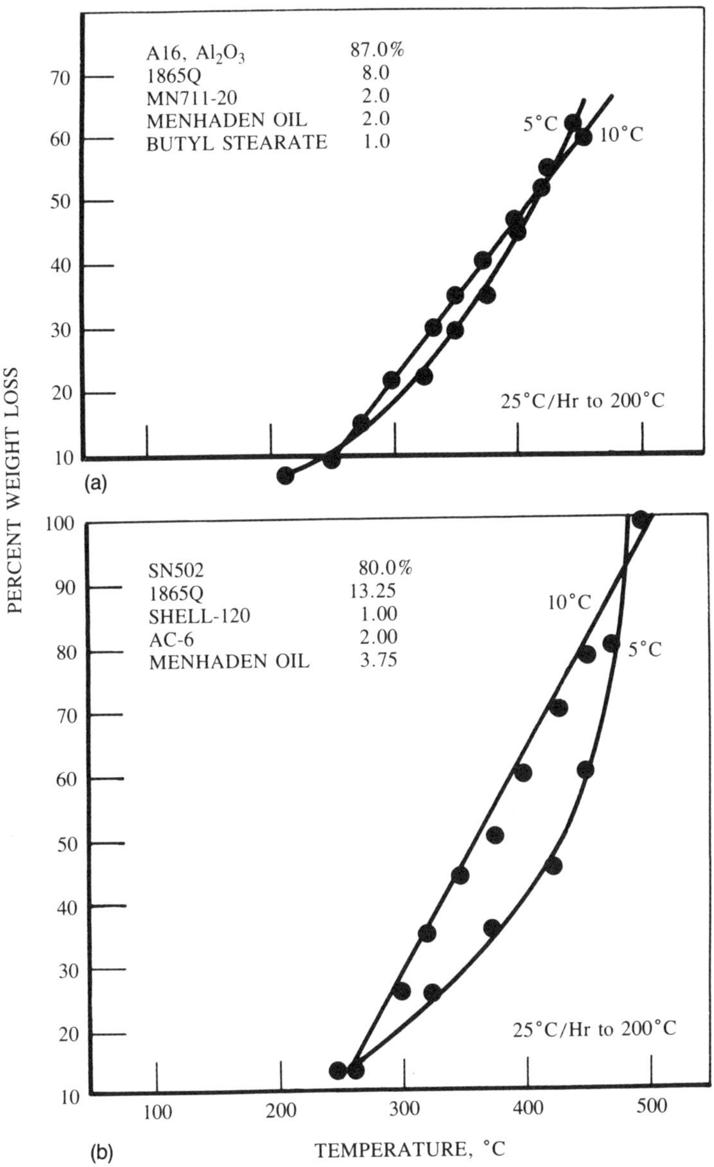

FIGURE 6-7 Percent weight loss as a function of temperature for two different mixes: (a) a mix of A16-SG alumina, and (b) a mix of SN502 silicon nitride. From Battelle Columbus, OH.

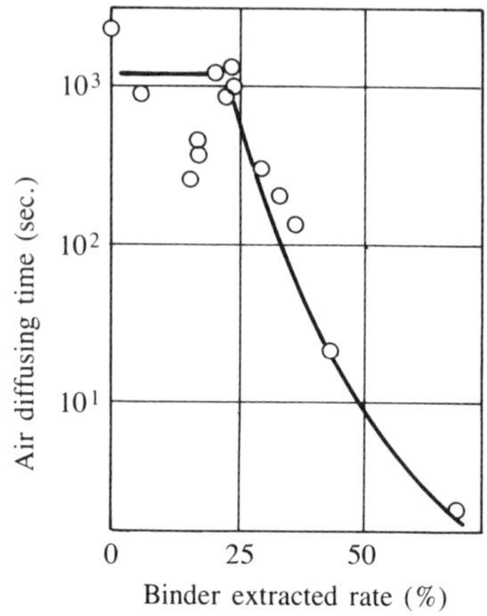

FIGURE 6-8 Air diffusion time as a function of the rate of binder extraction.

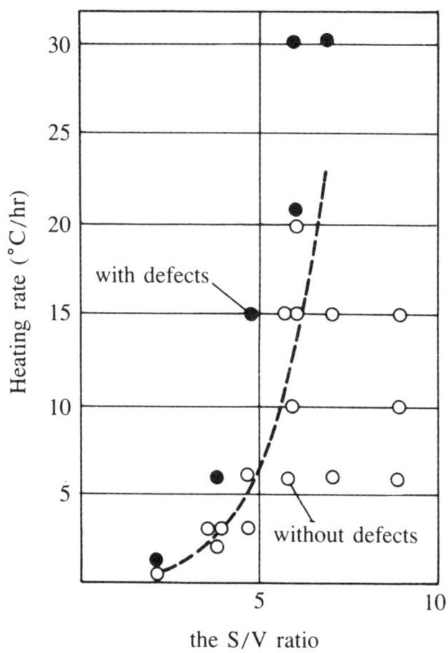

FIGURE 6-9 The development of flaws as a function of the surface-to-volume ratio of the part for various heating rates.

vaporized binder. It also is possible that the flow rate of air or gas could similarly affect the weight loss behavior. Therefore, before an optimum dewaxing schedule is established preliminary trials should be carried out with as many variations of size and shape as practicable.

DEWAXING UNDER VACUUM

Vacuum extraction can be used to dewax injection molded parts. It offers the advantages of eliminating the exothermic reaction between binder components and air or oxygen, allowing the controlled diffusion of volatile species, and minimizing a buildup of internal gas pressure. Figure 6-10 shows that the dewaxing rate can be increased in some binder formulations that contain components with low vapor pressures, such as long-chain

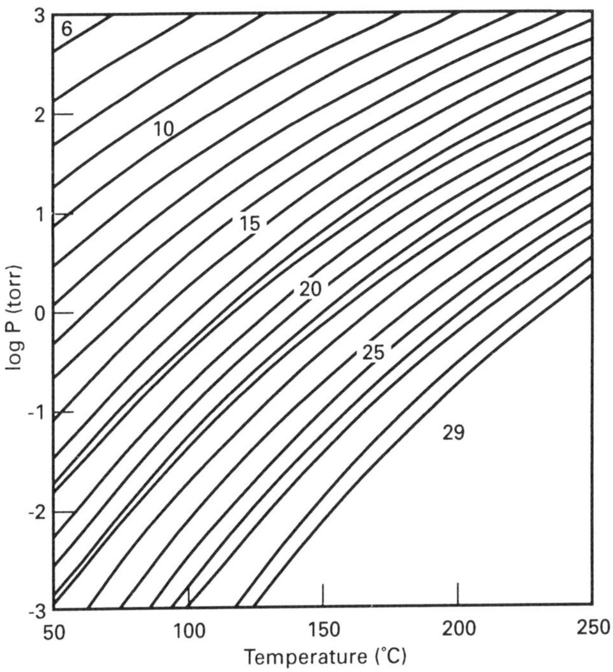

FIGURE 6-10 The vapor pressure of straight-chain paraffin hydrocarbons (C6–C29). After Finn (1991).

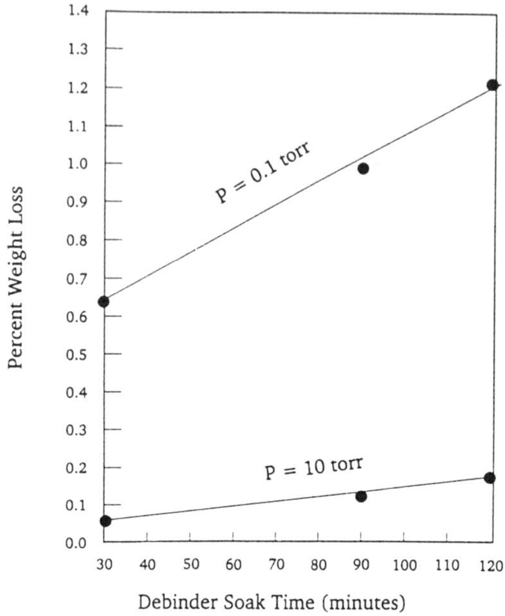

FIGURE 6-11 Lowering the pressure to a submicron level reduces the binder removal time. After Finn (1991).

(C6–C29) paraffin hydrocarbons (Finn 1991). The rate of binder removal has been found to be a function of both pressure and the shape of the parts (Kennedy and Finn 1990). Lowering the pressure to a submicron level, as shown in Figure 6-11, reduces the binder removal time.

There are, however, some limitations to vacuum extraction that are associated with poor heat transfer and lack of temperature control plus a relatively sluggish breakdown of the binder components. These factors are particularly critical with binder formulations that contain polymers, such as polyethylene, with average molecular weights above 10,000.

A low ambient pressure or vacuum is considered optimal for reducing the debinding time under permeation control. A relationship has been developed to estimate the debinding time under permeation control (German 1987):

$$t = \frac{22.5\, H^2(1 - E)^2 PG}{[E^3 D^2 F(P^2 - P_o^2)]} \tag{6-6}$$

where H is the thickness of the part, E is the fractional porosity, P is the pressure at the binder-vapor interface, G is the viscosity of the vapor, D is the particle diameter, F is the ratio of the volume of the solid binder to the volume of the vapor at the pressure P, and P_o is the external gas pressure.

Equation (6-6) suggests that the debinding time depends directly on the viscosity of the vapor and the square of the wall section thickness, and varies inversely with the square of the particle size and the drop in pressure.

Several simplifying assumptions are incorporated into this model, including unidirectional flow, isothermal conditions, and a planar liquid-vapor interface. What the model lacks is a consideration of the internal forces created in parts as a result of entrapped vapors—in other words, what is happening behind the planar interface.

PYROLYSIS

The thermal degradation in flowing air of nonoxide injection molded ceramics could have serious consequences because of their possible oxidation. Silicon carbide (SiC) is more likely to become oxidized than silicon nitride (Si_3N_4) or silicon. Silicon carbide cannot tolerate an oxidizing environment, particularly in the presence of the excess carbon necessary for densification.

For this reason it is important to understand the nature of the thermal degradation process of ceramic materials in atmospheres other than air. The relationship of the weight loss to temperature for polycarbosilane in argon is shown in Figure 6-12. As can be seen, the thermal decomposition of polycarbosilane takes place in two distinct stages: from 100°C to 620°C and from 620°C to 850°C. The weight losses that occur during these two temperature ranges indicate that selective volatilization of functional groups is occurring.

The thermal degradation of milled SN502 silicon nitride mixes in air and in nitrogen shown in Figure 6-13 indicates a marked difference in these two atmospheres. A maximum temperature exists for both mixes in air but not in nitrogen. In nitrogen, thermal decomposition starts slowly around 200°C as in air but progresses rapidly from 475°C upward. For this reason, in nitrogen the temperature becomes extremely critical above 475°C, and a slow heating rate is required, less than 0.5 K/min, to prevent the injection molded part from failing (Wei and German 1988).

In Figure 6-14, the thermogravimetric analysis of water soluble methylcellulose in air shows an initial weight loss of 3%, followed by no further loss until 250°C to 350°C. Above 350°C there is further oxidative degradation, with 100% weight loss by 500°C (Sarkar and Greminger 1983). In a nitrogen atmosphere there is an initial weight loss of 5% at 50°C due to the

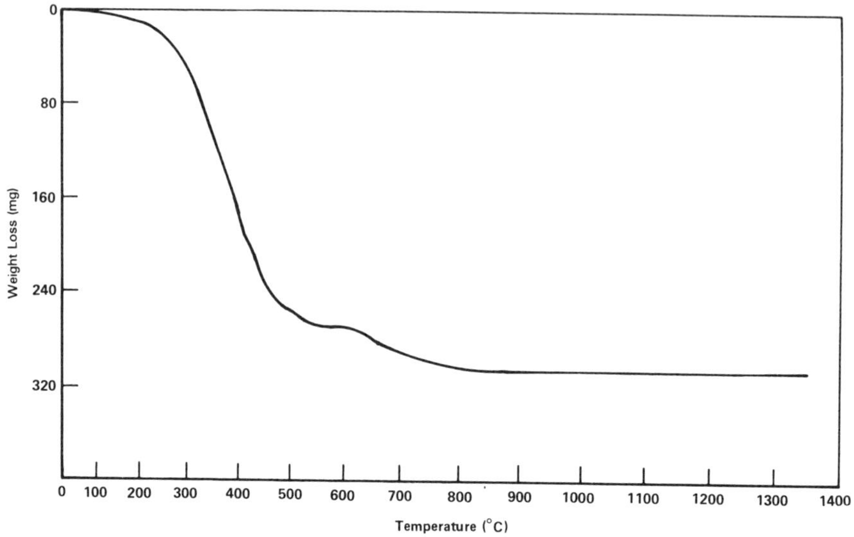

FIGURE 6-12 The thermogravimetric analysis of polycarbosilane in an argon atmosphere, showing the relationship of weight loss to temperature.

vaporization of water. There is no further weight loss until about 200°C, when pyrolytic degradation starts. This results in a further weight loss of 82% between 280°C and 410°C. Another 2% is gradually lost between 410°C and 800°C, leaving a carbon residue of about 10%.

During the early stages of decomposition in air, changes are occurring such as breaking of the main polymer chains, cross-linking between chains,

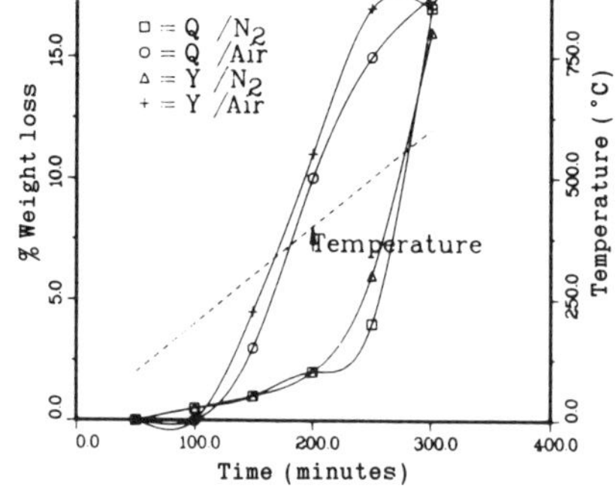

FIGURE 6-13 The weight loss of milled SN502 silicon nitride in air and in nitrogen.

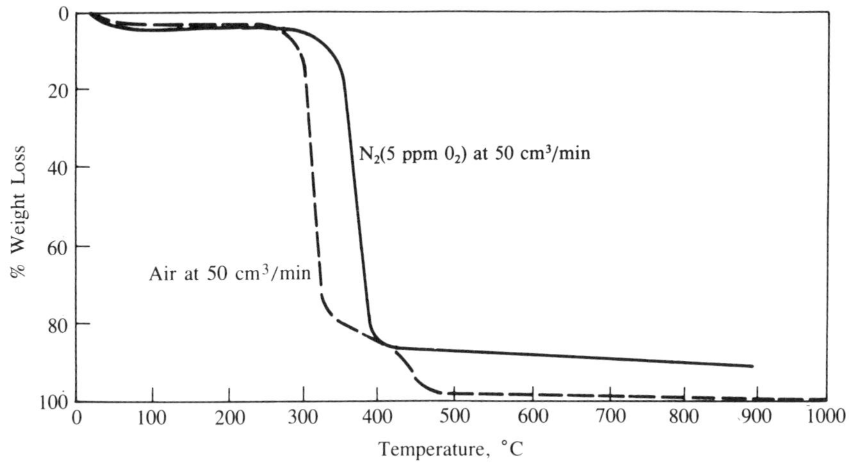

FIGURE 6-14 The thermogravimetric analysis of methylcellulose in air and in nitrogen.

and cyclization. The breakdown of the polymer chains forms some free radicals that can reform into monomers or react inter- or intramolecularly. In the absence of oxygen, such as in a nitrogen atmosphere, depolymerization processes are hindered. Furthermore, any cross-linked and cyclic molecules that form are extremely difficult to break down owing to their carbonization (Scheiffele and Sacks 1988; Shin et al 1988; Sun et al. 1988).

It has been suggested that pyrolysis depends on both chemical and physical factors (Cima et al. 1989). Chemical components influence the rate at which volatile substances are produced and require a system design that minimizes residual carbon in the part. The physical factors involve the relationship between binder removal and the limitations of mass and heat transfer. In pyrolytic debinding, the chemical components are the critical factors.

DRAINING THROUGH A POROUS BED

A powder bed that extracts the binder via capillary action has several advantages. It provides additional support, better thermal uniformity, a reduction in gas partial pressure gradients at the surface of the part, and homogeneous wicking out of the binder throughout the part. A model for binder removal by wicking along with a relation between debinding time and process parameters has been proposed (German 1987):

$$t = 4.5\left[\frac{(1 - E_c)^2 G H^2 D_w}{E_c^{\,3} W D_c (D_c - D_w)}\right] \tag{6-7}$$

where E_c is the fractional porosity of the part, G is the viscosity, H is the part's thickness, D_w is the diameter of the particles composing the wick, W is the binder–vapor surface energy, and D_c is the diameter of the particles composing the part.

Equation (6-7) indicates that rapid debinding would be favored by large part porosity, low binder viscosity, thin part dimensions, small wick particles, high binder surface energy, and large part particles. This concept has been extended by others (Cima et al. 1988, 1989; Cima and Lewis 1990).

DEWAXING UNDER HIGH PRESSURE

Dewaxed parts of good quality can be produced at high heating rates by using air or nonoxidizing gas above atmospheric pressure. This is so because higher gas pressures favor diffusion and suppress pressure gradients. A schematic diagram of a pressure dewaxing unit is shown in Figure 6-15. The ceramic parts are placed in the wire basket on top of the spool piece. This ensures that they will be positioned in the heated portion of the dewaxing unit just below a thermocouple suspended above it. A metered sweep gas is injected into the dewaxing unit through the distributor plate in the spool piece. On exiting the heated portion, the gases can be cooled, depressurized by a back-pressure-regulator valve, metered, and vented. The total system pressure is automatically controlled to the desired set pressure.

The data given in Table 6-9 indicate that the measured weight loss is between 11.7% and 14.0% of the weight of the molded part. This corresponds to between 90% and 100% removal of the binder. Figure 6-16 shows a conceptualized pressure debinding cycle (Inoue et al. 1986).

Figure 6-17a and Figure 6-17b illustrate the interaction between three processing variables: the pressure of the binder removal atmosphere, the rate of temperature rise, and the complexity of the part's geometry (Swank et al. 1986). The analysis, which does not provide any specific information about either heating rate or pressure, indicates that maximum binder removal occurs at high pressures and low heating rates. In addition, as indicated in Figure 6-17c, there is a strong interaction between pressure and part size.

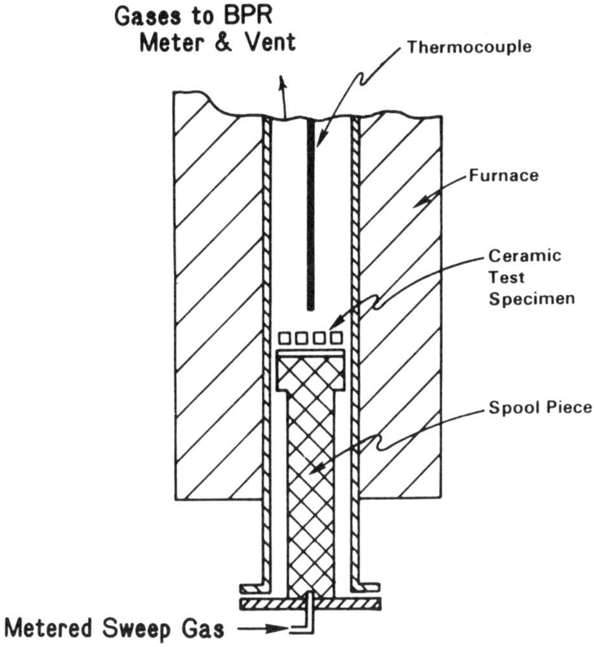

FIGURE 6-15 Diagram of a pressure dewaxing unit. BPR = back pressure regulator valve.

TABLE 6-9. Weight loss at different dewaxing pressures.

	Conditions			
Atmosphere	Pressure (mPa)	Heating rate (°C/hr)	Av. wt. loss (%)	Remarks
Air	0.034	10	14.0	I.C.[a]; Se.C.[b]
Air	0.034	5	13.3	Few I.C.
Air	0.034	50	13.5	Few I.C.
Air	0.345	50	13.5	Large I.C.
Air	0.345	10	13.4	No flaws
N_2	0.034	10	13.4	I.C.
N_2	0.034	5	13.0	Few I.C.
N_2	0.034	50	13.2	Large I.C.
N_2	0.345	50	12.8	Large I.C.
N_2	0.345	10	13.4	Few I.C.
H_2	0.345	5	12.5	Few I.C.
H_2	0.345	10	12.1	Few I.C.
H_2	0.345	50	11.7	Large I.C.

a. Internal cracks.
b. Seam cracks.

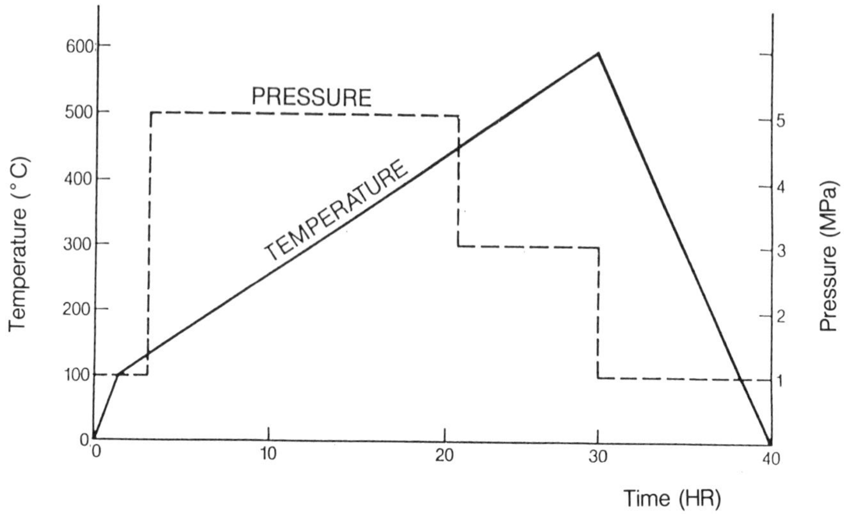

FIGURE 6-16 A conceptualized pressure debinding cycle.

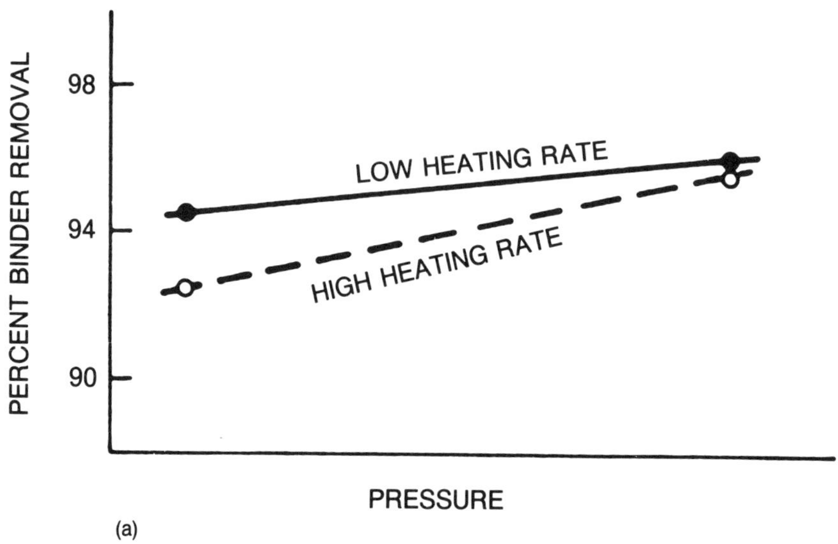

FIGURE 6-17 The relation between the pressure of the binder removal atmosphere and different variables associated with binder removal: (a) percent of binder removal as a function of pressure at two different heating rates; (b) number of cracks that develop as a function of pressure for components of two different sizes; and (c) number of cracks that develop as a function of pressure at two different heating rates.

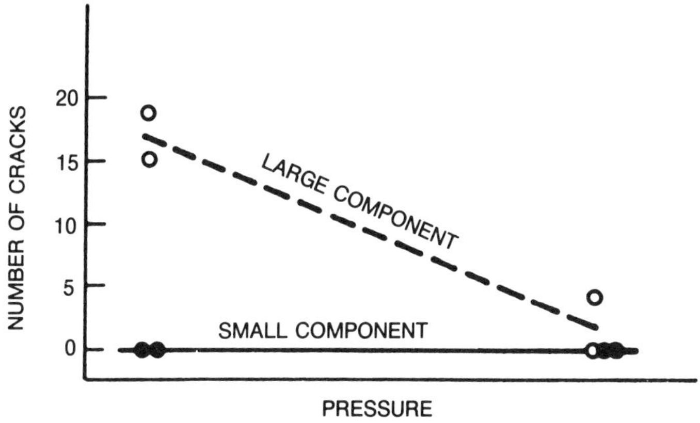

(b)

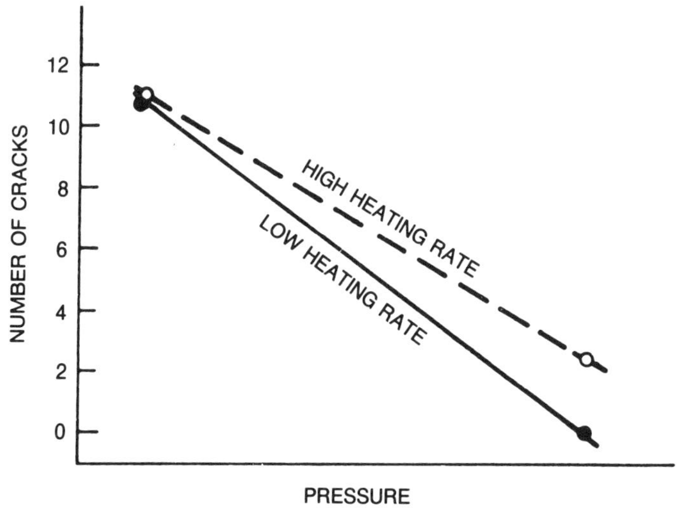

(c)

FIGURE 6-17 *contd*

THE MECHANISMS OF THERMAL DEWAXING

During thermal degradation, some of the binder ingredients at some temperature can liquefy, some can begin to volatilize, and others can remain very viscous. All of these processes are likely to occur simultaneously. For such complex overall reactions, thermodynamic calculations cannot predict how a polymeric binder will decompose thermally. Therefore it is necessary to investigate both degradation and transport mechanisms.

Degradation

It has been found that when polymers undergo thermal degradation, their molecular weight decreases. This is an indication that breaks occur in the polymer backbones even before any weight loss is detected. The reactions vary with a polymer's chemical structure and with the conditions at which oxidation occurs. Nevertheless, some generalizations about reaction mechanisms are possible and could be applied to the oxidation of hydrocarbon polymers.

As summarized in Table 6-10, the oxidation of these polymers takes place through a series of chemical reactions. In the initial step, the energy from heating causes radicals to be formed within the polymer. Initiation via reaction (1) probably contributes very little to polymer oxidation because it assumes that the original polymer, RH, is a chemically pure substance. Steps (2) and (3), in which propagation occurs, represent the chain reaction in which the first radical generated reacts rapidly with oxygen to form a peroxy radical, which then abstracts hydrogen from another polymer molecule to form a molecule of hydroperoxide and a second radical. This second radical then undergoes reactions (2) and (3). A chain reaction ensues that can be repeated hundreds of time before its termination.

As some preoxidation is unavoidable during the mixing and molding steps, traces of peroxides usually are present. Polymer oxidation probably is initiated by the radicals formed from their decomposition. The hydroperoxides formed in reaction (3) in Table 6-10 are unstable. They also similarly decompose into radicals, which, by abstracting hydrogen from the polymer molecules, initiate new oxidative chains and chain branching, the combined result of reactions (4) and (5).

The regular occurrence of branching along a polymer chain enhances the oxidation rate. The process terminates either when radicals couple, thus dissipating, or when they disproportionate into inert species. The autocata-

TABLE 6-10. Oxidation scheme for polymers.

1)	RH[a] + IN[b]	→	R*	Initiation
2)	R* + O	→	ROO*	
3)	ROO* + RH	→	ROOH + R*	Propagation
4)	ROOH	→	RO* + HO*	
5)	RO* + RH	→	ROH + R*	Chain branching
6)	HO* + RH	→	HOH + R*	
7)	ROO* (RO* etc.)	→	Inert products	Termination

a. Polymer.
b. Initiator.

lytic phase of polymer oxidation during which chain branching occurs is followed by a period during which the oxidation rate appears to be steady. In time the reaction starts to decelerate, and the oxidation rate slows, probably because of the exhaustion of reactive sites on the polymer chains. Although chemical reactivity is important, the physical structure of the polymer also influences the amount of oxygen that reacts before the rate slows down. The effect of polymer morphology on the oxidation rate is different for different polymers.

Transport

After the temperature of the ceramic part has been raised enough to melt the binder, the part resembles a porous solid. Removing the binder at a desirable rate then involves the mass transport of a gas in a porous solid. In such systems, there are two possible mechanisms of fluid flow: forced flow, in which a pressure differential acts as a driving force, or diffusion, in which the driving force depends on the concentration or the difference between the partial pressures of the binder ingredients.

Temperature, atmosphere, and pressure affect the two transport mechanisms in different ways. The most obvious effect of temperature is an increase in the partial pressure of the binder components. At temperatures below the boiling point of the binder components, the increased partial pressure produces a concentration gradient that favors diffusion. At and above the boiling point, pressure gradients occur that result in increased force flow. At higher temperatures, decomposition reactions occur that produce additional gases and further increase the pressure.

Diffusive transport can be further subdivided into bulk diffusion and Knudsen or capillary diffusion. The bulk diffusion coefficient of gases increases with increasing temperature, but the increase in the diffusion rate is relatively small compared with the increased pressure gradients from the gases produced by the volatilization and dehydrogenation of the binder components. These pressure gradients result in force flow. Stresses created within the part by this force flow can result in cracking and other internal defects.

If the pore sizes are large compared to the mean free molecular path of the gaseous binder molecules, bulk diffusion takes place. Therefore, large particle sizes (20–75 μm in diameter) and a high degree of porosity in the part (20–30%) would favor bulk diffusion.

If the pore sizes are comparable to the mean free molecular path of the vaporized binder molecules, collisions with the walls of the pores occur, and capillary diffusion dominates. This type of diffusion appreciably delays the flow of gas through the capillary.

If the major transport mechanism is capillary diffusion, the effect of the atmosphere will be relatively insignificant and limited mainly to changes in the composition. Although bulk diffusion rates are related to the molecular size and weight of the organic species in the binder, the dominant mechanism governing capillary diffusion is the collisions of the molecules of the volatilized organic binder with the walls of the pores. Although gaseous diffusion at the surface of the part would be increased by using lighter gases such as hydrogen, it is not considered the limiting mechanism for transfer of the binder.

Changes in the composition of the volatilized binder most likely will be determined by whether the atmosphere used is oxidizing or reducing. A hydrogen atmosphere would suppress dehydrogenation reactions, but it is unlikely that any formation of hydrogenated gaseous species would occur at the debinding temperature (about 600°C). The major effect of the atmosphere is to minimize the generation of hot spots from any combustion of the organic binder.

The use of overpressure elevates the boiling point of the binder components but does not affect the partial pressure. It also tends to shift the thermodynamic equilibrium away from any decomposition reactions that produce gaseous products which increase the pressure gradients. Both of these effects favor diffusion over forced transport. Overpressure also tends to reduce the molecular mean free path. If the pore sizes are sufficiently large, this could promote the conversion from capillary to bulk diffusion within the ceramic part.

It has been proposed that the dewaxing process is controlled by permeation in which laminar flow controls the rate of release of volatiles from an injection molded part (German 1987). This mechanism is most effective when pore sizes are relatively large and the vapor pressure is high. The experimental evidence supporting the permeation mechanism assumes the formation of connected open porosity (channel pores) after the rate of binder removal exceeds about 25 volume percent (Wada and Oyama 1989).

An important variable in the dewaxing process is the removal time, that is, the rate of removal. The implication of the transport mechanisms discussed above is that binder removal is a slow process. For bulk or capillary diffusion to be an effective transport mechanism it is necessary to consider the rate of binder removal as a function of time. The volatilization of a mix of silicon nitride, polyethylene, butyl oleate, and castor oil with time at different temperatures is shown in Figure 6-18. In Figure 6-19, an inverse relationship is indicated between the dewaxing rate and the amount of binder removed.

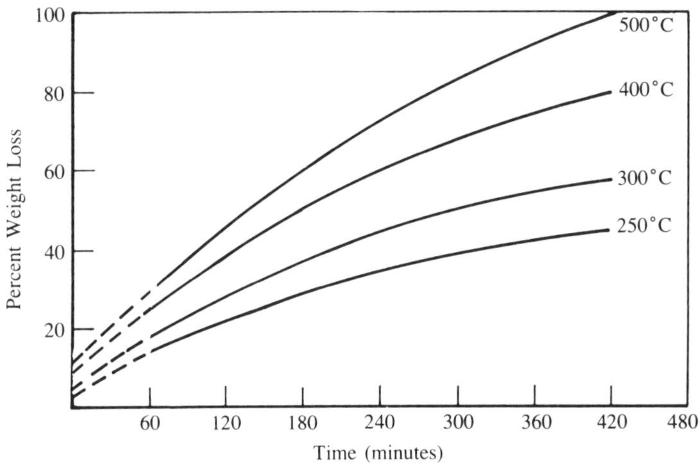

FIGURE 6-18 The weight loss for a silicon nitride mix as a function of time at four different temperatures.

The rate of removal strongly depends on the size of the part, specifically the thickness of its thickest section. An analysis of the effect of various parameters on the dewaxing time concluded that there is a direct dependence between the dewaxing time and the square of the section thickness (German 1987).

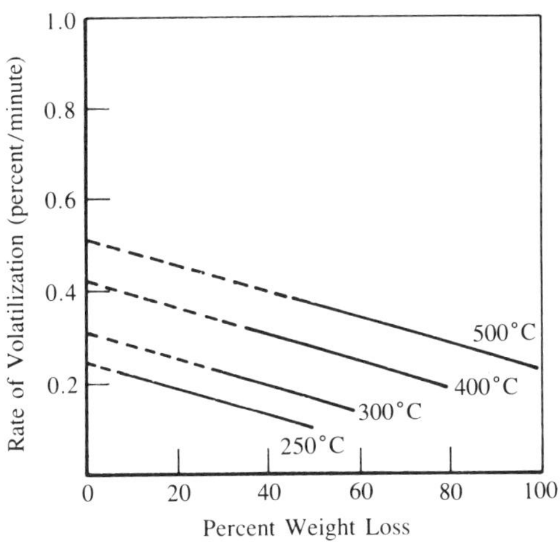

FIGURE 6-19 Thermal degradation rates for a silicon nitride mix at four different temperatures.

ALTERNATIVE DEWAXING PROCESSES

One alternative approach for dewaxing utilizes the selective solubility of the binder components in organic solvents. If an appropriate solvent is selected, a binder can be completely or partially removed without damaging the part. Among the organic solvents that can be used are ethylene dichloride, trichloroethane, hexane, toluene, pentane, heptane, methylene chloride, and acetone. The chief advantage of solvent extraction for dewaxing is the high solubility of the binder components in these organic solvents. They have properties very similar to those of gases, such as low viscosity, high rates of diffusion, and low or no surface tension.

Advantages claimed for solvent extraction are faster removal of the binder and greater economy because less costly equipment is required. An additional potential advantage is that the process can be automated by continuously conveying the parts through a solvent bath and a dryer. The solvent then can be purified by distillation and recycled.

In solvent extraction, the part is immersed in or washed with a solvent for a period of time that can be anywhere from 10 minutes to several hours depending on the size and thickness of the part and the extraction temperature. After the part is taken out of the solvent, any retained solvent is evaporated in air either at room temperature or by heating. Once the excess solvent is eliminated any components of the binder that are insoluble in the organic solvents are removed by further heating. Immersing parts in a solvent over an extended period of time, which depends on their size and the temperature, could cause either slumping of the parts or loss of powder due to excessive removal of the binder.

Even though the use of a solvent with a high vapor pressure, such as the chlorofluorohydrocarbon Freon, has been suggested for binder extraction (Weich 1983, 1986), this now is believed to be destructive to the ozone layer and thus environmentally unacceptable and should not be done. More desirable are organic solvents, such as hexane and heptane, that have a high solubility for paraffin wax because of their similar chemical structures, a low viscosity, and a wide evaporation range of from 2°C to 98°C (Lin and German 1989). The diffusion coefficient of the binder in these solvents is expected to be higher than that for other solvents.

The solvent extraction process can be thought of as proceeding via four steps: diffusion of the solvent into the binder phase, dissolution of the binder in the solvent, interdiffusion of the solvent and the binder, and diffusion of the solvent–binder solution to the surface of the part. The solubility parameters can be used to manipulate the first two steps. In practice, close values of the solubility parameters for the solvent and the binder are selected.

Assuming that the removal of the solvent–binder solution is rapid, a temperature-dependent interdiffusion coefficient, D_x, for the binder and the solvent has been derived (Lin and German 1989):

$$D_x = D_o \exp(-E/kT) \tag{6-8}$$

where D_o is the pre-exponential frequency factor, E is the activation energy, k is the Boltzmann constant, and T is the temperature in degrees Kelvin.

The fraction, F, of soluble binder remaining in the part, C_a, can be expressed as either:

$$F = \frac{C_a}{C_i} = \exp\left(\frac{-D_x t \pi^2}{4L^2}\right) \tag{6-9}$$

or:

$$-\ln(F) = \frac{D_x t \pi^2}{4\,L^2} \tag{6-10}$$

where L is the thickness of the part, C_i is the initial composition of the soluble binder, and t is the extraction time.

By substituting equation (6-8) in equation (6-10), one gets the following:

$$-\ln(F) = D_o t \pi^2 \exp\left(\frac{-E/kT}{4\,L^2}\right) \tag{6-11}$$

and:

$$\ln[-\ln(F)] = \ln\left(\frac{D_o t \pi^2}{4\,L^2}\right) - (E/k)(1/T) \tag{6-12}$$

The interdiffusion coefficient, D_x, as a function of temperature can be obtained from linear regression of a plot of $\ln[-\ln(F)]$ as a function of $1/T$, shown in Figure 6-20:

$$D_x = 1.52 \exp(-4208/T) \tag{6-13}$$

where the units of D_x are cm^2/sec.

Supercritical fluids, which are intermediate between gases and liquids, also can be used for solvent extraction. Certain substances, such as carbon

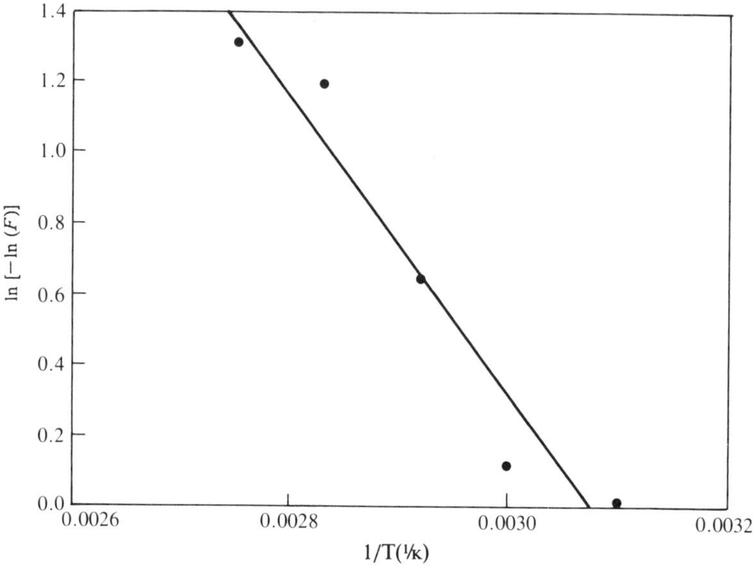

FIGURE 6-20 A plot that can be used to determine the interdiffusion coefficient as a function of temperature.

dioxide, become supercritical fluids above but near their critical temperature and above their critical pressure (the temperature and the pressure at which their liquid and vapor phases are in equilibrium). As supercritical fluids they are better solvents than ordinary liquids at relatively low temperatures. A supercritical solvent also can be a mixture of two or more substances. The critical temperature of such a mixture is generally (but not always) between the critical temperatures of its components.

If a supercritical solvent is near or above its critical temperature, its diffusion rate will be much higher than that of an ordinary liquid at room temperature, and its resistance to flow (or viscosity) will be much lower than that of an ordinary liquid at room temperature and atmospheric pressure.

Hydrodynamic theory indicates that the diffusion coefficient, D_x, is inversely proportional to the radius of the solvent molecule, α, and the viscosity of the solvent, η (Reid et al. 1977):

$$D_x = \frac{kT}{6\pi\eta\alpha} \tag{6-14}$$

where k is the Boltzmann constant, and T is the absolute temperature.

Furthermore, there is no capillary attraction because the density is lower than for ordinary liquids, and there is no surface tension. This is advantageous for dewaxing because capillary attraction causes rearrangements in the packed particles, which result in microdefects, such as density variations and internal cracks, in a part. Another advantage is that binder removal is two to three times faster than with thermal dewaxing processes because no gases are generated within the part.

Carbon dioxide is the preferred solvent for supercritical fluid extraction. As shown in Figure 6-21, a phase diagram of carbon dioxide, it becomes a supercritical fluid at 31 °C and 72.8 atmospheres of pressure. Figure 6-22 is a schematic diagram of the equipment for supercritical fluid extraction (Miyamoto, et al. 1990). The degree of extraction is shown as a function of time in Figure 6-23. Other solvents, such as hexane and heptane, also can be used effectively. The selection of the solvent is based on the solubilities of the binder components, which vary widely from solvent to solvent.

The major limitation of solvent extraction is the solubility of the solvent in certain organic polymers such as waxes, which results in swelling of the polymer network. The rate of solvent absorption is in large part diffusion-controlled and is related to the size of the solvent molecule. During the initial stages of solvent extraction, absorption occurs below the glass transition temperature of the organic polymers.

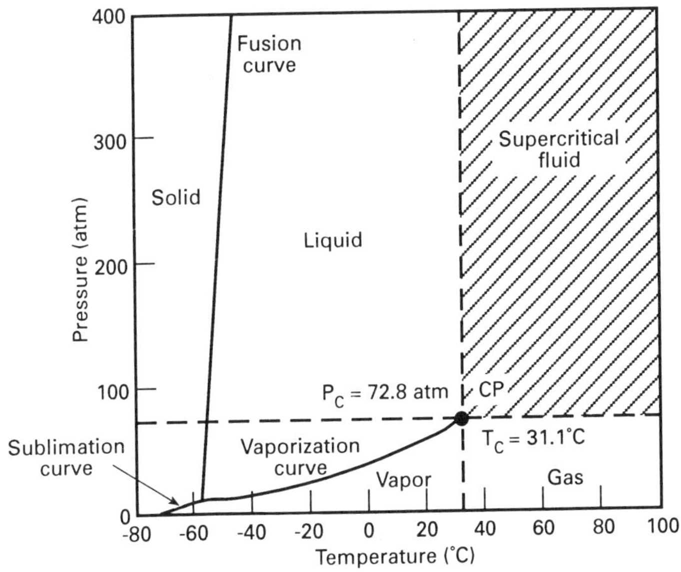

FIGURE 6-21 Phase diagram of carbon dioxide.

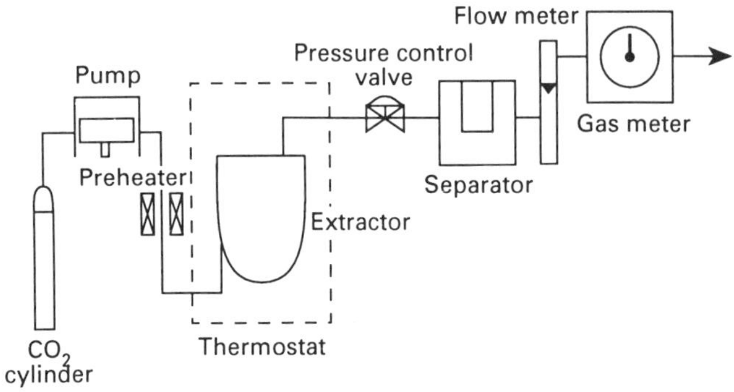

FIGURE 6-22 Schematic of apparatus for supercritical extraction. After Miyamoto et al. (1989).

DEFECTS IN DEWAXED PARTS

Four major defects can occur in dewaxed parts: skin formation, cracks, knit lines, and delaminations. However, they are not necessarily caused by the dewaxing process. Defects also are caused by the incompatibility of components in the binder formulation, inhomogeneous mixing, poor mold design, and inadequate control of the molding conditions.

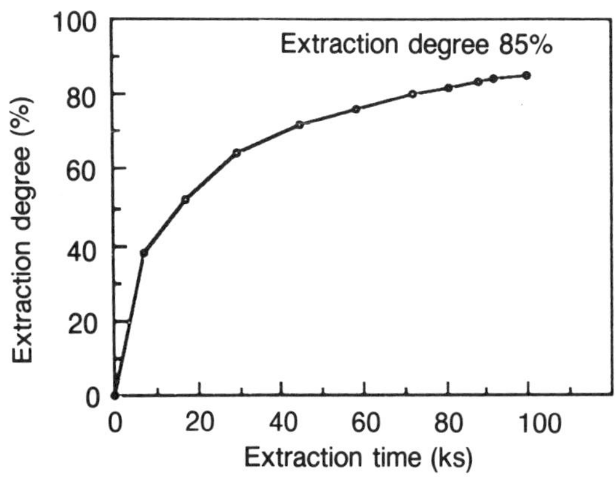

FIGURE 6-23 Degree of extraction as a function of time.

Skin Formation

The chemical compatibility of different binder ingredients and the ceramic powder has the greatest effect on the quality of the final molded part, as discussed in Chapter 3. Therefore, to prevent the formation of a skin on the molded part it is essential to understand the chemical interactions between the binder ingredients as well as the compatibility of the binder ingredients with various ceramic powders. A skin can form on the surface of the part during the molding process for several reasons: phase separation under high shear; excessive restriction in the mold, such as inserts; a narrow gate size relative to the size of the part; and excessive cavity pressure with a mold temperature that is too low or too high. For example, a high cavity pressure relative to the size of the part can force fine ceramic particles to move to the surface. When, in addition, the mold temperature is low for a wax-type binder formulation (less than 50°C), a hard shell or skin is formed. As noted earlier, skin also can form during dewaxing owing to the migration of fine particles to the surface via capillary pressure.

Knit Lines

Problems with mold design associated with the formation of knit lines are discussed in Chapter 5. Suggested changes in mold design to overcome this problem include alterations in gate size, location, and shape, and possibly also using a heated runner and bushing.

Crack Formation

The formation of cracks is one of the most intractable problems with injection molded parts. The more complex the part, the worse the problem. A typical crack is illustrated in Figure 6-24, which shows a crack originating from the weld line. Complete elimination of cracks that start from the weld line necessitates a radical change in mold design.

Other types of cracks, which frequently appear after dewaxing parts such as a combustion prechamber or an airfoil, are shown in Figure 6-25a. The cracks in the combustion prechamber were probably due to the insert in the combustion prechamber mold, and those in the airfoil, shown in Figure 6-25b, probably originated underneath the platform of the airfoil mold. It is a reasonable assumption that these cracks were caused by localized stress concentration from several possible sources: uneven packing during molding, the part geometry, or nonuniform shrinkage during both dewaxing and cooling down from the sintering temperature.

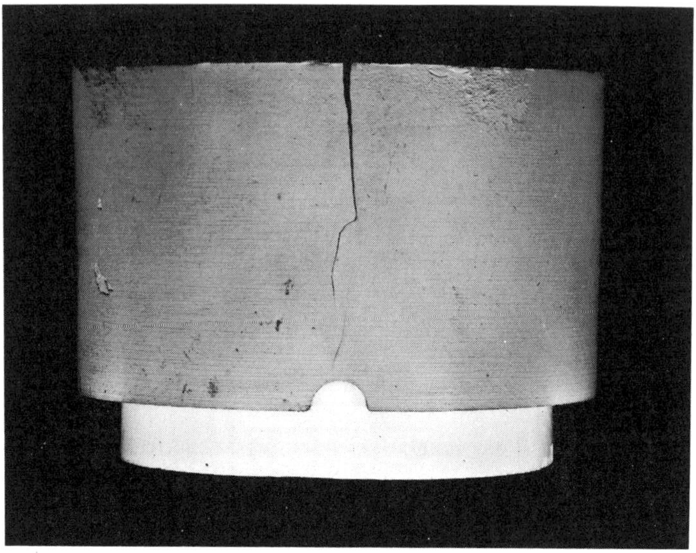

FIGURE 6-24 A crack originating from a weld line.

Several approaches can be used to overcome this problem: redesigning the insert in the combustion prechamber mold, enlarging the radius underneath the platform of the airfoil mold, and reducing the hold pressure during the packing and cooling stages of the molding process. Shrinkage or expansion can be minimized by increasing the filler loading in the mix and maintaining a mold temperature that allows relieving the stress, for example, 50°C for a wax-type binder formulation. Stress cracks that originate from adhesion of the part to the mold can be reduced by applying a mold-release agent before each molding cycle.

Cracks also can arise from density variations within the part, which can be due to physical characteristics of the powder particles such as their size, size distribution, shape, and surface area, and also their packing behavior. The distribution of the binder in the molded part also can contribute to density variations, as a consequence of the rearrangement of pores and particles during the debinding process (Cima et al. 1989). Density variations are the most likely source of the microcracks that appear in injection molded parts after dewaxing.

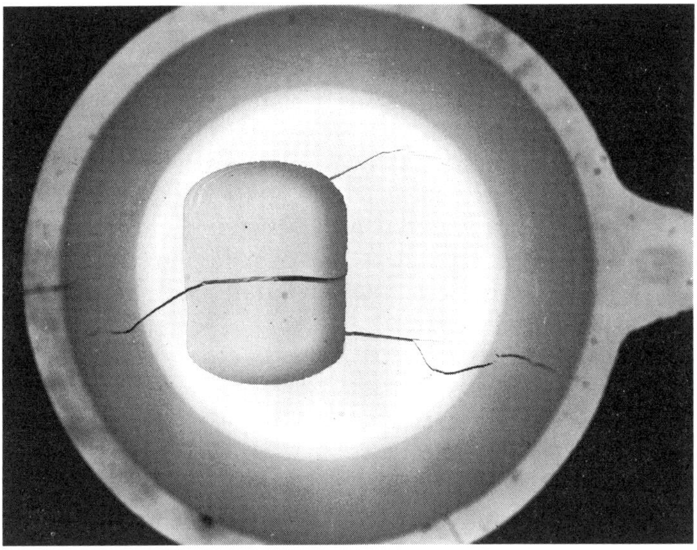

(a)

(b)

FIGURE 6-25 Cracks in two different parts after dewaxing: (a) a crack in a combustion chamber; and (b) a crack in an airfoil. From Battelle Columbus, OH.

Delamination or Internal Cracks

One of the most challenging problems to overcome is delamination in the dewaxed parts. Approaches for solving this problem have been suggested by experiments that determine the temperature at which the delamination or internal cracks begin to appear. In one such experiment, the heating rate is held at 10°C/hr and interrupted at 100, 200, 300, 350, and 450°C. In addition, the samples are placed either freestanding or bedded in alumina powder (100 and 325 mesh). The use of a bedding material provides support and improves both thermal uniformity and wicking out of the binder through capillary action.

Whether or not bedding material is used, internal cracks or delaminations start to appear between 100°C and 200°C, intensifying between 300°C and 350°C, as shown in Figure 6-26. The intensification coincides with the onset of pyrolysis, around 300°C, which is consistent with mass spectrometry data. From 100°C to 350°C, a part can undergo a number of simultaneous changes. Some of the binder ingredients may liquefy, and some may start to volatilize, whereas others may remain viscous. Delamin-

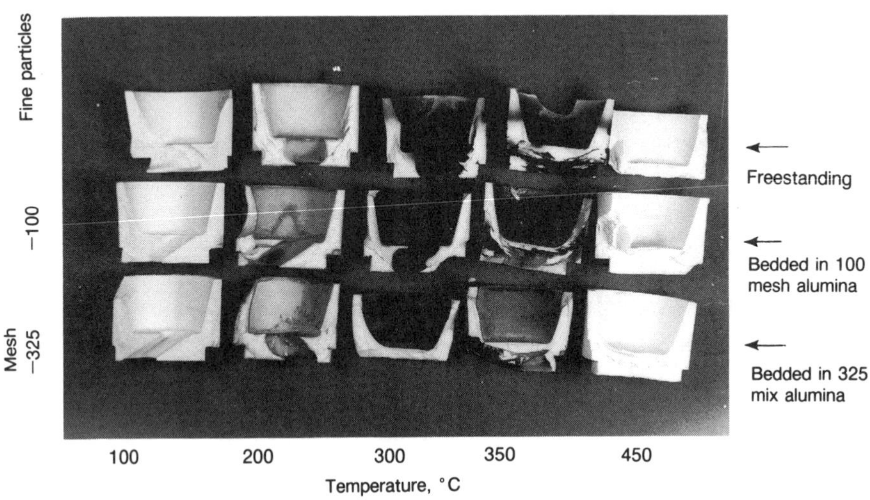

FIGURE 6-26 Delamination or internal cracks in A16-SG alumina at different temperatures.

ation could be initiated by the simultaneous effects of phase separation and the internal pressure of the gases produced during dewaxing (H_2, O_2, CO_2, and CO).

It also has been suggested that rearrangement of ceramic particles in the part's interior during binder removal, which results in denser packing, causes internal cracks or delamination (Quackenbush et al. 1982). A possible explanation for this is that as the interior binder is lost, capillary forces draw the ceramic particles together into denser packing, but the resulting decrease in volume is resisted by the part's surface, which remains expanded, and thus tension cracks can develop. Properties of the ceramic powder, such as particle size, as well as the level of powder loading and the binder system used, are among the factors affecting the extent to which this flaw-generating mechanism can destroy a part.

Other sources of delaminations in the molded part, which also should be taken into consideration, are the knit lines that can form during the mold-filling process. If such defects are introduced during molding, they cannot be corrected during dewaxing.

Entrapped air also is a potential source of delamination. Therefore, during the molding operation entrapped air must be removed from the mix, and displaced air removed from the mold cavity. Increasing the pressure and lowering the heating rate both have been suggested as ways to reduce the amount of internal cracking (Swank et al. 1986).

Other Flaws

Besides these major flaws, other defects occasionally are seen in the dewaxed parts. Blisters, pinholes, and crazing are due to inhomogeneities in the mix or uncontrolled heating during dewaxing. Speckles generally are caused by the presence of foreign inclusions in the binder, ceramic powder, mixer, injection molding machine, or mold. Slumping is attributed to a variety of causes, among them too much binder, the effect of gravity, an uncontrolled heating rate, and possibly even the size and shape of the particles of the ceramic powder.

References

Ando, T. 1986. Production Method for Ceramic Parts, Jap. Patent 83682 531–533.

Bandyopadhyay, G., K. W. French, L. J. Bowen, and J. T. Neil. 1986. Method for Fabricating of Large Injection Molded Shapes, Eur. Patent 0196600.

Barone, M. R. and J. C. Ulicny. 1990. *J. Am. Ceram. Soc.*, 73(11), 3323–3333.

Barone, M. R., J. C. Ulicny, R. R. Hengst, and J. P. Pollinger. 1989. Removal of Organic Binder in Ceramic Powder Compact, *Ceram. Trans.*, Vol. 1A, *Ceram. Powder Sci.*, II, ed. G. Messing et al., Am. Ceram. Soc., 578–583.

Calvert, P. and M. J. Cima. 1988. Theoretical Models for Binder Burnout, MIT Ceramic Processing Research Laboratory Report 91, Materials Processing Center, MIT, Cambridge, MA.

Cima, M. J. and J. A. Lewis. 1990. Diffusion of Dialkyl Phthalate in Plasticized PVB, *J. Am. Ceram. Soc.*, 73(9), 2702–2707.

Cima, M. J., M. Dudziak, and J. A. Lewis. 1988. Observation of Poly(Vinyl butyral)-Dibutyl Phthalate Binder Capillary Migration, *J. Am. Ceram. Soc.*, 72(6), 1087–1090.

Cima, M. J., J. A. Lewis and A. D. Devoej. 1989. Binder Distribution in Ceramic Greenware during Thermolysis, *J. Am. Ceram. Soc.*, 72(7), 1192–1199.

Finn C. W. 1991. Vacuum Binder Removal and Collection, *Internat. J. Powder Met.*, 27(2), 127–132.

German, R. M., 1987. Theory of Thermal Debinding, *J. Powder Met., Am. Powder Met. Inst.*, 23(4), 237–245.

Gilissen, R. E. M., A. G. Smolders, and J. F. C. Cooymans. 1985. Belgium Patent 903308.

Ikejima, M. and T. Imai. 1984. Production Method of Ceramic Bodies, Jap. Patent 141462 337–342.

Inoue, M., T. Sakai, and T. Katagiri. 1986. Reduction of Dewaxing Time by Pressurized Atmosphere in the Ceramics Injection Molding Process, *Yogyo Kyokai Shi* (*J. Japan Ceram. Soc.*), 94(1), 78–80.

Johnsson, A., E. Carlstrom, L. Hermansson, and R. Carlsson. 1983. Minimisation of the Extraction Time for Injection Molded Ceramics. *Proc. Brit. Ceram. Soc.*, 33, 139–147.

Kato, S. 1985. Debinding Method for Ceramic Moldings. Jap. Patent 53147 249–254.

Kennedy, S. W. and C. W. P. Finn. 1990. Metal Injection Molded Part Debinding and Sintering Vacuum Furnaces, *Powder Met. Sci. Tech.*, 1(3), 37.

Lin, S. T. and R. M. German. 1989. Extraction Debinding of Injection Molded Parts by Condensed Solvent, *Powder Met. Internat.*, 21(5), 19.

Miyamoto, A., Nishio, and Miyashite. 1990. Slip Injection Molding and Supercritical Extraction—SIMSE, Austceram '90, Perth, Western Australia, 26–31, Aug. 1990.

Mutsuddy, B. C. 1987. A Review of Binder Removal Processes for Ceramics Injection Molded Parts (CIM), paper presented at the Conference on Emerging Technologies in Materials, AIChE, Minneapolis, MN.

Quackenbush, C. L., K. French, and J. T. Neil 1982. Fabrication of Sinterable Silicon Nitride by Injection Molding, *Ceram. Eng.-Sci., Proc.*, 3, 1–2, 20.

Reid, R. C., J. M. Prausnitz, and T. K. Sherwood 1977. *The Properties of Gases and Liquids*, 3rd edn., McGraw-Hill Book Co., New York.

Sakai, T., M. Inoue, Y. Kihara, and Y. Kawabata 1985. Debinding Method of Ceramic Green Moldings, Jap. Patents. 215577 461–463 and 215578 465–467.

Sarkar, N. and G. K. Greminger, Jr. 1983. Methylcellulose Polymers as Multifunctional Processing Aids in Ceramics, *Am. Ceram. Soc. Bull.*, 62(11), 1280–1284/88.

Scheiffele, G. W. and M. D. Sacks. 1988. Pyrolysis of Poly(Vinyl Butyral) Binders II: Effects of Processing Variables, *Ceram. Trans. Ceram. Powder Sci., Am. Ceram. Soc.*, 1(A), 559-566.

Schreiber, R., O. Vogt, and R. Wosnitzka 1980. Heat Treatment of Ceramic Moldings, DDR Patent 142330.

Shin, W. K., M. D. Sacks, G. W. Scheiffele, Y. N. Sun, and J. W. Williams. 1988. Pyrolysis of Poly(Vinyl Butyral) Binders I: Degradation Mechanisms, *Ceram. Trans. Ceram. Powder Sci., Am. Ceram. Soc.*, 1(A), 549-558.

Sun, Y. N., M. D. Sacks, and J. W. Williams 1988. Pyrolysis Behavior of Acrylic Polymers and Acrylic Polymer/Ceramic Mixtures, *Ceram. Trans. Ceram. Powder Sci., Am. Ceram. Soc.*, 1(A), 538-548.

Swank, L. R., J. A. Mangels, J. C. Caverly, and R. K. Govila 1986. Life Prediction Methodology Final Report, Ford Motor, Contract DAAG 46-77-C-0028, US AMTL, 1-69.

Tanikawa, W. and N. Kurata. 1985. Debinding Materials for Ceramic Moldings, Their Production Method, Their Use, and Debinding Apparatus in Which a Full Profit Is Made of Their Properties, Jap. Patent 251173 587-593.

Verweij, H. and W. H. M. Bruggink. 1988. Apparatus and Method for Removal of Organic Binder from a Green Powder Compact, Dutch Patent PHN 12387.

Wada, S. and Y. Oyama, 1989. Thermal Extraction of Binder Components from Injection Molded Bodies, Toyota Central Res. and Dev. Labs., Nagakute, Aichi, 480-11, Japan.

Wei, T. S. and R. M. German. 1989. Injection Molded Tungsten Heavy Alloy, *Powder Met. Internat.*, 24(4), 327-335.

Weich, R. E. Jr. 1983, 1986. Method of Removing Binder from a Green Body, U.S. Patents. 4,404,166 and 4,305,756.

7

Sintering and product evaluation

Sintering is treated here as one of the factors used in evaluating the quality of parts produced by injection molding. The key parameters in evaluating sintering data are density and shrinkage. In addition to the sintering of nonoxide ceramics, the nitriding process for making reaction-bonded silicon nitride and pressure sintering processes for covalently bonded ceramics are similarly discussed. The conditions described are applicable to the initial pilot-plant stage and, with certain adjustments, to full-scale production.

In this chapter and its appendices, the sintering process and a process profile are described; density and shrinkage data are presented; and significant properties in product evaluation such as strength and fracture toughness are delineated.

THE SINTERING PROCESS

Alumina

High-performance alumina ceramics were developed in the mid-1930s with the incorporation of significant quantities of alpha alumina (α-Al_2O_3) in spark plug insulators. Since then the applications of engineering grade alumina ceramics have grown spectacularly.

Sintering Aids for Alumina

The technical feasibility of sintering alumina powder to very high density (greater than 99%) by the addition of a small amount (about 1.0 wt%) of magnesia (MgO) was first suggested in 1948 (Ryshkewitch 1948). Similar results have been reported using fine grained alpha alumina (0.5–0.7 μm) containing 0.5 to 1.0 wt% magnesia and sintered at 1750°C for 5 to 10 minutes (Pavlushkin 1956). The mechanism of sintering alumina powder and the role of magnesia have been extensively investigated. It is now believed that magnesia prevents discontinuous grain growth by a solid solution effect (Coble 1961), and also increases the sintering rate.

The most important effect of doping with magnesia is the ability of the magnesia solute to reduce grain boundary mobility via a solid solution pinning mechanism despite the presence of various impurities. Consequences of this pinning action are a reduced tendency for separation at the pore and grain boundaries and for nonuniform distribution of the liquid phase, and protection against abnormal grain growth.

The bulk of alumina ceramics used are 90 to 98% pure, with silica and silicates, particularly clays, the most common additives. A glassy phase, which can have a very complex composition, is produced when silica (SiO_2) or clays react with alumina around 1400°C.

The addition of a variety of metal oxides[1] reduces the firing temperature and modifies the properties of the fired parts. Furthermore, the use of ultrafine powder (less than 0.05 μm to 1.0 μm) with a high specific surface area plus an agglomerate-free green compact with about 69% relative density reduces the sintering temperature to less than 1200°C, and produces a fine-grained, uniform microstructure in the sintered part (Kumagai and Messing 1984, 1985; Yen and Sacks 1988).

Reducing Porosity

It is particularly important to eliminate the residual porosity in a compacted alumina part. The factors that determine the relationship of pore removal to secondary recrystallization are related to the factors that determine the reactivity and the initial state of the alumina powder, such as the size of the particles and their distribution, the degree of compaction that can be attained, the size of the pores in the green compact, the ratio of the pore size to the grain size, and the coordination number of the pores (the number of pores surrounding each pore). The distribution of pores and their location with respect to the grain boundaries in the green compact affect the length of sintering time required to attain full density.

[1]These additives include the oxides of barium, strontium, cerium, manganese, nickel, cobalt, tantalum, thorium, clay-lithia-magnesia, zirconia-silica, calcia-magnesia-silica, zirconia-strontia, alkaline earth oxides, and silica-strontia-magnesia.

There is a persistent compressive stress on a pore in a fine-grained matrix, which is the net result of the appearance of the grain boundary area and the disappearance of surface area during shrinkage. During sintering, this compressive stress is a positive and increasing quantity, with a rate that depends on the rates of grain growth and pore shrinkage (Lin and German 1988).

In the final stage of sintering, after normal grain growth, the pores are located at the junction of four grains. Another possibility is that the pores collect within the centers of the grains, often with a clear demarcation around their boundaries resulting from secondary recrystallization. If large numbers of closed pores form within the grains, the rate of densification is markedly reduced (Bruch 1962).

The rate of sintering or the removal of residual porosity in alumina compacts is affected by the sintering atmosphere (Walker 1955). For example, the sintering rate at 1960°C is faster in hydrogen, oxygen, or a vacuum than in air; and between 1600°C and 1960°C, the rate of grain growth is minimized in dry hydrogen. However, it increases in other atmospheres, for example, in carbon monoxide, argon, and air, in increasing order.

Porosity can be completely eliminated by controlling discontinuous grain growth and by sintering in hydrogen, oxygen, or a vacuum (Coble 1962); but sintering in helium, argon, nitrogen, or air does not completely eliminate porosity. The different results obtained in the various atmospheres are attributed to differences in the solubilities of these gases in alumina. However, it has been suggested that near-theoretical density (over 99%) can be achieved in sintering fine powder (0.3 μm) regardless of the sintering atmosphere (Paek et al. 1988).

Fast firing (heating rapidly at a rate of about 10^5°C/hr) has been found to produce high-density alumina (Harmer et al. 1979). The success of densifying alumina by rapid heating has been explained on the basis of the relative values of the enthalpies of activation of the coarsening and densification processes (Brook 1982). In the simplest case, these values are assumed to be associated with the values of the surface and lattice coefficients, D_s and D_l, respectively. By heating the ceramic powder compact rapidly through the low-temperature region, where the coarsening mechanism dominates, to the region where the densification mechanism dominates, a dense product is obtained in a very short dwell time (about 5–10 minutes) at the peak temperature.

Non-oxide Ceramics

The non-oxide ceramics are covalently-bonded refractory compounds of the light elements boron, silicon, carbon, and nitrogen—such as silicon nitride (Si_3N_4), silicon carbide (SiC), and boron nitride (BN). In general, these materials cannot be readily sintered. This has led to the development of specialized methods to sinter them. It has been suggested that the densification of these non-oxide ceramics is hindered by the presence of strong directional covalent bonds. These contribute to a low concentration and mobility of lattice defects, and thus inhibit diffusion within their lattices. An alternative explanation is that surface diffusion or vapor transport could be the primary transport mechanisms preventing their densification.

Pressureless Sintering of Silicon Nitride

A combination of approaches have been used to successfully densify silicon nitride (Gazza and Katz 1978). These include improving the packing of the finely powdered material, using sintering aids, introducing lattice defects, reducing the ratio of the free energy of the grain boundaries to the free energy of the surface (γ_{gb}/γ_{sv}), and minimizing competitive processes that hinder densification.

In the early 1970s, it was shown that sialon-type materials, containing silicon, aluminum, oxygen, and nitrogen, could be readily sintered without applying pressure (Jack 1973). It also was shown that silicon nitride containing 5 wt% magnesium oxide (MgO) as a sintering aid could be sintered in nitrogen at atmospheric pressure (0.1 MPa) to 90% of theoretical density. A loss in weight of about 10% was attributed to the decomposition of the silicon nitride (Terwilliger and Lange 1975). With the addition of the spinel, $Al_2O_3.MgO$, silicon nitride with greater than 96% of theoretical density was produced at less than 0.1 MPa (1 atm) of N_2 pressure. By embedding the parts in loose silicon nitride powder both the decomposition of the silicon nitride and the volatilization of the magnesium oxide could be prevented at these same sintering conditions (Giachello et al. 1979).

When a mixture of Si_3N_4-Al_2O_3-AlN was sintered in nitrogen at 0.1 MPa, it densified to 98% of theoretical density (Stewart et al. 1977). The mechanism of "transient liquid phase sintering" has been proposed for this densification. It is suggested that initially a liquid forms that accelerates densification. Subsequently, this liquid crystallizes, producing dense beta silicon nitride (Layden 1976).[2] When silicon nitride is synthesized from starting materials ($3Si + 2N_2 \rightarrow Si_3N_4$), the total surface energy is reduced

---

[2]There are two crystallographic forms of silicon nitride—alpha and beta. It is believed that the alpha and beta forms are stoichiometric polymorphs, where the alpha form has a higher free energy of formation. Nitridation below the melting point of silicon favors formation of the alpha phase. However, heat treatment of the alpha phase above 1500°C results in its considerable transformation to the beta phase.

by the completion of the chemical reaction and by the reduction of the total surface area of the particles. Both these effects also can contribute to densification.

When silicon nitride is sintered with combinations of either magnesia and beryllia (MgO–BeO) or magnesia, beryllia, and ceria (MgO–BeO–CeO_2), it is densified to about 97% of theoretical density (Oda et al. 1977). Beryllia would be expected to act in the same way as magnesia as an additive because beryllium silicate ($BeSiO_4$) has the same atomic arrangement as beta silicon nitride and is much more soluble both in beta and in beta-sialon than is magnesium silicate (Mg_2SiO_4, forsterite).

The effect of varying amounts (2, 4, 8, and 17%) of the sintering aid yttria (Y_2O_3), on silicon nitride powder has been investigated (Rowcliffe and Jorgensen 1977). Densities greater than 95% of theoretical density were achieved with various amounts of the additive yttria and sintering for 24 hours. As shown in Figure 7-1, the yttria initially reacts with silica on the surface and some nitride, forming a liquid, which allows densification by liquid phase sintering. As the reaction proceeds, this liquid combines with additional silicon nitride to form one or more refractory bonding phases.

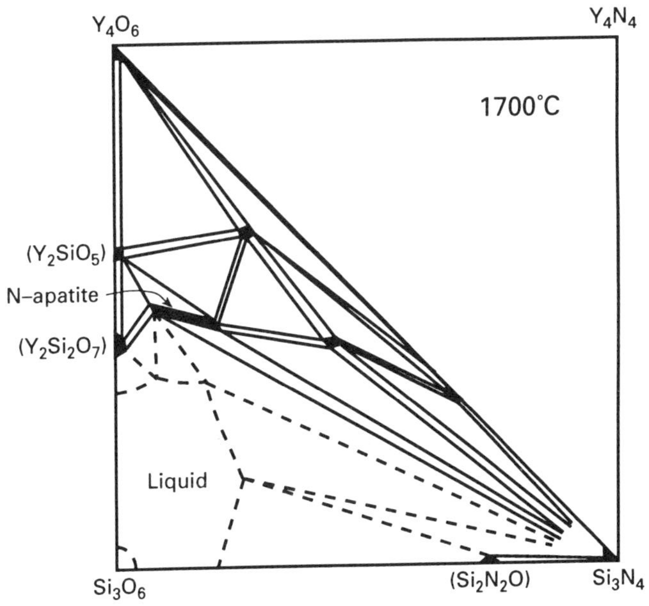

FIGURE 7-1 The Y–sialon phase relationship at 1700°C.

On cooling, any unreacted liquid present forms a glass that can be devitrified by heat treatment, thereby improving the product's properties (Tsuge et al. 1975). The addition of 10 wt% of a eutectic composition of yttria and alumina (Y_2O_3–Al_2O_3) as a sintering aid markedly improves sinterability (Venables et al. 1976). This improvement is attributed to a large volume of a low-viscosity liquid that apparently wets the solid completely and migrates rapidly throughout the microstructure. Within the Si_3N_4–Y_2O_3–Al_2O_3 triangle shown in Figure 7-2, there is a wide composition range, which expands with increasing temperature, where the products are only beta sialon and Y-sialon glass.

It has been shown that as the melting temperature of a single component oxide sintering aid increases (for example, CeO_2. Nd_2O_3, La_2O_3, Sm_2O_3, Y_2O_3), the sintering temperature must be increased in order to reach theoretical density (Hirosaki et al. 1988). When a mixed oxide additive such as Y_2O_3–Ln_2O_3 (Ln = Ce, Nd, La, Sm) is used, higher densification is achieved below 2000 °C because of the lower temperature at which a liquid forms.

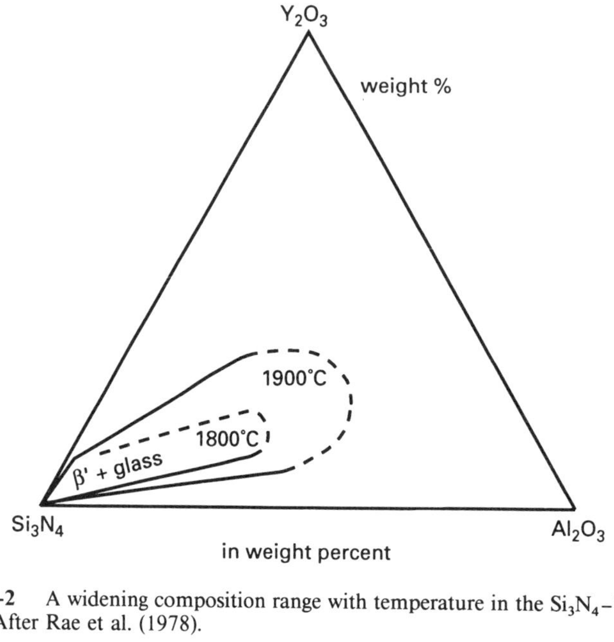

FIGURE 7-2 A widening composition range with temperature in the Si_3N_4–Y_2O_3–Al_2O_3 system. After Rae et al. (1978).

During sintering at temperatures above 1750°C, in the presence of a liquid phase, the transformation from the alpha to the beta phase takes place via a solution-precipitation mechanism. The growth of the beta phase during this transformation occurs by the preferential solution and reprecipitation on preexisting beta particles (Lange 1978). During their transformation, the morphology of the grains changes from equiaxed to fibrous. This is noteworthy in light of the higher fracture toughness and strength of the fibrous beta-phase microstructure (Drew and Lewis 1974).

The Role of High Nitrogen Pressure

The use of high nitrogen pressure (1–70 MPa) to suppress thermal decomposition and thus increase the ability to sinter at high temperatures has been investigated (Gazza and Katz 1978). Nitrogen pressure decreases the partial pressure of silicon vapor (P_{Si}) to below the vapor pressure in equilibrium with liquid silicon and also prevents the spontaneous decomposition of silicon nitride into liquid silicon and nitrogen. If, for example, a silicon nitride sample is kept at 2000°C in a nitrogen environment in which the pressure (P_{N_2}) is 0.5 MPa (5 atm), the silicon nitride will evaporate. The rate of evaporation will be controlled either by the rate of silicon vapor transported away from the sample or by the rate of silicon nitride decomposition.

If a closed system is used with the nitrogen pressure maintained at 1.0 MPa (10 atm), and the temperature is gradually increased, the vapor pressure of the silicon will increase (along the vertical line in Figure 7-3) until the boundary is reached where the liquid and the solid coexist (indicated on Figure 7-3 by the middle arrow), that is, where liquid silicon forms. If further exposed to nitrogen, all the silicon nitride would isothermally decompose into liquid silicon and gaseous nitrogen. The rate of decomposition would be controlled by the heat flow into the system.

For actual sintering experiments, the thermodynamics would depart from the ideal calculations used for constructing Figure 7-3. For example, the presence of additives, which are somewhat soluble in silicon nitride, would alter the coexistence boundaries and change the regions of stability. For this reason, it is more desirable to operate as close to the liquid–solid–vapor coexistence boundary as possible while keeping decomposition to a minimum. Therefore, in order to define the optimum sintering parameters, it is necessary to experiment with the specific additive system for each starting material.

There are a number of examples of the successful densification of silicon nitride by using high nitrogen pressure (1–50 MPa) with a variety of sintering aids. At 1.0 MPa nitrogen pressure with magnesia plus other

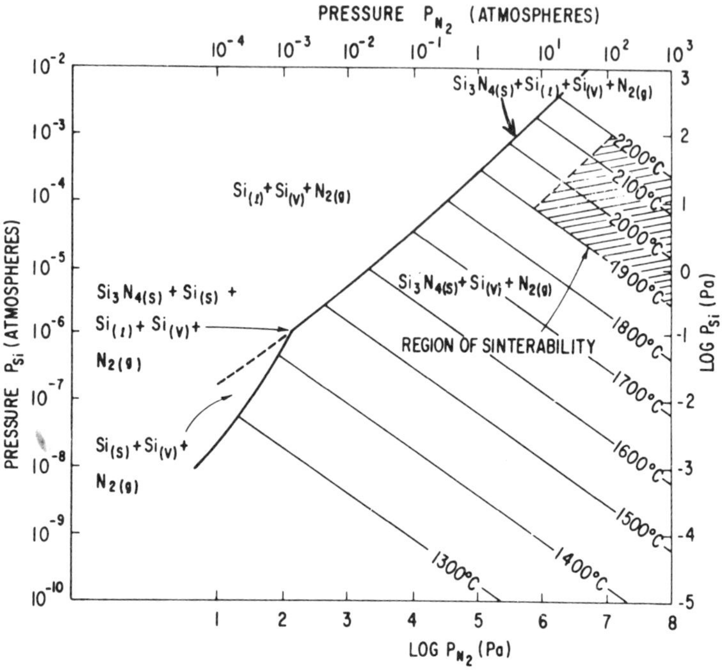

FIGURE 7-3 The calculated partial pressure of silicon at elevated temperatures and nitrogen pressure. After Gazza and Katz (1978).

additives such as alumina and yttria, densification to 98% of theoretical was achieved (Mitomo 1976); using 2.0 MPa nitrogen pressure plus 20 wt% ceria, 90 to 95% of theoretical density was achieved (Priest et al. 1977); and at 5.0 to 8.0 MPa nitrogen pressure with the sintering additives Be_3N_2 plus Mg_2Si or Be_2SiN_2 plus Mg_3N_2, sintered above 1800°C, 99% of theoretical density was achieved (Greskovich et al. 1979).

Another approach for achieving close to theoretical density is to sinter the part in two stages, first at a low pressure and then at a high pressure. In one demonstration of dual pressure sintering, the part first was sintered to the "closed-pore" stage (about 92–94% of theoretical density) at 0.1 MPa nitrogen, followed by sintering at 2.0 MPa. The density was increased to close to 98% of theoretical after sintering at the higher nitrogen pressure (Gazza and Katz 1980). However, if the samples were not sintered to the closed pore stage at the lower pressure, they could not be densified further by sintering at a higher nitrogen pressure because a pressure differential was not produced between the external pressure and the gas pressure within the pores. The use of dual pressure sintering to achieve 98% of theoretical

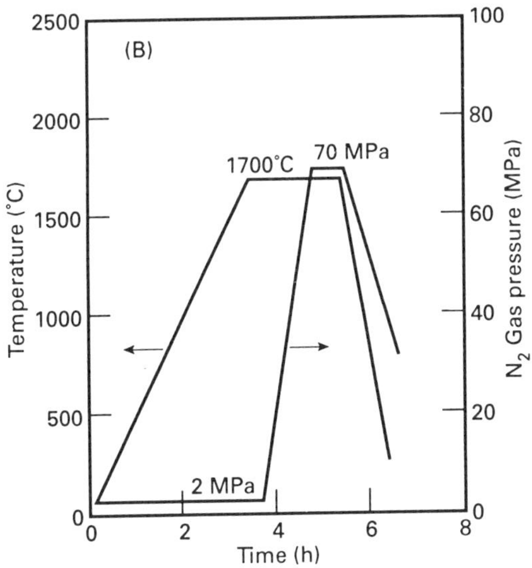

FIGURE 7-4 The use of dual pressure sintering to achieve 98% of theoretical density in a mixture of 10% Y_2O_3 and 5% Al_2O_3 as sintering aids.

density in a silicon nitride sample containing a mixture of 10 wt% yttria and 5 wt% alumina as sintering aids is shown in Figure 7-4 (Hirosaki and Okada 1988).

Reaction-bonded Silicon Nitride

Reaction-bonded silicon nitride refers to a material produced by heating silicon,[3] which has been formed into a desired shape, in an atmosphere of nitrogen or nitrogen plus hydrogen. The reaction is carried out in a cold-wall type furnace or in a furnace lined with refractory material, using heating elements of graphite or silicon carbide. The equation for the exothermic reaction is:

$$3Si_{(s)} + 2N_{2(g)} = Si_3N_{4(s)} \tag{7-1}$$

---

[3]Silicon nitride also has been produced from raw materials such as silica and kaolin, in addition to silicon. However, to date only silicon has been used to produce reaction-bonded silicon nitride.

The product consists of both the alpha and the beta phases of silicon nitride. The inability to form a single-phase material is attributed to the difficulty of controlling the kinetics of the type of gas/solid reaction involved in producing a solid product such as silicon nitride. One proposed model for the reaction that takes place during the formation of silicon nitride from silicon under normal nitriding conditions is that the alpha silicon nitride is formed via the dominant process in which volatilized silicon undergoes a vapor-phase reaction with nitrogen. Beta silicon nitride is formed via a secondary process involving the nucleation of silicon nitride on solid silicon, and its subsequent growth following the surface diffusion of silicon to the reaction site (Moulson 1979).

In another model, the silica/nitrogen reaction is considered as the dominant one in the nitriding process (Morgan 1977; Lindley et al. 1979). This is supported by some experimental evidence indicating that silica rapidly volatilizes off the surface of silicon particles in the presence of iron. Although the role of iron in this reaction is not fully understood, its influence in accelerating the nitridation process is recognized. The amount of beta silicon nitride formed can be correlated both to the iron concentration and to the presence of a liquid at the reaction temperature. A high nitriding temperature also favors the growth of the beta phase.

A large number of materials have been found to accelerate nitridation, such as calcium fluoride, (CaF_2), barium fluoride (BaF_2), copper, cobalt nickel, chromium, manganese, and vanadium. Aluminum also plays an important role. It has been observed that a high ratio of the beta to the alpha phase is promoted by aluminum (Atkinson et al. 1976). A possible explanation is that the aluminum reduces any silica present. Nitridation also is accelerated when the the silicon particles are 1 μm or less in diameter (Herbell et al. 1980; Cannon et al. 1982). The surface area undoubtedly is the controlling factor.

The kinetics of the nitriding process measured as a function of temperature between 1250°C and 1370°C and nitrogen pressure between 2.5 kPa and 100 kPa involve several stages (Atkinson et al. 1976). In the initial stage, silicon nitride nuclei form on the surface of the silicon. These nuclei then grow by what is thought to be a reaction between chemisorbed nitrogen and silicon vapor, which has arrived at the reaction site via a combination of surface diffusion and an evaporation condensation process. During this nucleation and growth stage the kinetics are linear with time. As the nitrided layer extends over the surface, the supply of silicon to the reaction sites decreases. Consequently, the reaction rate decreases until nitridation effectively ceases.

The texture of the silicon nitride formed, that is, whether it is coarse or fine-grained, also is related to the temperature and the nitrogen pressure. A fine-grained texture develops at temperatures less than 1400°C and pressures of about 1 MPa (Heinrich 1987). A high nitrogen pressure of about 50 MPa promotes the formation of beta silicon nitride, and results in the development of a fined-grained, homogeneous microstructure due to the growth of a large number of silicon nitride nuclei under high-pressure nitridation.

The nature of the nitriding gas also affects the quality of the product. Creep resistance, for example, can be improved significantly using a hydrogen/nitrogen nitriding atmosphere (Mangels 1978); but because the reaction is exothermic, the reaction rate is difficult to control.

The nitriding cycle is a two-step process with one temperature soak below and one above the melting point of silicon. Nitriding above the melting point produces poor microstructures in reaction-bonded silicon nitride because of the draining of molten silicon from the parts to the container (Evans and Davidge 1970; Godfrey 1975). A multistep nitriding cycle with a peak temperature below the melting point of silicon has been proposed as one solution (Messier and Wong 1974).

A nitrogen-demand nitriding cycle, in which the length of the cycle depends on both the atmosphere and the nitriding load and the temperature cycle is dictated by the rate of nitrogen consumption, has been proposed (Mangels 1978). The nitrogen demand cycle allows the reactants to control the rate of reaction. This is accomplished by monitoring the reaction rate at one temperature, and then allowing the temperature to increase.

Nitriding profiles with three different furnace loads based on the nitrogen demand cycle are shown in Figure 7-5. The curves indicate that the nitriding reaction starts at around 1000°C and reaches completion at about 1300°C. With the demand cycle, the real temperature–time profile shows a major shift from the programmed temperature input. In addition, the gas consumption curves indicate that the kinetics of the nitriding process are close to linear, and that the cycle time depends primarily on the amount of silicon in the furnace.

Sintered Reaction-bonded Silicon Nitride (SRBSN)

Reaction-bonded silicon nitride (RBSN) contains between 15 volume percent and 30 volume percent porosity. Consequently, the strength of RBSN is relatively low and its resistance to internal oxidation only moderate. By sintering RBSN at a high temperature, the residual porosity is eliminated forming sintered reaction-bonded silicon nitride (SRBSN). In the SRBSN process, the initial silicon preform is first nitrided as described in the previous section, then the nitrided preform is sintered between 1780°C and

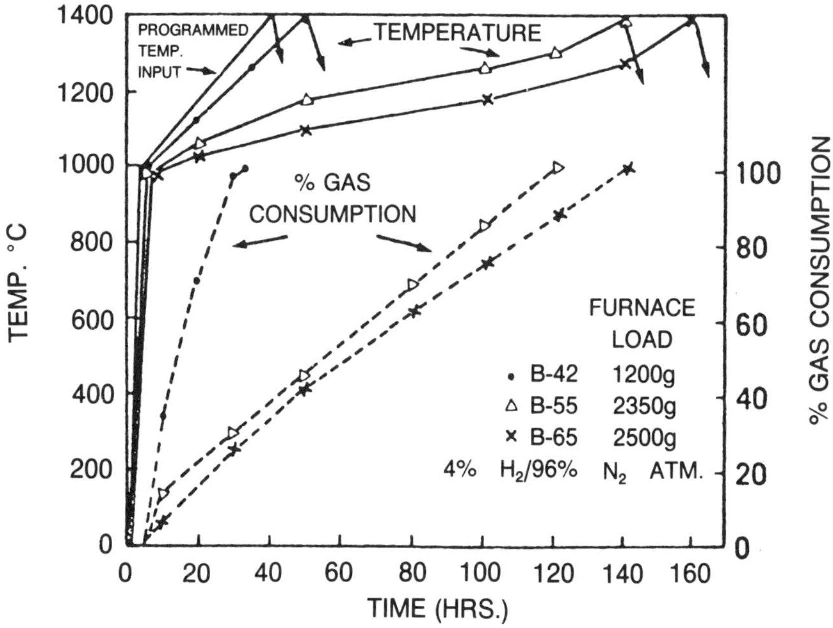

FIGURE 7-5 Nitriding profiles based on the nitrogen demand cycle. After Mangels (1978).

2000 °C in nitrogen at about 10 MPa pressure. Prior to nitridation, ytrria or magnesia are added to the silicon powder (Mangels 1986). Densities above 96% of theoretical have been achieved with this process.

The Nitriding Furnace

The preferred furnace design for making reaction-bonded silicon nitride is a cold-wall furnace with silicon carbide or graphite heating elements, shown in Figure 7-6. The heating cycle should be programmable and based on the variable conditions of temperature, vacuum, pressure, and nitrogen demand. Every stage of the heating process should start and end only when the programmed condition of each of the variables is fulfilled at the correct time. The furnace instrumentation should indicate the temperature, power, vacuum pressure, and nitrogen consumption. To guard against the possibility of a temperature overshoot (that is, an exothermic reaction), the furnace should be fitted with a safety trip, which is set 50 °C above the programmed set point temperature. It should also be fitted with a heat fuse designed to open at 1500 °C.

FIGURE 7-6 A cold-wall nitriding furnace. Courtesy Vacuum Industries.

Pressureless Sintering of Silicon Carbide

As with silicon nitride, the fineness of the silicon carbide (SiC) powder plays an important role in obtaining high sintered densities. Optimum sintering results can be achieved only with powders in the submicron range. Specific surface areas between 15 m^2/g and 21.5 m^2/g are considered the most favorable, for both alpha and beta silicon carbide (α-SiC: Coppola et al. 1978, and Bocker and Hausner 1977; β-SiC: Prochazka 1973, Greskovich and Rosolowski 1977, and Whalen and Winterbottom 1986).

The breakthrough in sintering silicon carbide came with the discovery of the positive effect of the simultaneous presence of boron and carbon in the silicon carbide powder. Liquid phase sintering of alpha silicon carbide was observed in the presence of a B_4C–SiC eutectic (Billington et al. 1962). However, the densification was limited to 90% of theoretical. By using as sintering aids a very small concentration (<1.0 wt%) of a boron compound, such as boron carbide plus carbon, beta silicon carbide is sintered to densi-

TABLE 7-1. Density of SiC with 1 wt% boron added in different boron compounds.

Addition	B (wt%)	Density (g/cm^3)
Amorphous B	0.5	3.01
Amorphous B	1.0	3.06
Crystalline B	1.0	3.04
B_4C	0.4	3.02
B_4C	0.8	3.05
LiBH	1.0	2.84
H_3BO_3	1.0	2.02

From Prochazka (1974b).

ties near theoretical (Prochazka 1973 and 1974a, Prochazka and Scanlan 1974). The sintering aids can be added during the synthesis of the silicon carbide powder or at a subsequent powder-processing stage.

The sintering of alpha silicon carbide using additions of boron and carbon or their compounds was first reported in 1974 and later confirmed (Bocker and Hausner 1978). Table 7-1 and Figure 7-7a and b show that the maximum effect on sintering occurs only when boron and carbon are both present in relatively small concentrations (Prochazka and Scanlan 1974). Table 7-2 shows that the introduction elemental carbon at higher concentrations (<0.7 wt/%) has no apparent effect on the final density (Prochazka 1974b). This result is attributed to the limited dispersion of carbon that follows mixing of powders and to the low carbon diffusivity in silicon carbide, which does not allow proper carbon redistribution over the silicon carbide surface.

Boron reduces the grain boundary energy by diffusing into the boundary surfaces, thereby creating the conditions for the growth of the boundary surfaces. Depending on the values of the diffusion coefficients, diffusion now can occur throughout the volume and at the grain boundaries. The effect of the boron is explained by the possibility of forming a ternary solid solution that also would effectively accelerate grain boundary diffusion (Prochazka 1974b).

Models for various grain assemblages show that for a given ratio of the grain boundary energy to the surface energy, in order for a material to be transported to the pores the individual grains must enclose the pores in as large a dihedral angle as possible. The minimum angle required depends on how the grains are arranged relative to each other. Powder packing and consequently green density in the powder compact are critical under these conditions (Stephenson and White 1967; Hoge and Pask 1973).

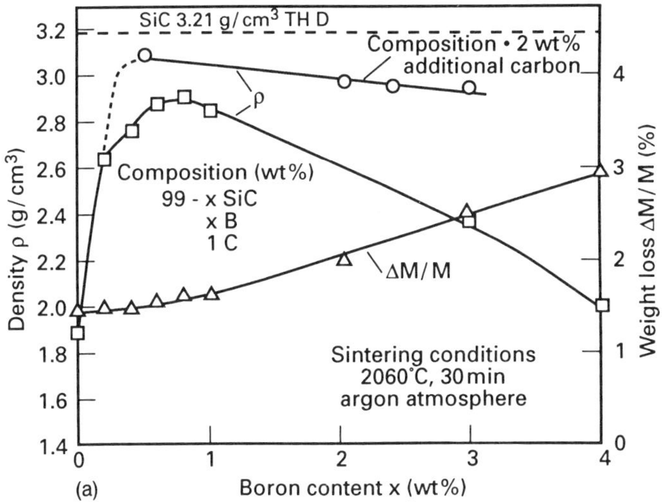

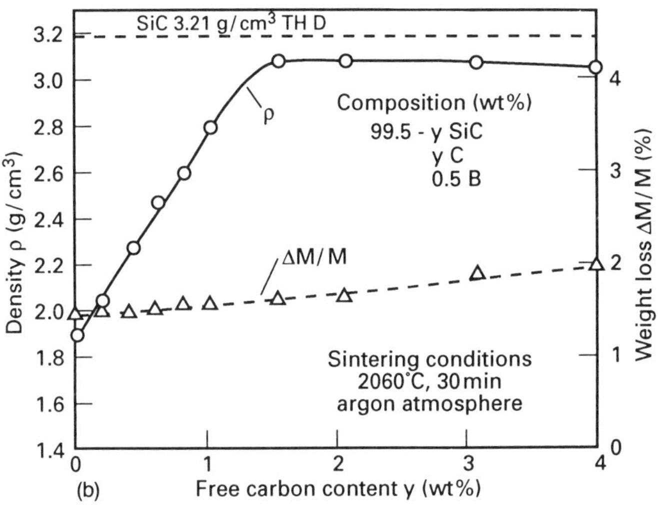

FIGURE 7-7 The effects of concentration of (a) boron and (b) carbon on the sintering of silicon carbide. After Böcker and Hausner (1978).

TABLE 7-2. The effect of adding carbon at a fixed content of boron.

Method	Addition to SiC (wt%) C	B	Density (g/cm^3)	Shrinkage (%)
C added as polymethyl-phenylene	0.0	0.36	1.96	3.5
	0.12	0.36	2.62	11.6
	0.25	0.36	3.08	16.7
	0.50	0.36	3.09	17.0
	0.80	0.36	3.09	17.0
C added as carbon black	0.25	0.36	2.66	11.5
	0.75	0.36	2.80	13.8
	1.50	0.36	2.75	13.8

From Prochazka (1974b).

With either continued sintering or decreasing porosity, the concentration of the oxygen remaining on the grain surfaces is increased. This can stop densification because the energy ratio shifts as a result of the increased oxygen content on the grain surfaces. Therefore, when sintering continues for longer than a few hours, pores can re-form.

The addition of carbon removes the oxygen adsorbed on the silicon carbide surface as silica. At the same time the cleaned silicon carbide surface is provided with high surface energy. Because of carbon's low diffusivity in silicon carbide, it must be uniformly distributed after pyrolysis by adding it in a polymer solution.

An explanation of the densification of silicon carbide, in addition to densification by volume or grain boundary diffusion in the solid state, involves a reaction sintering and liquid phase sintering mechanism (Lange and Gupta 1976). This is supported by the evidence of a second phase rich in boron that is present in the liquid state during sintering. In addition, a loss in weight has been observed that can only be explained by the dissociation of silicon carbide into silicon and carbon. Some of the silicon evaporates in the process, but could also react with boron carbide (B_4C) to produce silicon carbide and silicon boride (SiB_3), with a lowering of the liquidus temperature (Kieffer et al. 1972). The formation of a liquid phase when boron is added, and its possible role in the sintering process, also could be considered responsible for the exaggerated grain growth observed under certain conditions.

As the amount of liquid phase present is very small, it can be assumed that both sintering mechanisms take place and complement each other. A dense grain structure is obtained under conditions where a melt of silicon, boron, and carbon can serve to dissolve and precipitate silicon carbide while at the same time making possible a close-packed arrangement of the grains. A favorable ratio of the grain boundary energy to the surface energy is needed as a driving force in order to initiate these processes. In addition to boron, other possible sintering aids are beryllium, aluminum, iron, titanium, tungsten, and magnesium as elements or compounds (Bocker et al. 1979). Submicron alpha silicon carbide powder has been successfully sintered by the addition of elemental aluminum and certain of its compounds, such as AlN, Al_4C_3, $LiAlH_4$, plus an excess of carbon.

A mechanism has been proposed for the reactive liquid phase sintering (pressureless) of alpha silicon carbide plus rare earth oxides, such as yttrium oxide, in the presence of a gaseous species of aluminum oxide (Cordrey et al. 1990). The reactive liquid could be a eutectic liquid formed between aluminum oxide and yttrium oxide. However, the reactive liquid phase is fugitive in nature and at high temperatures (about 2000°C) probably is expelled from the compact as a vapor.

The Sintering Atmosphere, Temperature, and Time

Sintering usually is carried out in an atmosphere containing gaseous nitrogen, hydrogen, helium, or argon, or in a vacuum. If boron is used as the sintering aid, an atmosphere containing boron will favorably affect the sintering process. A boron-containing atmosphere is created by adding boron trichloride or other boron compounds that have a high vapor pressure at the sintering temperature.

When nitrogen is used as the sintering atmosphere, transformation of the beta phase to the alpha phase is inhibited, so excessive grain growth is prevented. But nitrogen reduces the sintering rate, necessitating a higher sintering temperature.

In general, the sintering interval at the appropriate temperature is very narrow; thus optimum sintered densities for a given powder are achieved after a relatively short sintering period. For example, to sinter beta silicon carbide, a temperature between 1900°C and 2100°C maintained for 30 to 60 minutes is sufficient. An excessively high temperature prevents densification by promoting the transformation from the beta to the alpha phase, which produces very coarse grains. Alpha silicon carbide is less sensitive to temperature than the beta, so that its optimum sintering temperature range is higher than that for the beta phase—from 2200°C to 2300°C.

The results of a statistical experimental approach to determine the processing parameters for injection molding complex shapes of beta silicon carbide are summarized in Table 7-3. The lower sintering temperature at both longer sintering times (12 minutes and 60 minutes) appears to result in

TABLE 7-3. The effect of temperature, time, and atmosphere on the sintered density and strength of injection molded beta silicon carbide.

Temperature		Atmosphere			Strength
°C	Time, minutes	Vacuum	Argon	Density, % of theoretical density	MPa
2080	6	Yes		97.25%	498.5
2080	12	Yes		97.95%	575.0
2100	6	Yes		97.52%	431.0
2100	12	Yes		97.57%	482.0
2120	30		Yes	92.93%	403.0
2120	60		Yes	92.45%	475.0
2135	30		Yes	92.59%	420.0
2135	60		Yes	92.57%	449.0

From Whalen (1987).

higher strength. The deleterious effect of the higher temperatures on the strength could be attributed to excessive grain growth, which is known to affect strength adversely, but there is clearly an increase in strength with sintering time at the lower temperatures from 6 to 12 minutes in a vacuum and from 30 to 60 minutes in argon.

The effect of temperature on density, however, is not so clear-cut. The sintering time at the sintering temperature apparently has little effect on the density in both cases although the density of the material sintered in a vacuum is relatively higher than in argon. Rapid heating, however, could assist in densification as well as produce a smaller grain size. Both vacancy diffusion and densification could increase at a faster rate than grain coarsening caused by diffusion. Therefore, fast firing with a hold at the peak temperature should be beneficial.

Industrial Sintering Approaches

In most cases, sintering temperatures vary between 1400°C and 1650°C. This is so largely because these temperatures can be attained relatively easily with current types of fuels and heating elements. Also furnaces can be built and trays (saggers) fabricated for these temperatures from readily available refractories.

Both silicon nitride and silicon carbide are sintered at much higher temperatures (1750–2200°C). Special furnaces such as vacuum or inert gas furnaces with high pressure capability are required. Water-cooled pressure vessels frequently are used along with molybdenum, tungsten, or graphite heating elements. The temperature and the gas flow are controlled with microprocessors.

The total sintering time can take from several hours to several days. Most of this time, however, is taken up in heating and gradually cooling the part. The time dwell at the highest temperature, where sintering is at a maximum, is usually just a fraction of the total sintering time. It is very important for the efficiency and the economy of the sintering process to take into consideration the rate of heating and cooling as well as the length of time for isothermal holding. Accelerating the sintering increases the throughput and reduces energy consumption.

During the initial stage of sintering, the part is gradually heated to the sintering temperature. The rate of heating should not exceed a critical limit beyond which the temperature gradient within the part would create internal stresses that could exceed the ultimate strength of the material. This initial stage also involves other processes such as polymorphic transformations, which might be an additional factor limiting the heating rate. The onset of formation of a melt and of new phases via chemical reactions also could occur in this stage.

The kinetics of the sintering process determine the time dwell at the maximum temperature. The sintering time can be shortened by raising the temperature, but this could cause an excessively large amount of grain growth, liquid formation, or gas evolution. Moreover, the sintering time cannot be reduced arbitrarily because the part, which has a relatively low thermal conductivity, must be heated uniformly.

The rate of cooling, like the rate of heating, is limited by the creation of stress. In the case of cooling, a tensile stress is created on the surface of the part, but the strength of the sintered part is considerably higher than in the initial sintering stage. Generally, special annealing procedures are not necessary even though controlled cooling is known to improve certain properties, particularly the mechanical strength, of ceramics such as silicon nitride and silicon carbide. Possible explanations include the relief of stresses in microregions composed of phases with various expansion coefficients, crystallization of the liquid, or the healing of some surface flaws.

The term "sintering interval," which is used in connection with optimum sintering conditions, refers to the temperature range within which the sintering is sufficiently rapid and the part is not deformed. In the presence of a liquid phase, it is determined by the temperature dependence of the liquid present and its viscosity. Phase diagrams provide a great deal of useful information. In the case of sintering where a liquid phase is not present, the upper limit of the sintering interval is determined by the temperature at which excessive grain growth occurs.

There is an exponential relation between the temperature and the requisite sintering time, within the sintering interval. This relation takes the following logarithmic form:

$$\log\frac{t_1}{t_2} = A\left[\frac{1}{T_1} - \frac{1}{T_2}\right] \tag{7-2}$$

where t_1 is the time to attain the requisite sintering time at temperature T_1, and t_2 is the time to attain the requisite sintering time at temperature T_2 (temperatures in K). The value of the constant, A, varies with the type of ceramic.

Heating and cooling are the factors that control the duration of the sintering process. The maximum heating rate that a part can sustain without damage controls the extent to which the sintering cycle can be shortened. Heat must be supplied to the part uniformly. Because of the problems encountered in determining and maintaining optimum heating conditions, the rate at which the temperature actually is increased in industrial furnaces generally is considerably below the theoretical value. In addition, the service life of the refractory materials used to line the furnace as well as the kiln furniture must be taken into account. Nevertheless, methods of sintering at the maximum possible rate, called fast firing, constantly are being investigated.

HOT ISOSTATIC PRESSING (HIP)

Hot isostatic pressing (HIP) is a method for consolidating compacted powder parts or preforms at elevated temperature under hydrostatic gas pressure. Figure 7-8 is a diagram of a typical HIP unit. Since 1955, when research on a process known as gas pressure bonding was initiated at Battelle Laboratories in Columbus, Ohio (Boyer 1982), progress has been rapid in the use of HIP technology for a wide range of powder metallurgical applications.

The potential for using pressures of 210 MPa or greater to isostatically press complex shapes, and to improve control of both the temperature and the atmosphere, has stimulated interest in applying HIP to ceramics. The advantages for ceramics include rapid and higher densification at relatively lower temperatures, a fine-grained microstructure and thus greater ability for microstructural control, and the capability of near-net shape forming.

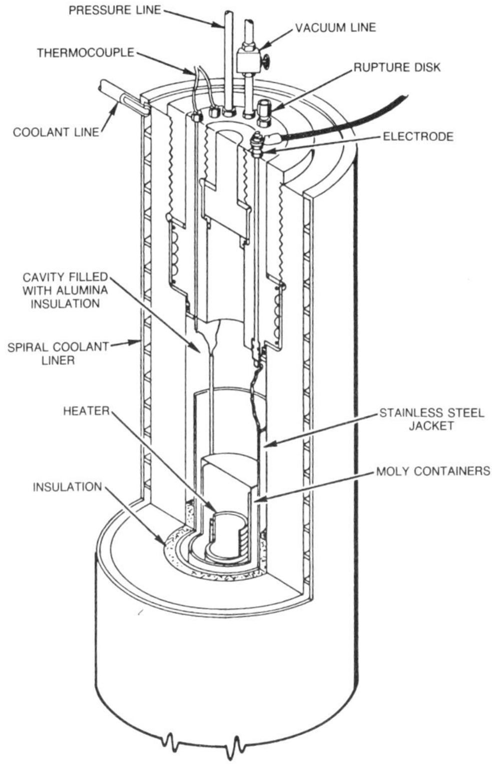

FIGURE 7-8 Schematic of a hot isostatic pressing unit.

In HIP processing, a compacted powder part (also called a powder compact) or a preform is placed in a deformable gas-impermeable container or "can" of metal or glass. Alternatively, the powder compact or preform first is sintered to the point where all the internal porosity is closed (for most ceramics, this state of closed porosity occurs at roughly 95% of theoretical density).

Next, the part is loaded into an autoclave in which a high-temperature furnace is installed. After bolting, the autoclave is evacuated (to about 2.0 Pa of pressure at close to 1000°C). Then it is back-filled with an inert gas (such as argon) at a pressure below the target pressure of about 210 MPa, and the temperature is raised. As the temperature of the confined gas increases, its pressure increases. The final, or target, pressure is set at the peak temperature.

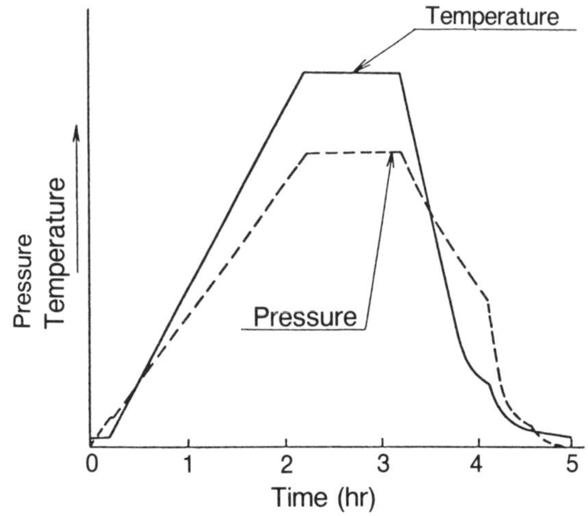

FIGURE 7-9 Pressure and temperature profiles in a hot isostatic pressing cycle.

After completion of the HIP cycle, the system is allowed to cool, and the pressure is released. Finally, the parts are removed from the autoclave and taken out of the can. Figure 7-9 is a diagram of a pressure and temperature cycle, and Figure 7-10 shows the key components of consolidation by HIP. The major developments and improvements in the equipment have been achieved on furnaces, compressors, and controls. The vessel, the gas handling, and the auxiliary equipment have changed very little.

The HIP of ceramics requires operating temperature above 1500°C. Therefore, the furnace structure and the materials used have been extensively investigated. It has been pointed out that the materials that are widely used for heating elements, such as molybdenum, tungsten, and tantalum, are subject to creep deformation at high temperatures and pressures (Fujikawa et al. 1982). These heating elements also become fragile after a number of HIP cycles, and could fail because of thermal shock as the workload is handled during the HIP process. For this reason, graphite is becoming the preferred material for heating elements at high temperatures. Platinum is used in some research and development units that can operate with a partial oxygen atmosphere, for example, 20% oxygen and 80% argon.

Even though HIP units are water-cooled, temperature gradients can exist between the walls of the equipment and the furnace. At high pressures, the presence of these gradients can create problems by causing the gaseous medium to densify. When density differences are present in the system,

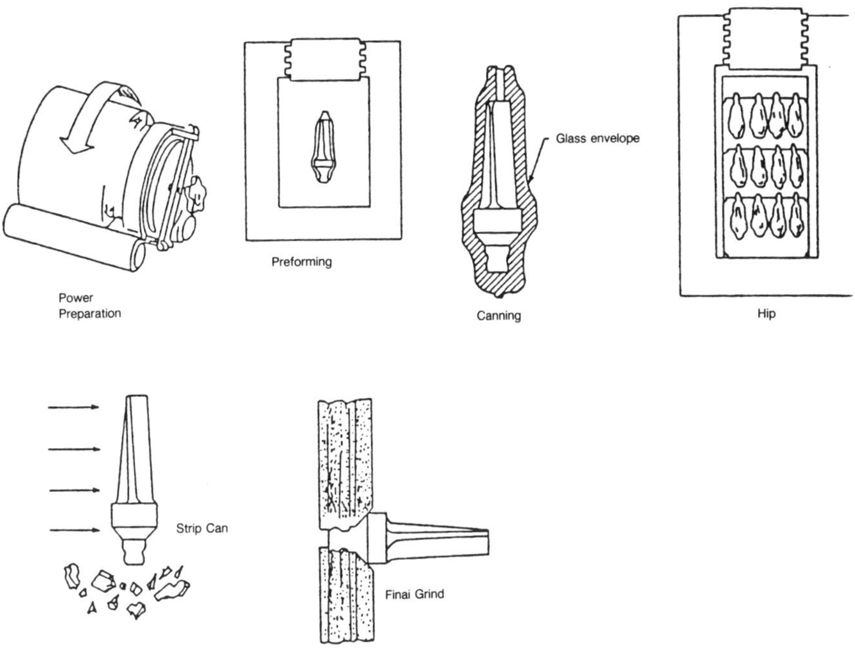

FIGURE 7-10 Key components of a hot isostatic pressing powder consolidation process.

considerable convective flow can occur, resulting in thermal fluctuations. To reduce the temperature gradients, an insulating mantle is placed over the furnace, or the open space in the autoclave is filled with a high-temperature non-reacting material such as flexible sheets of graphite.

Multiple thermocouples are used to monitor the temperature distribution in a HIP unit. A platinum–rhodium thermocouple can be used up to 1700°C. But above this temperature tungsten–rhenium, and more recently boron carbide–graphite (B_4C/C) thermocouples are being given consideration.

Argon is most widely used as the pressure transmitting medium; but in the silicon nitride process in which sintering is combined with HIP (sinter plus HIP), nitrogen has been used to suppress the dissociation of the silicon nitride (Fujikawa et al. 1982). However, nitrogen could damage the metal heating element.

A microprocessor-based control system now is a standard feature in HIP units. In addition, a dilatometer has been invented that can follow the shrinkage of a part under various pressure and heating rate conditions in real time (McCoy et al. 1985, Brun and Borom 1985). It can detect shrinkages of less than 0.01%. In production units such a dilatometer probably would not be feasible, but it has proved useful for investigating densification during the HIP cycle.

Metals or glasses can be used as encapsulating materials for complex ceramic preforms, the most commonly used metals being tantalum and molybdenum. The advantage of metals as encapsulants is that they deform easily at high temperatures and pressures. However, the processes used to remove the capsule after the HIP cycle is finished, by a machine or by dissolving it in acid, are costly. Furthermore, the metal capsule does not always completely conform to the ceramic's shape, causing the pressure to be transmitted nonuniformly. This behavior is most likely to occur in the fabrication of parts with complex geometries.

Unlike metal, the glass capsules that are used deform under temperature and pressure, and they closely and uniformly conform to the contours of the ceramic parts. The glass can be formed into a kind of envelope by standard glass-forming techniques, or glass frit can be coated on the surface of the ceramic part (Adlerborn and Larker 1978; Hintler et al. 1980; and Larker 1984). The advantages of glass frit are that large defects can be healed, and temperature gradients are minimized so that a final uniform density results. However, with any encapsulating method, a barrier material, such as boron nitride powder, must be used to prevent reaction between the encapsulating material and the ceramic part.

PRODUCT EVALUATION

The properties that give ceramics their desirable characteristics of strength, hardness, refractoriness, and chemical resistance also contribute to their brittleness and lack of toughness (or resistance to fracture). Their brittle nature always must be taken into consideration to successfully design ceramics for applications where they will be subjected to mechanical stress.

A design approach can be taken that is based on the concept of an allowable failure probability for a ceramic part subjected to specified load levels. This requires information about the mean failure strength of a specified surface area of the part under uniaxial stress conditions, as well as about the inherent variability of the material.

The Strength of a Ceramic

A characteristic of brittle fracture is that there is no prior deformation before failure. Brittle fracture usually occurs at stress levels well below those predicted theoretically from the inherent strength of a material, owing to atomic or molecular bonding. The low fracture strength of brittle materials is due to the presence of defects.

In 1920, it was proposed that microscopic cracks exist both within and on the surface of all materials (Griffith 1921). If the long dimension of a crack is perpendicular to the direction of an applied tensile stress, large stress concentrations develop at the crack tip, where the actual local stress can approach the theoretical strength of the material. By itself, a high local stress at the tip of a crack does not necessarily result in the propagation of the crack and ultimately brittle fracture. Rather, a crack will propagate only if there is a concomitant net reduction in energy. See Appendix 7-D for a detailed discussion of the relation of crack propagation to fracture strength in a brittle material.

The principal factors determining the strength of a ceramic are the physical characteristics of the flaws, such as size, shape, location, and distribution. Methods have been developed for calculating fracture stresses of flaws, including pores and machining defects (Evans and Tappin 1972). Also fracture stress and flaw size have been correlated for all types of flaws grouped together. It has been clearly shown that, in a ceramic, fracture stress is controlled by the characteristics of the flaws rather than by the average microstructure.

A ceramic's strength also depends on the grain size, and the critical flaw size is affected by changes in the grain size. The presence of inclusions (solid material present as a contaminant) also affects strength because of differences between the thermal, elastic, or chemical properties of the ceramic and the inclusion. A difference in thermal expansion could create cracks around the inclusion during cooling, whereas an elastic difference could introduce cracks when a load is applied. A chemical interaction could result in the formation of a new phase that might have inferior thermal and elastic properties. In addition, the influence of the environment on strength (during tests as well as in use) cannot be overlooked.

Clearly, the influence of flaws related to processing, fabrication, and densification on the strength of ceramics is complex. No straightforward approach exists that can be applied to control the variables in the manufacturing process to improve the strength of ceramics. This is particularly true for ceramic injection molding.

The Measurement of Strength

The measurement of strength is a basic procedure in ceramic fabrication. It often is used to compare different materials and to provide data for the design of complex engineering structures.

Under crushing loads (compressive strength), materials fracture at different stress levels from those under either stretching loads (tensile strength) or bending loads (modulus of rupture or flexural strength). Therefore, the measurement procedure must be specified. Other important strength-related properties of ceramics are the rate of permanent deformation at high temperature (creep rate) and the relationship between stress level and recoverable deformation (modulus of elasticity).

Modulus of Rupture or Flexural Strength

The modulus of rupture (MOR) is the fracture strength of a material under a bending load. The MOR measurement is made on a bar of material with a rectangular cross section, supported near its ends with a load applied to the central portion of the supported span. There are two methods for applying the load: three-point loading and four-point loading.

For three-point loading using rectangular bars, the MOR is given by:

$$\mathrm{MOR} = \frac{3PL}{2bd^2} \tag{7-3}$$

where P is the load required to break the bar, L is the span (distance between the outer supports), b is the width of the bar, and d is the depth of the bar.

For four-point loading using a rectangular bar, the MOR is given by:

$$\mathrm{MOR} = \frac{3Pa}{bd^2} \tag{7-4}$$

where a is the distance between the two inner load application points.

Because of the inherent variability of a brittle material, the mean strength of a set of samples depends both on the maximum stress in each sample and on the distribution of stress throughout the sample. In other words, the effective surface of the sample being tested must be known. Therefore, in a bend test, the shape and the dimensions of the sample as well as the test rig design are chosen so that deviations from the theoretical stress distribution are acceptably small (Timoshenko and Young 1962).

Other important parameters to control in MOR determinations are the rate of loading (particularly in tests made above room temperature), the ratio of the thickness of the span to the thickness of the sample (should not be less than 10), and the alignment of the sample. The sample must be carefully aligned in the holder so that it cannot twist while being loaded.

Four-point bend tests are preferable to three-point bend tests because four-point tests have larger effective surfaces, and thus local deviations from the theoretical stress distribution will have a proportionately smaller effect on the measured strength. Also, because the inner span of the four-point test is completely free from shear stresses, it experiences uniaxial stress.

TABLE 7-4. **Room temperature flexural strength of injection molded alumina.**

Type	Strength bend test 3-point (kg/cm^2)	4-point (MPa)
A16-SG Al_2O_3[a]	—	344
96 wt% Al_2O_3[b]	1470–1970	—

a. From Mutsuddy (1987).
b. From Moteki (1960).

Roller-type bend rigs are highly recommended to reduce frictional effects at support points.

It is now standard practice in the United States when preparing samples to follow the MIL-STD-1942 (Military Standard). Efforts also are being made to produce an international standard through a Technical Cooperation Program (Geiger 1990). Tables 7-4, 7-5, and 7-6 present some MOR data for different injection molded ceramics.

Tensile Strength

The tensile strength of a material is a measure of its resistance to fracture under a stretching or pulling load. In recent years, the tensile testing of

TABLE 7-5. **Room temperature flexural strength of injection molded silicon nitride.**

Type	Strength bend test 3-point (MPa)	4-point (MPa)	Weibull modulus
RBSN[a]	342		10.5
SRBSN[b]		698	
SRBSN (Y_2O_3)[a]		699	10.1
SSN[b]		593–673	
RBSN (4 wt% Y_2O_3/HIP)[c]		≈800	
SSN (4 wt% Y_2O_3/HIP)[c]		≈650	
SSN (6 wt% Y_2O_3/HIP)[d]		772	
SSN (6 wt% Y_2O_3/HIP)[e]		910	34.0

a. Mangels (1978/86).
b. Hunold et al. (1989).
c. Heinrich et al. (1988)
d. Helms et al. (1990).
e. Bandyopadhyay (1988).
RBSN: reaction-bonded silicon nitride. SSN: sintered silicon nitride. HIP: hot isostatic pressing.

TABLE 7-6. Room temperature flexural strength of injection molded silicon carbide.

Type	Strength bend test 4-point (MPa)	Weibull modulus
Alpha (B_4C + resin)/SSC[a]	402	9.5
Alpha (B_4C + resin)/SSC[b]	383	
Alpha (1.0 Al)/SSC[c]	404	
Alpha (B_4C + resin)[b]	518	

a. Wada et al. (1987).
b. Helms et al. (1990).
c. Hunold et al. (1989).
SSC: sintered silicon carbide.

ceramics has been receiving attention (Carroll et al. 1989), but a number of issues need to be resolved before this test method can be used on a routine basis. These issues include machining procedures and allowable tolerances; alignment requirements; and the optimal sample design, strain rate, and combination of sample, grip, and design.

Weibull Analysis

Engineering design is based on the concept of a fail-safe structure. The probability of zero failure is ensured by incorporating a safety factor that is sufficient to cover any uncertainties in the properties of the material and the load conditions. In the case of ceramics, the objective of total safety leads to inefficient design. This is particularly true in dynamic applications such as the turbocharging rotor.

There is a considerable amount of scatter in the results of strength measurements of ceramics. These measurements must be statistically analyzed in order to estimate reliability at a given stress so that a realistic safety factor can be incorporated. Weibull analysis, which is widely applied for this purpose, rests on two assumptions (Weibull 1951):

- The sample is assumed to be divided into volume elements that contain a large number of flaws, in each of which the stress is considered to be approximately uniform. To support this assumption the size of critical flaws is compared with the sample's dimensions.
- Failure occurs when a critical local stress is reached at any one of the flaws. This so-called weakest link hypothesis leads to conservative design in scaling up from small samples to large components.

See Appendix 7-E for a discussion of a variation of Weibull analysis.

Fracture Toughness

The fracture toughness of a material measures its resistance to crack propagation, and is one of the chief factors that determine the load that a component containing cracks can sustain before it fails. The concept of fracture toughness is important for interpreting brittle fracture from the viewpoint of failure under stress.

In most materials, a small amount of local plastic flow accompanies fracture. Linear elastic fracture mechanics assumes that, for a given material, the distribution of stress in the plastic zone at the crack tip is determined by the distribution of stress in the surrounding elastic region. This is proportional to the stress intensity factor (K), which depends on the configuration of the specimen and its loading. As the load increases, K increases proportionally, and is reflected in an increase in the actual local stress at the crack tip. When this local stress is sufficient to extend the crack, the corresponding critical value of K is referred to as the critical stress intensity factor or fracture toughness of the material, symbolized K_{IC}. In general, a material's resistance to crack propagation increases with its fracture toughness.

Several techniques can be used to measure the fracture toughness of ceramics, including indentation, single-edge notch beam, and chevron notch.

For example, the hardness and the fracture toughness for an injection molded silicon nitride sample have been calculated from measured values of the diagonal impressions and the radial crack lengths made by an indenter, as shown in Figure 7-11a (Mutsuddy 1987). A typical crack pattern associated with the indentation (Vickers) of a silicon nitride sample and the equation for K_{IC} derived from indentation fracture mechanics are shown in Figure 7-11b.

The calculated results for Si_3N_4 are as follows:

	Hardness (GPa)	Fracture toughness (K_{IC}) ($MPa\sqrt{m}$)
Mean	12.1	6.24
Standard deviation	0.33	1.12

Because the fracture toughness values were calculated by assuming a Young's modulus of $E = 289.6$ GPa, the K_{IC} values should be corrected for the modulus by using the following equation:

$$K_{IC_2} = K_{IC_1} (E_2/E_1)^{1/2} \tag{7-5}$$

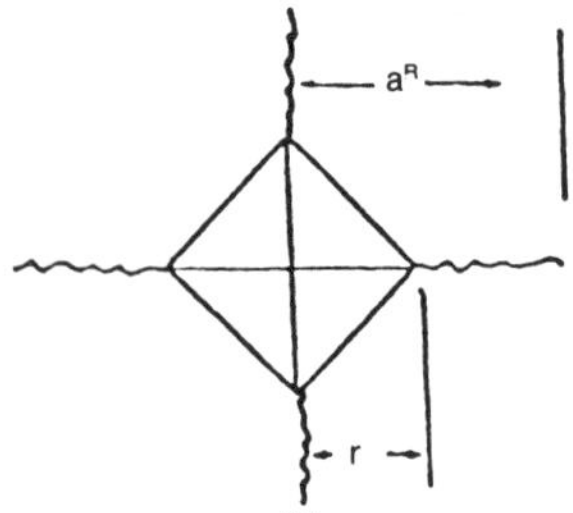

(a)

$$K_{IC} = \frac{0.028\, H r^{\frac{1}{2}} (E/H)^{\frac{1}{2}}}{(a^R/r)^{\frac{3}{2}}}$$

K_{IC} = Fracture Toughness

H = Hardness

E = Young's Modulus

a^R = Radial Crack Length

r = Plastic Indent Dimension

(a)

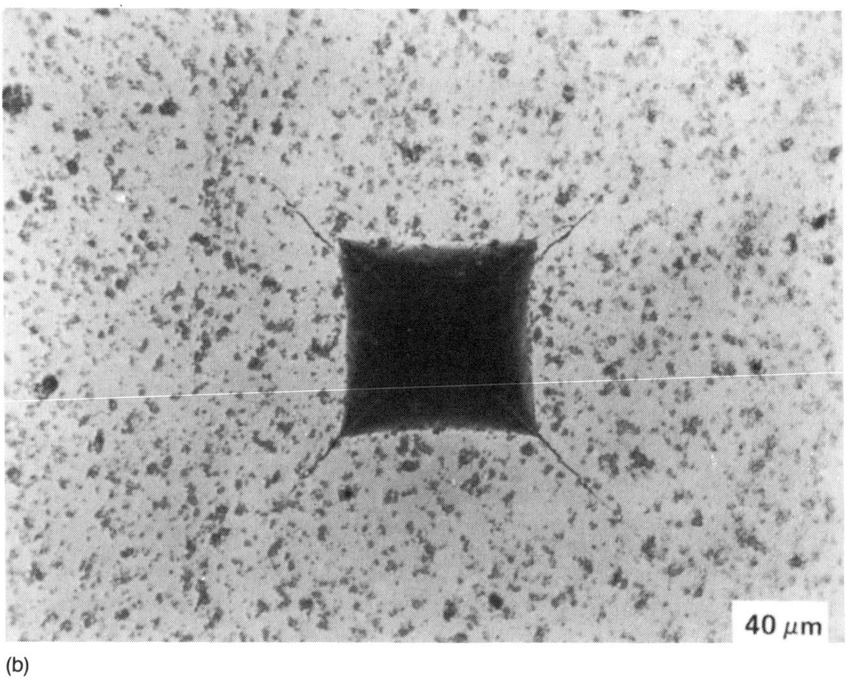

(b)

FIGURE 7-11 A fracture toughness measurement on sintered silicon nitride: (a) indentation parameters and (b) a typical crack pattern in silicon nitride (Vickers load = 15 kg).

Density

Probably the most often measured and reported property for all types of ceramics is density. Because the calculation of density requires that the weight of a sample or a component be divided by its volume, the measurement of volume constitutes an important part of any density determination. The density most commonly measured for a ceramic is the bulk density (*b*), which is based on the bulk volume. A common practice is to express a ceramic's density as the ratio of the bulk density to the true density (*t*). When reported in this way, it is called the percent theoretical density, calculated as shown in equation (7-6):

$$\% \text{ theoretical density} = \rho_b/\rho_t \times 100 \qquad (7\text{-}6)$$

Some typical densities of injection molded ceramics are given in Tables 7-7 through 7-10.

Shrinkage

Ceramic components undergo shrinkage at several steps in the manufacturing process. Major shrinkage occurs during sintering as pores are gradually reduced or eliminated. The linear shrinkage (the amount each dimension changes) is more commonly reported than the volume shrinkage (the amount that the volume of the entire component changes). However, both these measurements are directly related.

In ceramic injection molding, shrinkage is a critical factor. Among other variables, the volume of the powder loading in the mix has a direct bearing on the shrinkage, as shown in Figure 7-12. Nonuniform shrinkage caused by variations in the density and by the orientation of the particles can result in severe cracking and limit the ability to achieve close dimensional tolerances in the finished part.

Tables 7-7 through 7-10 present some typical shrinkage data for a number of injection molded ceramics.

TABLE 7-7. Density and linear shrinkage of A16-SG alumina.

Parts of binder	Density (g/cm^3)	Linear shrinkage (%)
18.86	3.9	16.07
19.94	3.89	16.65
21.01	3.88	17.17

From Tanaka et al. (1985).

TABLE 7-8. Density and linear shrinkage of injection molded 96 wt% alumina and 4 wt% talc.

Wt% binder	Density (g/cm^3)	Linear shrinkage (%)
18.3	3.20–3.62	15.3–15.5

From Moteki (1960).

TABLE 7-9. Density and linear shrinkage of injection molded silicon nitride.

Sintering aid(s) (wt %)	Type of silicon nitride: RBSN g/cm^3[d]	RBSN g/cm^3	SRBSN, g/cm^3	Sintered, %[e]	HIP g/cm^3
—	2.52–2.7a	—	—	—	—
$7Y_2O_3$[a]	—	3.08	—	—	—
$8Y_2O_3 + 4Al_2O_3$[b]	—	—	3.05–3.23	15–20	—
$7Y_2O_3$	—	—	—	—	3.27
PY 6[c]	—	—	—	—	3.25

a. Heinrich et al. (1988).
b. Mutsuddy (1987).
c. Helms et al. (1990).
d. Density.
e. Linear shrinkage.
RBSN: reaction-bonded silicon nitride. SRBSN: sintered reaction-bonded silicon nitride. HIP: hot isostatic pressing.

TABLE 7-10. Density and linear shrinkage of injection molded silicon carbide.

Grade	Sintered density (g/cm^3)	Linear shrinkage (%)	HIP density (g/cm^3)
Alpha (B_4C + C)[a]	3.13	—	—
Alpha (B_4C + C)[b]	3.12		
Alpha (1.0 Al)[c]	3.07	16–17	—
Alpha (1.0 Al)[c]	—	—	3.19
Alpha (B_4C + C)[b]	—	—	3.17

a. Wada et al. (1987).
b. Helms et al. (1990).
c. Hunold et al. (1989).

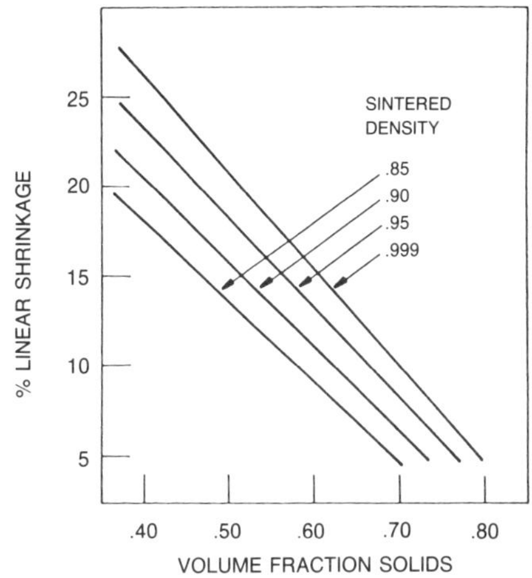

FIGURE 7-12 Linear shrinkage as a function of ceramic powder loading.

Nondestructive Evaluation

Nondestructive evaluation (NDE) of the part at every stage of the ceramic injection molding process would be desirable in process quality control. However, the science and the technology of NDE for ceramics are still developmental. Most of the advances in the NDE of ceramics have been made with sintered parts. NDE techniques for the starting powders, powder–binder mixes, and green (presintered) parts, although promising, are still in the early stages of development.

Neutron attenuation or neutron scattering techniques have potential because of the relatively high neutron cross section of the hydrogen-rich composition of the powder–binder mix. Nuclear magnetic resonance (NMR) is another technique sensitive to the presence of hydrogen in the binder. Changes in the spin (and magnetic moment) of proton-rich nuclei in the presence of external magnetic fields make it possible to detect material rich in protons.

Many NDE methods, particularly the various techniques of ultrasonics and radiography, have been investigated and applied to the detection and evaluation of flaws in sintered ceramics. For flaw characterization, ultrasonics requires using higher-than-conventional frequencies (25–100 MHz) and sophisticated signal processing for the analysis of velocity, attenuation spectra, and reflections from the flaws. Microradiographic techniques (both contact and those with microfocus X-ray tubes) have shown good resolution and sensitivity for small flaws. Tomographic techniques, such as computed tomography (CT), can be used for the three-dimensional localization of flaws and other variables, as shown in Figure 7-13a and b (Rokugawa et al. 1991; Sawicka and Sonnenberg 1992). Also photoacoustic microscopy shows promise as a high-sensitivity and high-resolution technique for detecting surface and subsurface flaws; but because it is relatively slow, its application probably will be limited largely to development and sampling.

Laser-assisted optical examinations are potentially applicable for rapid automated interpretation of surface-related flaws in sintered ceramics. Optical holography using lasers has shown promise for surface and subsurface detection of flaws and could have applications in materials development and in the statistical sampling of production lots. Various ultrasonic techniques, such as surface acoustic waves and critical-angle reflectivity, also have shown sensitivity to surface flaws.

The high rejection rates for injection molded ceramic parts, equivalent to 15 to 20% of total manufacturing costs, plus the demand for maximum achievable performance and reliability, make in-process NDE a necessity. Microfocus radiography is the only technique presently used for in-process inspection (Pujari et al. 1990; Pasto and Natansohn 1990). It is used for detecting agglomerates and variations in the packing density of powders, and for detecting inclusions (both metallic and nonmetallic) in binder ingredients. It also is used for detecting flaws such as voids, cracks, and density gradients in sintered parts. Radiography in real-time reveals voids and variations in the density of as-molded parts, and thus allows adjustments to be made in molding parameters as well as modifications in mold design. In the future, an adaptation of magnetic resonance imaging (MRI) could become a useful NDE method for probing the distribution of binder in ceramic mix formulations.

CONCLUDING REMARKS

Although at present structural ceramics have not experienced the growth originally predicted for them, there is good reason to believe that an enormous expansion in the market for ceramic parts will occur shortly after the turn of the century. This book has been written in the conviction that injection molding will become the major forming technology for producing cost-effective, reliable, and reproducible complex shaped ceramic parts in large quantities. Of less importance, however, is the process—high pressure, low pressure, or other—that will be used to pump a highly viscous slurry into a complex mold cavity.

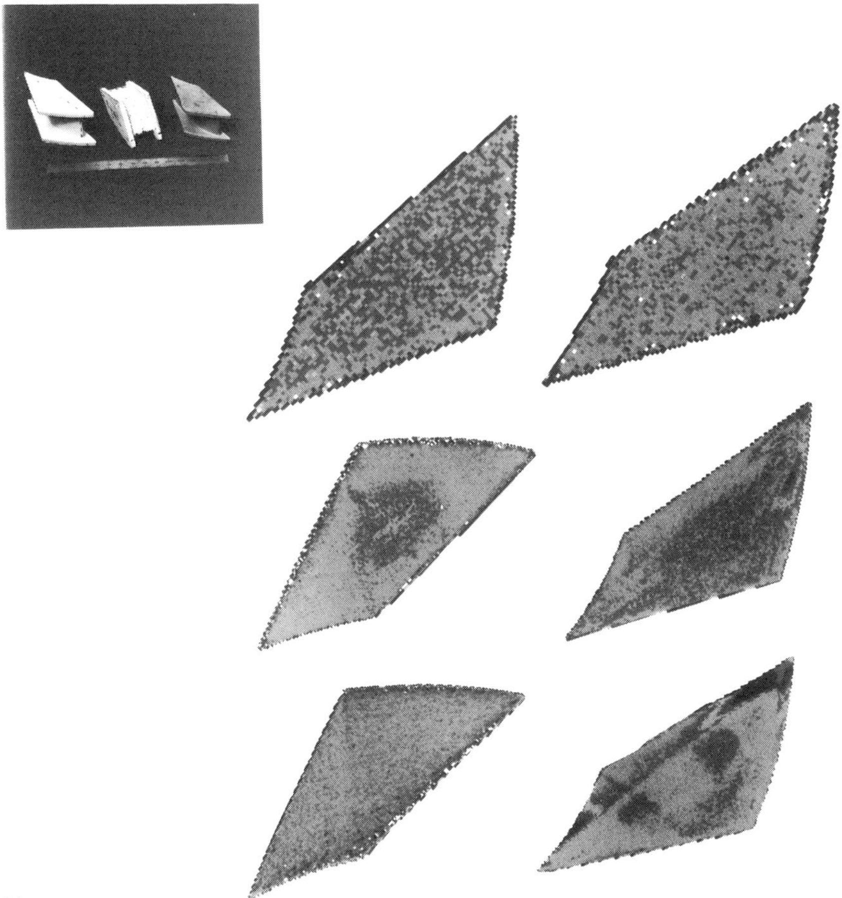

(a)

FIGURE 7-13 (a) Computed tomography (CT) images of two different ceramic nozzle plates (slip A and slip B) after three stages of fabrication: injection molding (top), sintering (middle) and hot isostatic pressing (HIPing; bottom). The ceramic nozzles are shown on the upper right. The nonuniformities in the material distribution shown to be present in both slips increase after sintering. However, after HIPing the uniformity of slip A is better than 0.4% while slip B is still nonuniform. (b) CT images of two ceramic cylinders (A and B) measured after injection molding (left) and after consecutive dewaxing (right). The examined objects are shown at the bottom. After dewaxing, an extended crack pattern is shown in A while B shows uniform distribution of the material with no cracking (the two dark lines are cracks that occurred in handling). Courtesy of AECL Research, Chalk River Laboratories, Canada and the Ceramics Department, NKK Corporation, Japan.

SLIP A

SLIP B

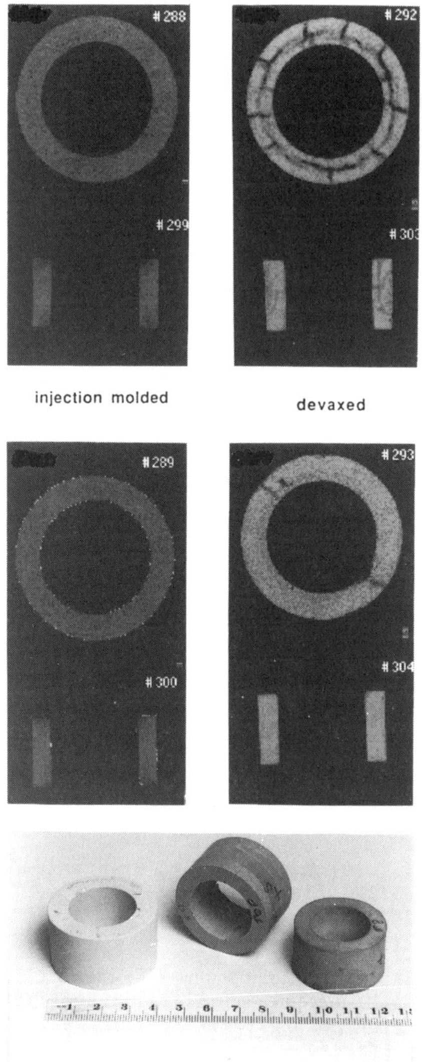

(b)

APPENDIX 7-A
THE MECHANISM OF SINTERING

In the sintering process, powder compacts are strengthened at elevated temperature. Sintering usually involves shrinkage and densification, that is, the reduction of porosity. A polycrystalline material is produced in which the particles are strongly mutually bonded. Progress during sintering generally is followed by measuring the strength, shrinkage, or density of the parts. Sintering can take place in the solid state or through the effect of a viscous liquid. Although the sintering of multicomponent systems can involve concomitant chemical changes, in principle it is a physical process.

In the earliest theoretical work, sintering was considered to be a viscous flow of crystal solids as a consequence of surface tension (Frenkel 1946). A satisfactory theory of sintering should explain how the centers of individual particles can approach each other closely enough for shrinkage to occur. To account for the details of movement on the atomic scale, the internal geometry of a powder compact must be taken into consideration.

Several processes are operational in directing the way material moves under the influence of chemical potential gradients created by the presence of curved surfaces within the compact. The vacancy concentration is related to the curvature of these surfaces.

The difference in vacancy concentration, ΔC, immediately below a surface of curvature radius r, with respect to a flat surface can be stated as:

$$\frac{\Delta C}{C_0} = \frac{2\gamma v}{rkT} \tag{7A-1}$$

where C_0 is the concentration at the flat surface, γ is the surface energy, v is the volume of a vacancy, and k is the Boltzmann constant.

The vacancy concentration is increased under a concave surface and decreased under a convex surface. This suggests that sintering probably involves two mechanisms of mass transport, which result in the elimination of pores (densification) and in strengthening the sintered part: (1) viscous or plastic flow caused by stresses in the proximity of curved surfaces, and (2) volume diffusion caused by differences in the vacancy concentration. In both mechanisms, the centers of the particles move closer together. Without this condition, densification would not be possible.

In addition, two other transport mechanisms can strengthen a part through formation of a neck, but without densification:

- Evaporation and condensation: In materials with a high vapor pressure, mass transfer can occur in the gaseous phase. The vapor pressure, which like the vacancy concentration, depends on the curvature of the surface, is higher above convex surfaces than above concave ones. Consequently, vapor from these surfaces condenses on concave points of contact. In this case, the centers of the particles do not approach, so there is no shrinkage. The volume of the pores stays the same; only their shape is altered.

- Surface diffusion: Mass transfer by surface diffusion probably is involved in the early stage of sintering when partial strengthening of the powder compacts occurs without any shrinkage.

For all mechanisms, the results of an analysis based on the model shown in Figure 7A-1 of a spherical particle with a radius r, set on a planar base, can be expressed as (Kuczynski 1949):

$$\rho = \left(\frac{x}{2r}\right)^n = \frac{zt}{r^m} \tag{7A-2}$$

where ρ is the radius of curvature at the contact point, z is a combined material-and-geometric constant, the exponents n and m characterize the mechanism of mass transport and are given in Table 7A-1, x is the neck diameter, and t is time.

The relative contribution of these processes varies with temperature, particle size, particle shape, and the stage of sintering. Although densification is a continuous process, it can be considered to consist of a number of discrete consecutive stages. A model proposed for the first stage of sintering, shown in Figure 7A-2, includes a grain boundary between two spherical particles, which form a neck by the transfer of material from the grain

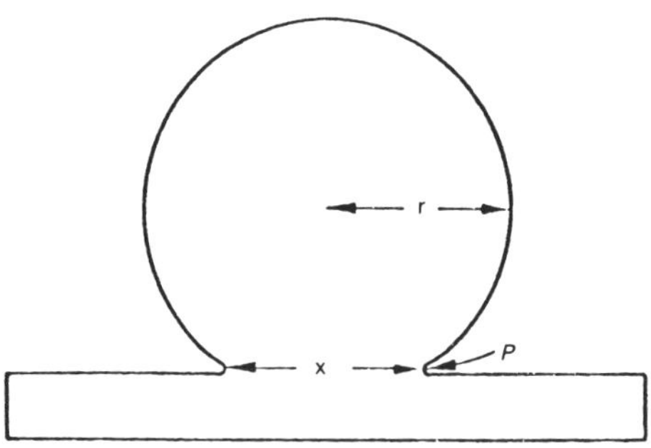

FIGURE 7A-1 A sperical particle set on a planar base.

TABLE 7A-1. Constants for equation (7A-2).

Mechanisms	n	m
Viscous or plastic flow	2	1
Evaporation/condensation	3	1
Lattice diffusion	5	3
Grain boundary diffusion	6	4
Surface diffusion	7	4

boundary region, that is, by the transfer of vacancies to the region (Kingery and Berg 1955). The equation relating the shrinkage to the neck size is:

$$\frac{\Delta l}{l_o} = \frac{y}{r} \approx \frac{x^2}{4r} \tag{7A-3}$$

Thus the fractional change in the length of the compact as a whole, $\Delta l/l_0$ is equal to the fractional shrinkage, y/r (where y is assumed to be equal to the radius of curvature (ρ) and r is the radius of the particle), and approximately equal to the square of the neck radius, x, divided by 4 times the radius of the particle.

It has also been proposed that pressure builds up in the system, which increases the packing density of the particles either because the particles slide past each other or because of the local fragmentation of the agglomerates. This has been described as the zero sintering stage (Wilkinson and Ashby 1975, 1978).

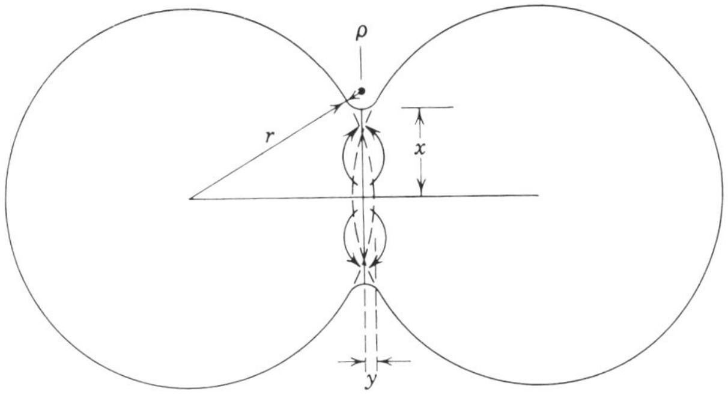

FIGURE 7A-2 A model proposed for the first stage of sintering. r is the radius of the particle, ρ is the radius of curvature, at the surface, x is the neck radius, and y is the radius of curvature of the neck.

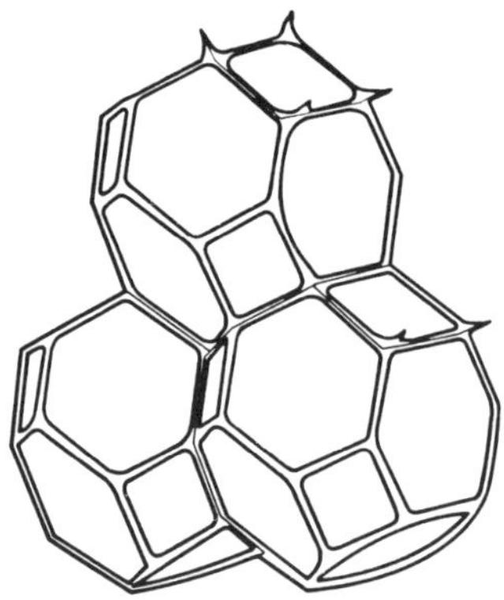

FIGURE 7A-3 A model of the second and the final stages of sintering represents the grains as body-centered stacked, truncated octahedra. After Shaw (1989).

A model of the second and the final stages of sintering, shown in Figure 7A-3, represents the grains as body-centered stacked, truncated octahedra. In the second stage of sintering cylindrical pores are located along the three-grain edges and in the final sintering stage, spherical pores are located at the four-grain corners, on grain boundaries, and within grains (Coble 1961).

Equations have been derived to express the rate of change of porosity, P, during the intermediate stage of sintering by considering the diffusion along the grain boundary, D_{gb}, or through the nearby lattice, D_l (Shaw 1989).

For grain boundary diffusion:

$$\frac{\mathrm{d}P}{\mathrm{d}t} = \frac{BD_{gb}\nu\gamma}{G^4kT}t \tag{7A-4}$$

For lattice diffusion:

$$\frac{\mathrm{d}P}{\mathrm{d}t} = \frac{AD_l\nu\gamma}{G^3kT}t \tag{7A-5}$$

where A and B are constants that depend on the mechanism of mass transfer. Values of A and B are typically in the range 3–4 for lattice and grain boundary diffusion.

Theoretically, it should be possible to distinguish between the two densification mechanisms by their dependency on grain size, with these two equations.

For a powder compact to continue to shrink until all the pores are eliminated, during the final sintering stage, it must correspond to the theoretical model in two key respects: the pores must remain at the grain boundaries, or the grain boundaries must be prevented from pulling away from the pores; and the pores must be, in effect, composed of vacancies (Coble 1961; Rosolowski and Greskovich 1973). In other words they must not contain a gas; or if they do, it must be able to diffuse away so that the pressure inside the pores does not increase as they shrink. A pressure increase would produce an internal force opposing their shrinkage.

As the pores become smaller, the force they can exert on the grain boundaries decreases, increasing the potential for the grain boundaries to separate from the pores. The more nonuniform the grain structure is, particularly if there is exaggerated grain growth, the earlier in the sintering process this is likely to occur. In this condition, a majority of the pores would be located within the grains.

Grain Growth

During the sintering process, density and grain growth take place concurrently. However, different processes of atom transport could be responsible for the two processes. A simple model for grain growth considers the movement of a single grain boundary in a pure dense material (Kingery et al. 1976). There is a free energy difference, ΔE across a curved grain boundary having a surface energy and principal radii of curvature r_1 and r_2:

$$\Delta E = \gamma V_m \left(\frac{1}{r_1} + \frac{1}{r_2} \right) \tag{7A-6}$$

The molar volume of atoms moving across the boundary is V_m. This free energy difference provides the driving force for the boundary to move toward its center of curvature.

The rate of boundary movement is proportional to the curvature and to the ability of the atoms to cross the grain boundary, D_{gb}, and inversely proportional to the average grain size, G. The rate of grain growth can then be expressed as:

$$\frac{dG}{dT} \approx \frac{D_{gb}}{G} \tag{7A-7}$$

and:

$$G^2 - G_o^2 \approx t \tag{7A-8}$$

Owing to factors that inhibit the movement of grain boundaries, the growth rate often is found to be less than that expressed in equation (7A-7) (Bennison and Harmer 1990). Some of these inhibiting factors are:

- The segregation of soluble impurities if the resulting effect reduces the free energy of the system.
- The inhibition of grain growth by the presence of second phase particles or other inclusions.
- The presence of pores, which can behave similarly to inclusions in retarding grain boundary movement. Unlike inclusions, however, the effect of a pore on boundary movement depends on its mobility relative to that of the boundary and the geometry of the system. Small pores move faster than large pores. During the final stage of sintering, small pores do not tend to exert a very large drag on the boundary movement unless there are too many present. On the other hand, the presence of many large pores can exert a large restraining force on the boundary movement.

The prediction of sintering behavior in actual situations is difficult even with these sintering models. This is so largely because of fluctuations in the level of impurities, variations in particle shapes and packing density, or any other departures from perfect uniformity.

The sintering diagram approach, shown in Figure 7A-4, is a useful way to link the controlling sintering mechanisms throughout the entire sintering process. Table 7A-2 lists the six distinct diffusion-controlled paths proposed for material transport (Ashby 1974). A useful tool for the assessment of the change in grain growth and densification behavior resulting from an additive or a variation in the process is provided by Figure 7A-5, a diagram of surface area as a function of density (German and Munir 1976; Shaw and Brook 1986).

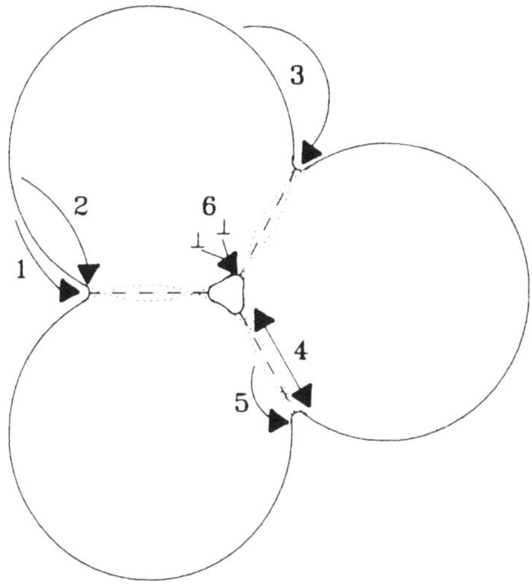

FIGURE 7A-4 Six different diffusion controlled paths. After Ashby (1974).

TABLE 7A-2. Alternate paths for atom transport during the initial stages of sintering.

Mechanism number	Transport path	Source of atoms	Sink of atoms
1	Surface diffusion	Surface	Neck
2	Lattice diffusion	Surface	Neck
3	Vapor	Surface	Neck
4	Boundary diffusion	Grain boundary	Neck
5	Lattice diffusion	Grain boundary	Neck
6	Lattice diffusion	Dislocations	Neck

From Ashby (1974).

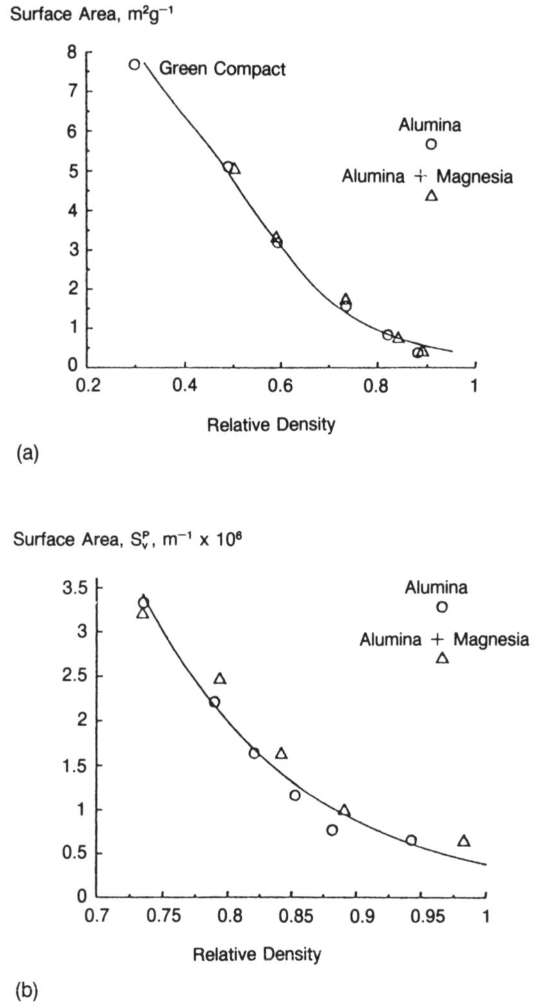

FIGURE 7A-5 Specific surface area as a function of relative density: (a) measured by BET in the early stages of sintering; (b) measured by stereology at higher densities.

APPENDIX 7-B
LIQUID PHASE SINTERING

When ceramics are sintered, there is usually a liquid phase present. Liquid phase sintering has three main characteristics: the initial consolidation of the powder compact by the liquid phase, which minimizes the porosity in the liquid; the formation of a skeleton structure that has the minimum energy configuration of the interfacial boundaries (where, in effect, the boundaries between the grains, or solid phase, and the liquid phase become arranged in a minimum energy configuration); and the transport of the solid phase through the liquid (Kingery 1950).

Figure 7B-1 shows a hypothetical densification curve for the three stages of liquid phase sintering. In the first stage, the liquid phase is formed, and the solid particles are rearranged, resulting in close packing (liquid flow). In the second stage, the fine particles dissolve in the liquid phase and are reprecipitated on the large particles (solution–precipitation). In the third stage, the powder compact is slowly consolidated as a consequence of the solid particles growing together to form a rigid skeleton (solid-phase sintering).

The driving force for rearrangement for various particle shapes and contact geometry is expressed as:

$$\frac{\mathrm{d}\,(\Delta\rho/\rho_0)}{\mathrm{d}t} = x_{(g)}\frac{\gamma_{lv}}{\eta r_s} \tag{7B-1}$$

where ρ is the relative density, t is the time, $x_{(g)}$ is a geometric constant, γ_{lv} is the surface tension, η is the viscosity of the liquid, and r_s is the radius of the particle. The value of $x_{(g)}$ decreases as the volume fraction of the solid and the relative density increase (Kwon 1991).

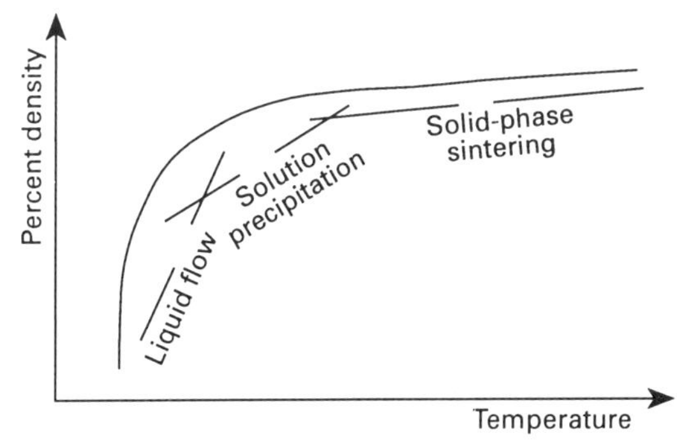

FIGURE 7B-1 Densification curve for three stages of liquid phase sintering.

Depending on the degree of packing and the physical characteristics of the particles, the volume of pores between the solid particles is from 25% to 50%. When 30% to 35% by volume of liquid phase is present, theoretical density can be attained as a consequence of the rearrangement process alone.

Solution precipitation is one of the processes that contribute to densification when there is only a relatively small amount of liquid phase present. In general, the solution and precipitation process results in grain growth due to the dissolving of fine particles that are soluble in the liquid phase, and the subsequent reprecipitation of the dissolved material on large particles.

This can happen in two possible ways: dissolving occurs over the entire surface of the particles (with subsequent reprecipitation of the dissolved material on the surface of other particles not in contact with them); or dissolving occurs at points that are in contact. At such points, the surface tension creates strong capillary pressure, which favors high solubility of the solid phase in the liquid phase. The solid material precipitates from the resulting supersaturated liquid phase onto the undissolved grains. This precipitation changes the shape of the particles, having the effect of fitting them closer together and thus promoting shrinkage.

Another condition for the occurrence of the solution and precipitation process is the degree of penetration of the liquid phase between the particles. The extent to which the liquid flows into the boundaries between the particles depends on the dihedral (two-sided) angle formed by the liquid phase at the boundary of two particles, shown in Figure 7B-2. Under equilibrium conditions this angle can be expressed as:

$$\sigma_{s_1} - \sigma_{s_2} = 2\ \sigma_{s-l}\cos(\phi/2) \tag{7B-2}$$

where $\sigma_{s_1} - \sigma_{s_2}$ is the interfacial energy at the boundary between two particles of the same phase, σ_{s-l} is the interfacial energy at the boundary between the solid and liquid phases, and ϕ is the dihedral angle. If σ_{s-l} and σ_{s_1} are roughly equal, the value of ϕ will be about 120 degrees, which indicates that the liquid will penetrate into the grain boundaries for a considerable distance.

Two different densification mechanisms can affect the densification rate for solution precipitation: diffusion through the liquid phase and control of reactions at the interface (Kwon and Messing 1989, 1991), as follows:

Diffusion through the liquid phase:

$$\frac{\mathrm{d}(\Delta\rho/\rho_o)}{\mathrm{d}t} = Y_{(g)}\frac{SD_bCV_m\gamma_{lv}}{kT}r_s^{-4} \tag{7B-3}$$

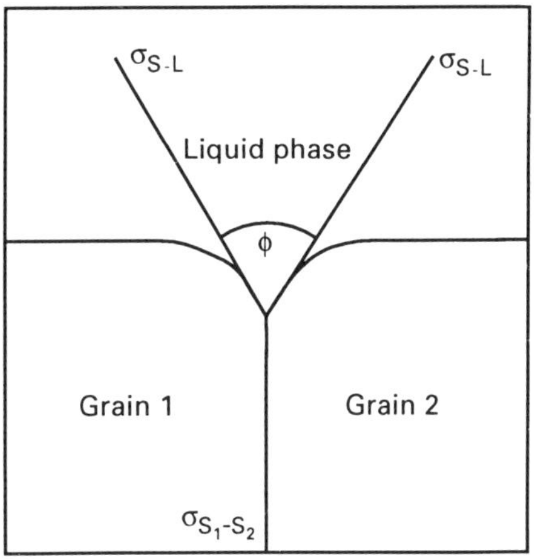

FIGURE 7B-2 The extent that the liquid flows into the boundaries between particles depends on the dihedral angle.

where $Y_{(g)}$ is a geometric constant that depends on the volumes of the solid and the liquid, the density, and the dihedral angle; S is the thickness of the liquid boundary; D_b is the diffusion constant of the grain boundary; C is the solubility of the solute; V_m is the molecular volume of the solute; k is the Boltzmann constant; and T is the absolute temperature.

Control of the interface reaction:

$$\frac{\mathrm{d}(\Delta\rho/\rho_o)}{\mathrm{d}t} = Z_{(g)}\frac{JCV_m\gamma_{lv}}{kT}r_s^{-2} \tag{7B-4}$$

where J is the interface constant, and $Z_{(g)}$ is the geometric constant. From both these expressions, it can be seen that the densification rate for solution-precipitation is strongly dependent on the particle radius, r_s.

As discussed, the extent to which the liquid penetrates between the solid particles depends on the value of the dihedral angle. At a certain ratio between the surface energies at the solid-solid and solid-liquid boundaries, which occurs when the interfacial energy does not satisfy the relationship expressed by equation 7B-2, the liquid does not flow into the boundaries between the particles.

In this case, neighboring particles of the solid phase can become welded together. The greater the number of welded particles, the more difficult it is for the liquid to flow. Occasionally, the number of such welded particles is so high that almost all of the particles in the solid phase grow together, creating their own skeleton with the liquid phase enclosed in the pores.

The liquid in the enclosed pores usually contains gaseous species from the sintering atmosphere, vaporized liquid, and decomposed solid. After pore closure, the driving force (*D.F.*) for densification can be expressed as (Kwon 1991):

$$D.F. = \frac{2\gamma_{lv}}{r_p} - P_p \tag{7B-5}$$

where P_p is the vapor pressure inside the pore. If the pore radius, r_p, and P_p remain small (that is, $D.F. > 0$), densification will continue. If r_p increases owing to the growth and coalescence of pores, and P_p increases because of the internal evolution of gases, the driving force will decrease. The result will be a reversal or loss of densification, that is, pore growth and bloating.

Whereas the dissolution of the solid phase in the liquid phase contributes to densification, the differential dissolution of particles that have different shapes and sizes results in an increase in the size of both pores and grains. This process is known as Ostwald ripening.

The relative importance of these mechanisms differs from system to system. It depends on the quantity and viscosity of the liquid phase formed, the particle sizes, the interfacial energies, and the solubility of the solid in the liquid.

APPENDIX 7-C
HIP DENSIFICATION MECHANISMS

Several mechanisms contribute to densification during hot isostatic pressing (HIP). These mechanisms vary with different types of ceramics and can depend on the stage of sintering as well as the applied pressure, temperature, and physical characteristics of the powder. Kinetically controlled processes that could be affected by the combinations of pressure and temperature available in HIP include the sintering mechanism, grain growth, and phase transformation.

Sintering

Pressure-enhanced densification consists of heating a mass of particulate material under externally applied pressure in order to achieve densification by interparticle bonding. The thermodynamic driving force for densification arises both from the reduction of surface energy and from the applied pressure. The kinetic response to the thermodynamic driving force can be complex. Among the mechanisms that are able to contribute to densification through their interaction with the processing parameters of applied pressure and temperature are stoichiometry, amount of porosity, pore size and location relative to grain boundaries, the material's surface energy and volume, grain-boundary self-diffusivities, and dislocations.

Possible mechanisms by which pressure-induced densification can take place during the initial stage of sintering include plastic yield, power-law creep, grain-boundary diffusion, lattice diffusion from the grain boundary, surface diffusion, lattice diffusion from the surface, and evaporation–condensation. Although in practice densification can be brought about only by the first four mechanisms, all of them can influence the rate of densification because they all result in neck growth (Wilkinson and Ashby 1975, 1978). Densification in the intermediate and final sintering stages results from dislocation motion and atomic diffusion to the pore surface, as well as from interface-controlled diffusion.

In order to quantitatively describe the kinetics of pressure-enhanced densification, constitutive or rate equations are required that define densification rates for given stages as well as for given mechanisms. Table 7C-1 gives examples of such equations (Ashby 1987). Constitutive equations used for pressure-induced densification are presented in the form of graphs illustrating domains with which given mechanisms control the rate of densification—for example, the plot shown in Figure 7C-1 of the relative density as a function of pressure, temperature, and time.

Grain Growth

Early work on the densification of ceramics by HIP has indicated that complete densification can be obtained without grain growth (Hodge 1964). But unless the appropriate HIP conditions are used, grain growth can occur. Becoming able to predict quantitatively the HIP conditions needed to achieve full density without grain growth would be an important step toward producing high-quality ceramic parts.

Phase Transformation

During HIP the grains of a ceramic powder compact are subjected to very high stresses. This is evidenced by the abnormally high dislocation densities seen in the microstructure of HIP'ed silicon nitride (Wills and Brockway 1981). It is anticipated that phase transitions not seen in conventionally processed ceramics will be observed as a result of the temperature and pressure regimes to which ceramic powder compacts are subjected during a HIP cycle.

TABLE 7C-1. **Constitutive or rate equations that define densification.***

	Stage 1	Stage 2
Initial Compaction	$D_i = \left\{ \frac{(1-D_o)}{1.3\sigma_y} P + D_o^3 \right\}^{1/3}$	$D_i = 1 - \exp\left\{ - \left[\frac{3}{2} \frac{P}{\sigma_y} \right] \right\}$
Volume Diffusion, pressure driven	$\dot{D} = 43\, C(1-D_o) \frac{D_v}{R^2} \frac{(P-P_o)\Omega}{kT}$	$\dot{D} = 270\,(1-D)^{1/2} \left[\frac{1-D}{6} \right]^{1/3} \frac{D_v}{R^2} \frac{(P-P_i)\Omega}{kT}$
Boundary Diffusion, pressure driven	$\dot{D} = 43\, C^2 \frac{\delta D_B}{R^3} \frac{(P-P_o)\Omega}{kT}$	$\dot{D} = 270\,(1-D)^{1/2} \frac{\delta D_B}{R^3} \frac{(P-P_i)\Omega}{kT}$
Power Law Creep	$\dot{D} = \frac{3.1}{C^{1/2}} D_{crp} D \left\{ \frac{C(P-P_o)}{3\,\sigma_o\, D^2} \right\}^n$	$\dot{D} = 1.5\, D_{crp} D(1-D) \left\{ \frac{1.5}{n} \frac{(P-P_i)}{\sigma_o} \frac{1}{(1-(1-D)^{1/n})} \right\}^n$

$$\text{where } D_v = D_{ov} \exp\left\{ -\frac{Q_v}{RT} \right\};\quad \delta D_B = \delta D_{oB} \exp\left\{ -\frac{Q_B}{RT} \right\};$$

$$D_{crp} = 10_{b-6} \exp\left\{ -\frac{Q_{crp}}{RT} \left[\frac{T_m}{T} - 2 \right] \right\};\quad C = \frac{1-D_o}{D-D_o};\quad P_i = \left[\frac{1-D_c}{1-D} \right] \frac{D}{D_c} P_o$$

*Nomenclature and Units

D	Relative density
D_o	Initial relative density (typically, 0.64)
D_i	Relative density from plastic yielding
D_c	Relative density at which pores close (around 0.95)
$\dot{D}$	Densification rate (/sec)
δD_B	Boundary diffusion coefficient times boundary thickness (m^3/sec)
D_v	Volume diffusion coefficient (m^2/sec)
k	The Boltzmann constant (J/K)
P	External pressure (MPa)
P_o	Outgassing pressure (MPa)
Q_B	Activation energy of boundary diffusion (J/mol)
Q_v	Activation energy of volume diffusion (J/mol)
Q_{crp}	Activation energy of creep diffusion (J/mol)
R	Particle radius (m)
T	Absolute temperature (K)
T_m	Absolute melting temperature (K)
σ_o, n	Creep parameters (s^{-1}, MPa)
Ω	Atomic or molecular volume (m^3)
σ_y	Yield strength (MPa)

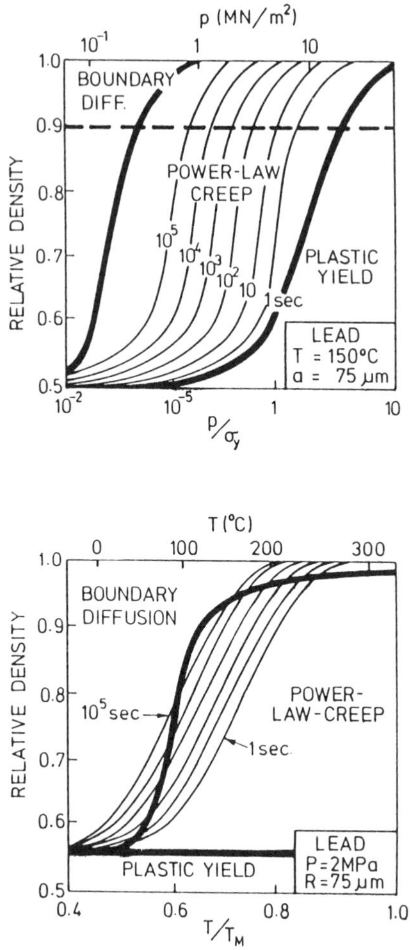

FIGURE 7C-1 Illustration of the domains within which given mechanisms control the rate of densification. Example: Relative density as a function of pressure, temperature, and time. After Ashby (1987).

APPENDIX 7-D THE RELATION OF CRACK PROPAGATION TO FRACTURE STRENGTH

If cracks are assumed to be elliptical in shape, as shown in Figure 7D-1, the highest stress (σ_m) at the tip of the crack can be calculated from the following relation:

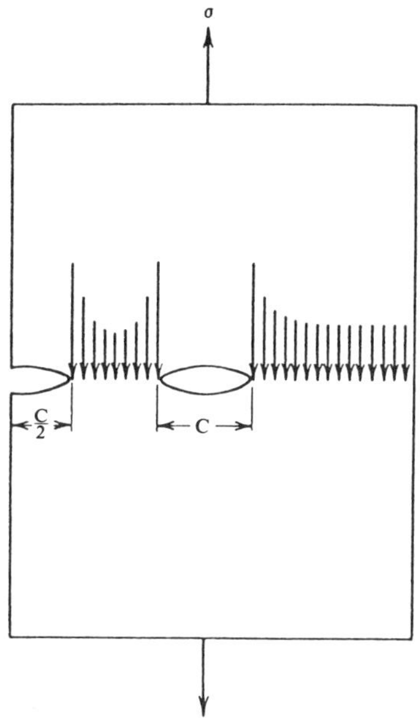

FIGURE 7D-1 Assuming cracks are elliptical in shape, the highest stress (σ_m) at the tip of the crack can be calculated by using expression (7-D1).

$$\sigma_m = \sigma\left[\frac{2C}{r}\right]^{1/2} \tag{7D-1}$$

where σ is the applied tensile stress, r is the radius of curvature at the crack tip, and C is the length of an interior crack or twice the full length of a surface crack.

Because the radius of curvature of the crack tip cannot be less than on the order of the diameter of an atom, it must remain finite; but it can attain the theoretical fracture strength of the material even though the bulk of the material is under a relatively low applied stress. If an internal crack of the type shown in Figure 7D-1 propagates, additional surface is created, which increases the total surface energy as the size of the crack increases. The total surface energy per unit thickness (U_s) of a flat elliptical crack through a thin plate is:

$$U_s = 2C\gamma \tag{7D-2}$$

where the factor 2 represents the two surfaces of the crack each with length C, and γ is the surface fracture energy. The surface fracture energy tends to restrict crack growth, but propagation of the crack can decrease the elastic strain energy. For a thin plate subjected to a fixed strain, the total elastic energy per unit thickness (U_e) due to the presence of an elliptical crack extending through the plate is:

$$U_e = \frac{\pi C^2 \sigma}{4E} \tag{7D-3}$$

where σ is the nominal tensile stress, and E is the elastic modulus of the plate. Equation (7D-3) represents the difference between the elastic energy for a plate containing a crack and one without a crack. The total energy (U) associated with the crack is the sum of equations (7D-2) and (7D-3). The critical stress (σ_F) required to destabilize a crack of a given size can be determined by differentiating $U = (U_s + U_e)$ with respect to C and setting the derivative equal to zero. The relationship then becomes:

$$\sigma_F = \left[\frac{4\gamma E}{\pi C}\right]^{1/2} \tag{7D-4}$$

Equation (7D-4) applies strictly to the case of a flat elliptical crack through a thin plate (Figure 7D-1). For other configurations, the numerical factor changes but is still on the order of unity. The fracture strength then can be determined by:

$$\sigma_F = \left[\frac{\gamma E}{C}\right]^{1/2} \tag{7D-5}$$

From either equation (7D-4) or equation (7D-5), the conclusion can be drawn that if a distribution of crack sizes exists within (or on the surface of) a brittle solid, the fracture strength will be determined by the most serious flaw, that is, the longest crack.

APPENDIX 7-E
WEIBULL ANALYSIS

According to Weibull theory the probability of survival, P_s, of a sample subjected to a stress, σ, can be stated as:

$$P_s = \exp\left[-\int v\left(\frac{\sigma - \sigma_u}{\sigma_0}\right)^m dv\right] \tag{7E-1}$$

where the integration is over the whole component, σ is the stress in a small element of volume dv, and σ_u, m, and σ_0 are material constants with the following significance: σ_u is the stress at which the material is absolutely safe, that is, $P_s = 1$ for $\sigma \leq \sigma_u$, and m is a dimensionless parameter that determines the variability of the quantity $\sigma_F - \sigma_u$ (σ_F is the failure stress). The variability of the failure stress thus depends on both m and σ_u. The quantity σ_0 is a scaling factor. For a given σ_u and m, and for a given sample volume, the mean strength of the material $\bar{\sigma}_F$ increases as σ_0 increases. Quantity σ_0 is not strictly a material parameter because it depends on the volume of the sample. When $\sigma_u = 0$, the Weibull equation (7E-1) becomes:

$$P_s = \exp\left[-\int v\left(\frac{\sigma}{\sigma_0}\right)^m dv\right] \tag{7E-2}$$

Because σ_0 is a complex quantity, the above form of the equation (7E-2) is not normally used. However, if a normalized stress variable (obtained by dividing each observed strength value by the mean value) is used, the basic Weibull function can be presented as:

$$P_s = \exp - \left[\beta\left(\frac{1}{m}\right)!\right]^m \tag{7E-3}$$

where β is the mean normalized variable and the exclamation mark denotes a factorial function.

There are a number of methods, both graphical and numerical, for estimating σ_0, σ_u, and m from experimental data. If n test results are available, and these results are ranked in order of increasing strength, then, if the rth failure stress is σ_r, the survival probability corresponding to σ_r can be estimated from r and n. Two plotting positions can be used:

$$\text{Mean rank position:} \quad P = 1 - \frac{r}{n+1} \tag{7E-4}$$

$$\text{Median rank position:} \quad P = 1 - \frac{r - \frac{1}{2}}{n} \tag{7E-5}$$

The following plotting procedure is used.

- Values of the reduced strength, β, are obtained by dividing each strength value by the mean.
- The n values obtained are ranked in order of increasing β.
- A value of $P = 1 - \dfrac{r - \frac{1}{2}}{n}$

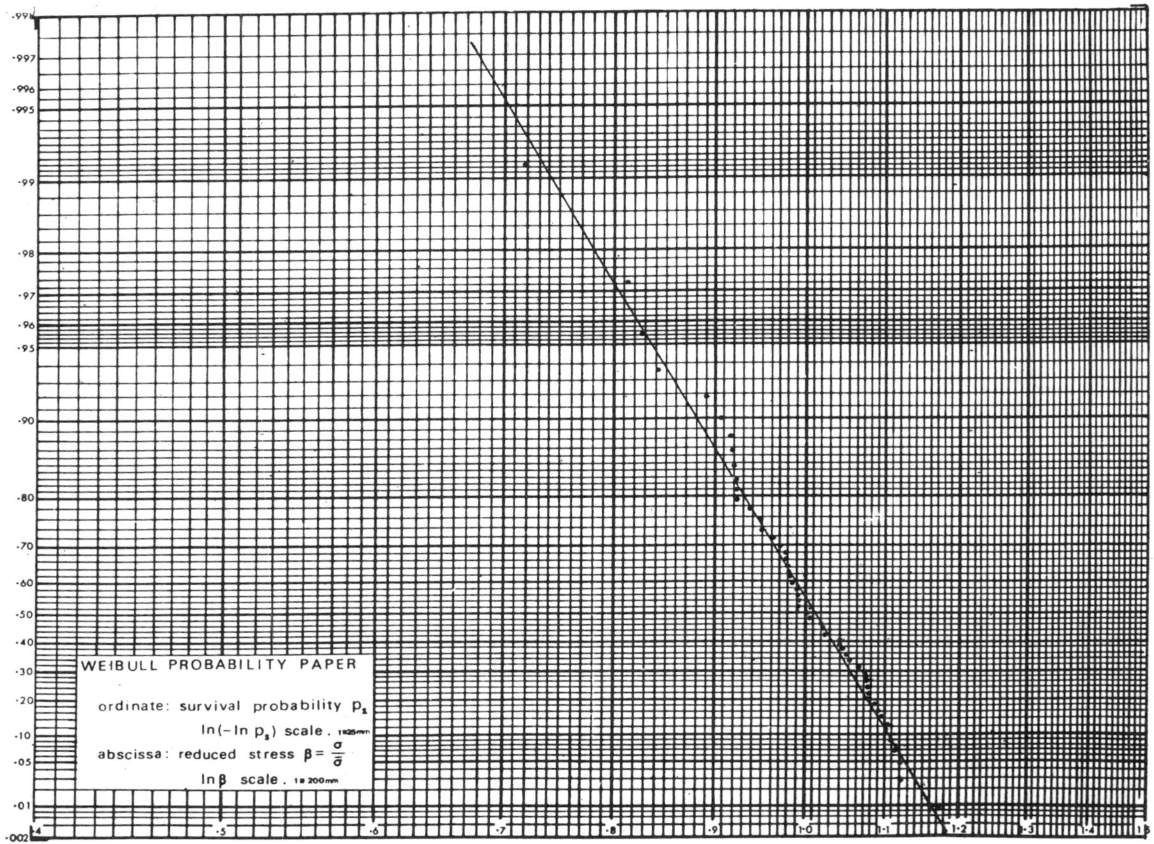

FIGURE 7E-1 The m-value is estimated as the slope of the plot, ln $(1/P_s)$ versus ln β.

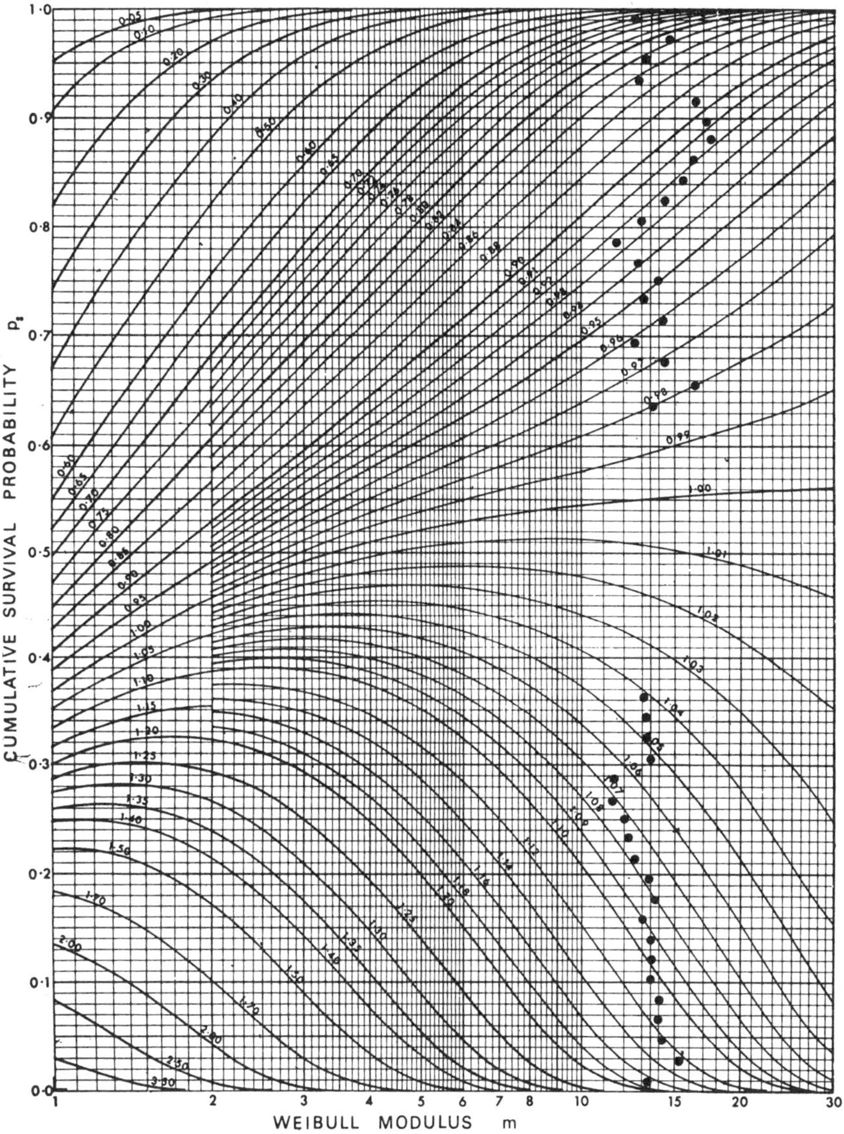

Curves of $p_s = \exp-\left[\beta\left(\frac{1}{m}\right)!\right]^m$ for the values of β shown. $\beta = \frac{\sigma}{\bar{\sigma}} = \frac{\text{applied stress}}{\text{mean strength}}$

FIGURE 7E-2 The *m*-value is read directly from the straight-line plot drawn parallel to the probability axis.

In the first method, the Weibull constant, m, is estimated as the slope of the plot of ln $(1/P_s)$ against ln β, shown in Figure 7E-1. In the second method, the data points should fall on a straight line parallel to the probability axis, that is drawn through the data points, and m is read directly from the plot, as shown in Figure 7E-2.

References

Adlerborn, J. and H. Larker. 1978. U. S. Patent 4,112,143.

Ashby, M. F. 1974. A First Report on Sintering Diagrams, *Acta Metall.*, 22, 275.

Ashby, M.F. 1987. The Modelling of Hot Isostatic Pressing, *Proc. Inter. Conf. on HIP*, 15–17 June 1987, Luleå, Sweden. Ed. Tore Garvare, Centek Publishers, Sweden.

Atkinson, A., A. J. Moulson, and E. W. Roberts. 1976. Nitridation of High Purity Silicon, *J. Am. Ceram. Soc.*, 59(7–8), 285–289.

Bandyopadhyay, G. et al. 1988. Fabrication and Evaluation of Silicon Nitride Heat Engine Components, Proc. 3rd Internat. Symp. on Ceramic Materials and Components for Engines, Las Vegas, NV.

Bennison, S. J. and M. P. Harmer. 1990. A History of the Role of MgO in the Sintering of Alpha Alumina, *Sintering of Advanced Ceramics, Cermam. Trans.*, ed. C. A. Handwerker, J. E. Blendell, and W. Kaysser, 7, 13–49.

Billington, W. R., J. Chown, and A. E. S White. 1962. The Sintering of Silicon Carbide, in *Special Ceramics*, ed. P. Popper, Academic Press, London, 19–34.

Bocker, W. and H. Hausner. 1977. Observations on the Sintering Characteristics of Submicron Silicon Carbide Powders, *Science of Ceramics*, 9, 168.

Bocker, W. and H. Hausner. 1978. The Influence of Boron and Carbon Additions on the Microstructure of Sintered Alpha Silicon Carbide, *Powder Metall. Int.*, 10(2), 87–89.

Bocker, W., H. Landfermann, and H. Hausner. 1979. Sintering of Alpha Silicon Carbide with Additions of Aluminium, *Powder Metall. Internat.* 11(2), 83–85.

Boyer, C. B. 1982. Evolution and Future of HIP Systems, 2nd International Conference on Isostatic Pressing, 21–23 Sept. 1982, at Stratford-upon-Avon, England, 7.2–7.25:MPR Publishing Services Ltd.

Brook, R. J. 1982. Fabrication Principles for the Production of Ceramics with Superior Mechanical Properties, *Proc. Brit. Ceram. Soc.*, 32, 7.

Bruch, C. A. 1962. Sintering Kinetics for the High Density Alumina Process, *Am. Ceram. Soc. Bull.*, 41(12), 799–860.

Brun, M. K. and M. P. Borom. 1985. Design and Construction of a HIP-Dilatometer, *Am. Ceram. Soc. Bull.*, 64(5), 719–723.

Cannon, W. R. et al. 1982. Sinterable Ceramic Powders from Laser Driven Reactions: Part 1: Process Description and Modeling, *J. Am. Ceram. Soc.*, 65(7), 324–330.

Carroll, D. F., S. M. Wiederhorn, and D. E. Roberts. 1989. Technique for Tensile Creep Testing of Ceramics, *J. Am. Ceram. Soc.* 72(9), 1610–1940.

Coble, R. L. 1961. Sintering Crystalline Solids, 1, Intermediate and Final State Diffusion Models, *J. Appl. Phys.*, 32, 787.

Coble, R. L. 1962. Sintering Alumina: Effect of Atmosphere, *J. Am. Ceram. Soc.*, 45, 123-127.

Coppola, J. A., L. N. Hailey, and C. H. McMurtry . 1978. Process for Producing Sintered Silicon Carbide Ceramic Body, U.S. Patent 4,124,667.

Cordrey, L., D. E. Niesz, and D. J. Shanefield. 1990. Sintering of Silicon Carbide with Rare Earth Oxide Additives, *Sintering of Advanced Ceramics, Ceram. Trans.*, ed. C. A. Handwerker, J. E. Blendell, and W. Kaysser, 7, 618-636.

Drew, P. and M. H. Lewis. 1974. Microstructures of Silicon Nitride Ceramics during Hot Pressing Transformations, *J. Mater. Sci.*, 9, 261-269.

Evans, A. G. and R. W. Davidge. 1970. *J. Mater. Sci.*, 5, 314.

Evans, A. G. and G. Tapping. 1972. *Proc. Brit. Ceram. Soc.*, 20, 275-297.

Frenkel, J. 1946. *Zhurn. eksp. Theor. Phiziki*, 16, 29.

Fujikawa, T., M. Moritoki, and K. Homma. 1982. The High Temperature HIP Furnace and Its Applications in Densification of Ceramics, 2nd International Conference on Isostatic Pressing, 12.3-12.5.

Gazza, G. E. and R. N. Katz. 1978. Development of Advanced Sinterable Si_3N_4, DOE Heat Engine Highway Vehicle Systems A.E.S Program, DOE/AMMRC Interagency Agreement EC-76-A-1017.

Gazza,G. E. and R. N. Katz. 1980. Development of Sinterable Si_3N_4, Automotive Tech. Dev., DOE/AMMRC IGA EC-77-A-01-1017.

Geiger, G. 1990. Advancements in Mechanical Testing of Advanced Ceramics, *Bull. Am. Ceram. Soc.*, 69(11), 1794-1800.

General Electric Company. 1977. U.S. Patent 2,734,425.

German, R. M. and Z. A. Munir. 1976. Surface Area Reduction during Isothermal Sintering, *J. Am. Ceram. Soc.*, 59, 379.

Giachello, A., P. C. Martinengo, and G. Tommasini. 1979. Sintering of Silicon Nitride in a Powder Bed, *J. Mat. Sci.*, 14, 2825-2830.

Godfrey, D. J. 1975. *Proc. Brit. Ceram. Soc.*, 25, 325.

Greskovich, C. D, and J. H. Rosolowski. 1977. Polycrystalline Silicon Articles by Sintering, U.S. Patent 4,040,849.

Greskovich, C., S. Prochazka, and J. H. Rosolowski. 1979. Basic Research on the Technical Development of Sintered Ceramics, Tech. Pert. AFML-TR-76-179, Cont. F33615-76-C-5033, General Electric, Schnectady, NY.

Griffith, A. A. 1921. The Phenomena of Rupture and Flow in Solids, *Phil. Trans. Roy. Soc. London*, 221A, 163-198.

Helms, H. E. et al. 1990. Advanced Turbine Technology Application Project, Ann. Auto. Tech., U.S. Dept. of Energy, Dev. Contractors' Coordination Meeting, Dearborn, MI.

Harmer, M., E. W. Roberts, and R. J. Brook. 1979. Rapid Sintering of Pure and Doped αAl_2O_3, *Trans. J. Brit. Ceram. Soc.*, 78(1), 22-25.

Heinrich, J. 1987. Nitridation of Silicon under High Pressure, *Advanced Ceramic Materials*, 2(3A), 239-242.

Herbell, T. P., T. K. Glasgow, and N. J. Shaw. 1980. Reaction Bonded Silicon Nitride Prepared from Wet Attrition Milled Silicon, NASA TM-81428.

Heinrich, J., E. Backer, and M. Bohmer. 1988. Hot Isostatic Pressing of Si_3N Powder Compacts and Reaction-Bonded Si_3N_4, *J. Am. Ceram. Soc.*, 71(1), C-28-C-31.

Hintler, W., K. Schweitzen, and A. Rossman. 1980. U.S. Patent 4,242,294.

Hirosaki, N., A. Okada, and K. Matoba. 1988. Sintering of Si_3N_4 with the Addition of Rare-Earth Oxides, *J. Am. Ceram. Soc.* 71(3), C-144–C-147.

Hirosaki, N. and A. Okada. 1988. Effects of N_2 Gas Pressure Sintering on the Surface and Strength of Si_3N_4, *Advanced Ceramic Materials*, 3(5), 515–516.

Hodge, E. S. 1964. Elevated Temperature Compaction of Metals and Ceramics by Gas Pressures, *Powder Metall.*, 1(14), 168–201.

Hoge, C. E. and J. A. Pask. 1973. Thermodynamics of Solid-State Sintering, 3rd Internat. Meeting on Sintering, Herceg Navy.

Hunold, K., J. Greim, and A. Lipp. 1989. Injection Molded Ceramic Rotors—Comparison of SiC and Si_3N_4, *Powder Metall. Inter.*, 21(4), 17–23.

Jack, K. H. 1973. Nitrogen Ceramics, *Trans. & J. Brit. Ceram. Soc.*, 72(8), 376–386.

Kieffer, R. et al. 1972. Investigation on the System Boron–Carbon–Silicon, *Ber. Dt. Keram. Ges.*, 49, 41–46.

Kingery, W. D. 1950. Densification during Sintering in the Presence of a Liquid Phase, *J. Appl. Phys.*, 30, 301–306. Kingery, W. D., and M. J. Berg. 1955. Study of Initial Stages of Sintering Solids by Viscous Flow, *J. Appl. Phys.*, 26(10), 1205–1212.

Kingery, W. D., H. K. Bowen, and D. R. Uhlmann. 1976. *Introduction to Ceramics*, John Wiley & Sons, New York.

Kuczynski, G. C. 1949. Self-Difussion in Sintering of Metallic Particles, *Trans. AIME*, 185, 169–178.

Kumagai, M. and G. L. Messing. 1984. Enhanced Densification of a Boehmite Sol-Gel by Alpha Alumina Seeding, *J. Am. Ceram. Soc.*, 67, 11, C-230–C-231.

Kumagai, M. and G. L. Messing. 1985. Controlled Transformation and Sintering of a Boehmite Sol-Gel by Alpha Alumina Seeding, *J. Am. Ceram. Soc.*, 68(9), 500–505.

Kwon, O. H. 1991. Liquid Phase Sintering, *Engineered Materials Handbook, ASM*, 4, 285–290.

Kwon, O. H. and G. L. Messing. 1989. Kinetic Analysis of Solution-Precipitation during Liquid Phase Sintering of Alumina, *J. Am. Ceram. Soc.*, 73(3), 275–281.

Kwon, O. H. and G. L. Messing. 1991. A Theoretical Analysis of Solution-Precipitation Controlled Densification during Liquid Phase Sintering, *Acta. Metall.*, 39(9), 2059–2068.

Lange, F. F. 1978. *Materials Science Research*, 11,
Processing of Crystalline Ceramics, Plenum Press, New York, 597.

Lange, F. F., and T. K. Gupta. 1976. *Sintering of SiC with Boron Compounds, J. Am. Ceram. Soc.*, 59, 537–538.

Larker, H. 1984. Dense Ceramic Parts Hot Pressed to Shape by HIP, *Mater. Sci. Res.*, 17, 571–582.

Layden, G. K. 1976. Process Development for Pressureless Sintering of SiAlON Ceramic Components, Report R75-91072-4, United Tech. Res. Cen.

Lin, S. T. and R. M. German. 1988. Compressive Stress for Large Pore Removal in Sintering, *J. Am. Ceram. Soc.*, 71(10), C- 432–433.

Lindley, M. W, et al. 1979. J. Mater. Sci., 14, 70.

Mangels, J. A. 1978. Development of Injection Molded Reaction Bonded Si_3N_4, in

Ceramics for High Performance Applications—II, eds. J. J Burke, E. N. Lenoe, and R. N. Katz, Brook Hill Pub. Co., Chestnut Hill, MA, 113-130.

Mangels, J. A. 1986. Sintered Reaction-Bonded Silicon Nitride, Proc. 5th Ann. Conf. on Composites and Advanced Ceramic Materials, 589-593.

McCoy, K. J., L. E. Muttart, and R. R. Wills, 1985. Continuous Monitoring of Volumetric Changes in Ceramic Powder Compacts During Hot Pressing, *Amer. Ceram. Bull.*, 64(9), 1240-1244.

Messier, D. R. and P. Wong. 1974. *Ceramics for High Performance Applications*, ed. J. J. Burke, A. E. Gorum, and R. N. Katz, Brook Hill, Chestnut Hill Publ. Co., MA, 181.

Mitomo, M. 1976. Pressure Sintering of Si_3N_4, *J. Mater. Sci.*, 11 1103-1107.

Morgan, P. E. D. 1977. In *Nitrogen Ceramics*, ed. F. L. Riley, Noordhoff, Leyden, 23.

Moteki, A. 1960. Properties on High Alumina Porcelain Produced by Injection Molding, Studies on the Injection Molding of Ceramics, III, *J. Ceram. Assoc. Japan*, 68(2), 23-33.

Moulson, A. J. 1979. Reaction-Bonded Silicon Nitride: Its Formation and Properties, *J. Mater. Sci.*, 14, 1017-1051.

Mutsuddy, B. 1987. Mechanical Properties of Injection Molded Ceramics, *Powder Metall. Internat.*, 19(2), 43-45.

Oda, M., N. Kaneno, and N. Yamamoto. 1977. Pressureless Sintered Silicon Nitride, in *Nitrogen Ceramics*, ed. F. L. Reley, Noordhoff, Leyden.

Paek, Y. K., K. Y. Eun, and S. J. L. Kang. 1988. Effect of Sintering Atmosphere on Densification of MgO Doped Al_2O_3, *J. Ceram. Soc.*, 71(8), C-380-C-382.

Pasto, A. E. and S. Natansohn. 1990. Improved Processing Methods for High Reliability Silicon Nitride, Proc. Ann. Auto. Tech. Dev. Contractors' Coordination Meeting, Dearborn, MI, 22-25 Oct., 143-160.

Pavlushkin, N. M. 1956. Production of Superstrong Corundum, *Glass and Ceramics*, 13(11), 479-484.

Priest, H. F., G. L. Priest, and G. E. Gazza. 1977. Sintering of Si_3N_4 Under High Nitrogen Pressure, *J. Am. Ceram. Soc.*, 60(1), 81.

Prochazka, S. 1973. Sintering of Silicon Carbide, Proc. of the Conference on Ceramics for High-Performance Applications, G. E Report No. 74 CRD067, Hyannis, MA.

Prochazka, S. 1974a. Sintering of Silicon Carbide, in *Ceramics for High Performance Applications*, ed. J. J. Burke, A. E. Gorum, and R. N. Katz, Brook Hill Publ. Co., Chestnut Hill, MA, 239.

Prochazka, S. 1974b. The Role of Boron and Carbon in the Sintering of Silicon Carbide, *Special Ceramics*, 6, 171-181.

Prochazka, S., and R. M. Scanlan. 1974. Effect of Boron and Carbon on Sintering of SiC, *J. Am. Ceram. Soc.*, 58 (1-2), 72.

Pujari, V. K., K. E. Amin, and P. H. Tewari. 1990. Development of Improved Processing and Evaluation of Silicon Nitride, Proc. Ann. Auto. Tech. Dev. Contractors' Coordination Meeting, Dearborn, MI, 22-25 Oct., 133-141.

Rokugawa, S., S. Usul and B. D. Sawicka. 1991. Nondestructive Evaluation of Ceramics Components by x-ray Computed Tomography, Report AISCL-10461, Atomic Energy of Canada Limited (Chalk River).

Rosolowski, J. H. and C. Greskovich. 1973. Analysis of Pore Shrinkage by Volume Diffusion during Final Stage Sintering, *J. Appl. Phys.*, 44, 1441.

Rowcliff, D. J., and P. J. Jorgensen. 1977. Proc. of the Workshop on Ceramics for Advanced Heat Engines, Contract 77011OUC 95a-ERDA Report.

Ryshkewitch, E. 1948. *Oxydkeramik der Einstoff-systeme*, Springer-Verlag, Berlin.

Sawicka, B. D. and D. K. Sonnenberg. 1992. Computed Tomography Examination of NKK Ceramic Objects: Part 3, Report ET-T-18, Atomic Energy of Canada Limited (Chalk River).

Shaw, N. J. 1989. Densification and Coarsening during Solid State Sintering of Ceramics; A Review of the Models. 1—Densification, *Powder Metall. Internat.* 21(3), 16–21.

Shaw, N. J. and R. J. Brook. 1986. Structure and Grain Coarsening during the Sintering of Alumina, *J. Am. Ceram. Soc.*, 69, 107.

Stephenson, I. M., and J. White. 1967. Factors Controlling the Microstructure in Two-Phase and Three-Phase Systems, *Trans. Brit. Ceram. Soc.*, 66(9).

Stewart, R. W., R. R. Wills, and J. M. Wimmer, 1977. Fabrication of Reaction Sintered Silicon Nitride, *J. Am. Ceram. Soc.*, 60(1), 65–67.

Tanaka, T., K. Saitoh, M. Satoh, M. Miyazaki, and Y. Kanako. 1985. Injection Molding of Alumina, Yogyo-Kyokai-Shi, 93(9), 572–576.

Terwilliger, G. R. and F. F. Lange. 1975. Pressureless Sintering of Si_3N_4, *J. Mater. Sci.*, 10(7), 1169–1174.

Timoshenko, S. P. and D. H. Young. 1962. *Elements of Strength of Materials*, 4th, edn., D. Van Nostrand, Princeton, NJ.

Tsuge, A. et al. 1975. High Strength Silicon Nitride Ceramics, *Toshiba Review*, 30, 580–584.

Venables, J. D., D. K. McNamara, and R. G. Lye. 1976. High Temperature Silicon Nitride Prepared with Eutectic Flux Additions, in *Nitrogen Ceramics*, Proc. N.A.T.O. Advanced Study Inst., ed. F. L. Riley.

Wada, S. et al. 1987. Injection Molding of SiC Turbocharger Rotor, Tokyo Internat. Gas Turbine Congress, I-143–I-146.

Walker, R. F. 1955. Mechanism of Material Transport during Sintering, *J. Am. Ceram. Soc.*, 38(12), 466.

Weibull, W. 1951. A Statistical Distribution Function of Wide Applicability, *J. Appl. Mech.*, p. 293.

Whalen, T. J. and W. L Winterbottom. 1986. Improved Silicon Carbide for the Advanced Heat Engine, NASA Contractor Report 179477.

Wilkinson, D. S. and M. F. Ashby. 1975. The Development of Pressure Sintering Maps, in *Materials Science Research*, 10, *Sintering and Catalysis*, ed. G. C. Kuczynski, Plenum, New York, 473–492.

Wilkinson, D. S., and M. F. Ashby. 1978. Mechanism Mapping of Sintering under Applied Pressure, *Sci. Sintering*, 10(2), 67–76.

Wills, R. R., and M. C. Brockway. 1981. Hot Isostatic Pressing of Silicon Based Ceramics, Technical Report AFWAL-TR-80-4193, Battelle Columbus Lab.

Yen, T-S., and M. D. Sacks. 1988. Low Temperature Sintering of Aluminum Oxide, *J. Am. Ceram. Soc.*, 71(10), 841–844.

Glossary

Absorption The capacity of a solid to receive and retain a liquid or gas by capillary, osmotic, or solvent action.

Adherence The stress required to cause separation of one material from another at their interface.

Adsorption The capacity of a substance to accept and retain on its surface a layer of a gas or liquid.

Agglomerate A collection of particles held together by relatively weak cohesive forces. The forces can be due to weak chemical bonding or to an electrostatic surface charge generated during the processing operation.

Aggregate A dense mass of particles held together by strong intermolecular or atomic cohesive forces. It is stable to normal handling and ordinary mixing techniques including high-speed mixing and ultrasonics.

Amorphous Lacking a detectable crystallinity; a disordered structure although solid in appearance.

Anneal To prevent or remove stresses in ceramics by controlled heat treatment.

Apparent density The mass of a material divided by its apparent volume, including closed pores.

Apparent porosity The relation of the open pore space to the bulk volume, expressed in percent.

Average molecular weight Indicates the chain length of the most typical chain in a given polymer, which is neither the longest nor the shortest chain.

Average particle size A single value representing the entire particle size distribution.

Back pressure control Determines the hydraulic back pressure applied to the injection ram during the plastication phase of the injection cycle.

Ball milling A method of grinding and mixing material with or without liquid, in a rotating cylinder or conical mill partially filled with grinding media.

Barrel A cylinder housing in which the screw rotates, which includes a replaceable liner, if used, or an integrally formed special surface material.

Bedding The materials, their placement, and configuration as designed to support a ceramic part.

BET (Brunauer–Emmett–Teller) An instrumental method for determining the surface area of a solid sample, by measuring monomolecular nitrogen gas adsorption at the temperature of liquid nitrogen.

Binder An organic or inorganic substance added to a ceramic powder to provide it with plasticity, flowablity, and green or dry strength.

Bingham fluid or solid Materials in which no flow or deformation occurs until a certain threshhold shear stress is applied, at which point the shear stress becomes a linear function of the shear rate. The characteristics of this linear function are the viscosity (slope) and the yield value (shear stress intercept).

Blister A defect consisting of a bubble that forms during thermal treatment of a ceramic part.

Boost cut-off Determines the machine parameter used to terminate injection velocity control. Hydraulic pressure, ram cushion, or cavity pressure can be selected as the activating parameter.

Bulk density The weight of a ceramic part divided by the volume of material it contains less the volume of its open pores.

Capillary action The phenomenon of intrusion of a liquid into interconnected small voids, pores, and channels in a solid as the result of surface tension.

Catalyst An agent that initiates or accelerates the speed of a reaction between substances but does not undergo any change itself.

Cation A positively charged atom or radical (chemically bonded cluster of atoms). A cation is attracted to the cathode in electrochemical systems.

Cavity Depression in the female half of a mold that helps to define the space to be filled by the molded material.

Cavity packing Begins when the cavity is filled and the material entering the gate is increasing the density of the part; it ends when the gate finally freezes.

Ceramics Pertaining to products manufactured from inorganic nonmetallic substances that are subjected to a high temperature during manufacture or use.

Chemisorption The binding of an adsorbate to the surface of a solid by forces that have energy levels similar to those of a chemical bond.

Clamp force The force that keeps the mold closed or clamped together during the injection cycle. The clamp force balances the injection pressure. The clamping mechanism can be either hydraulic or toggle.

Clamp stroke The maximum distance the moving platen will travel. It is desirable that this motion be fully adjustable to permit the minimum stroke required by the mold.

Closed porosity The volume fraction of all pores within a solid mass that are closed off by the surrounding dense solid and, therefore, are inaccessible to each other and to the extant surface; they thus are not detectable by gas or liquid penetration.

Colloid Colloids are particles that, when dispersed in a fluid, are fine enough so that Brownian motion maintains them in suspension without settling. In general, colloidal particles are submicron in size.

Combustion Rapid oxidation.

Comminution The process of reduction of particle size, usually, but not necessarily, by grinding or milling.

Compaction (1) The act of forcing particulate or granular material together (consolidation) under pressure to yield a relatively dense mass or formed part. (2) The preparation of a compact or part produced by the compression of a powder, generally while confined in a die, with or without the inclusion of lubricants, binders and the like.

Connected porosity The volume fraction of all pores, voids, and channels within a solid mass that are interconnected with each other and extend to the surface, and thus are measurable by gas or liquid penetration.

Contamination Undesirable impurities or foreign material present in a powder or part.

Core Movable sections in a mold that permit the molding of parts with undercuts; also referred to as core pins. Also passageways within a mold for circulation of temperature-control media.

Core pin Pin used to mold a hole.

Crack A fracture line, but not complete separation.

Cracking A defect in the dewaxed or sintered part consisting of fractures or separation.

Crazing The cracking that occurs in dewaxed or fired ceramics as a result of critical tensile stresses.

Crushing The process of reducing the size of lump material by mechanical forces.

Crystallization The progressive process in which crystals are first nucleated (initiated) and then grow in size within a host medium that supplies their atoms. The host can be a gas, a liquid, a glass, or another crystalline phase.

Cure Changing the physical properties of a material by chemical reaction, which may be condensation or polymerization; usually accomplished by the action of heat and catalysts, alone or in combination, with or without pressure.

Cushion control Determines the ram position at which boost cut-off occurs; or the ram position reference if boost cut-off is done with hydraulic or cavity pressure.

Cycle The total series of operations in the molding process, or part of a process, usually defined as the temporal difference between any step and the subsequent repetition of that step.

De-airing Removal of entrapped air from the mix.

Deflocculant An electrolyte adsorbed on colloidal particles in suspension, which charges the particles creating repulsive forces that maintain the particles in dispersed state, thus reducing the viscosity of the suspension.

Deflocculate The dispersion of a powder suspension so that it has little tendency to settle or coalesce.

Dehydration Removal of moisture, whether free or chemically combined.

Dilatancy or Dilatant fluid In general refers to materials that exhibit an increase in apparent viscosity with increasing rates of shear.

Draft The degree of taper of the side walls or clearances; the angle utilized to simplify demolding.

Ejection device Parts are ejected from the mold by knock-out pins, sleeves, stripper rings or stripper plates, either singly or in combination.

Ejector pin A rod, pin, or sleeve that pushes a molding off a force or out of a cavity of a mold. It is attached to an ejector bar or plate that can be actuated by the ejector rod(s) of the press or by an auxiliary hydraulic or air cylinder.

Endothermic reaction A reaction that absorbs heat.

Entrance angle The maximum angle at which the mixture enters the land area of the mold.

Exaggerated grain growth (secondary recrystallization) A phenomenon occasionally observed in ceramics in which a few grains grow very rapidly in an uncontrolled fashion. Exaggerated grains separated from pores commonly appear in some regions of the microstructure when the processing is not carefully controlled. The terms exaggerated grain growth and discontinuous grain growth are sometimes used interchangeably.

Exothermic reaction A reaction that liberates heat.

Fan gate A shallow gate somewhat wider than the runner from which it extends.

Feed section A separate section, located at the upstream end of the barrel that contains the feed opening into the barrel.

Female The indented half of a mold; the side of the mold containing the cavity.

Flash Extra material attached to a molding along the parting line; it must be removed before sintering the parts.

Flexural strength A mechanical property of solid material that indicates its ability to withstand a flexural or transverse load.

Flight landing The surface at the radial extremity of the flight constituting the periphery or outside diameter of the screw.

Flocculate A grouping of primary particles or agglomerates in which the bonding is weaker than in the agglomerate structures.

Flow A qualitative description of the fluidity of a ceramic mix during the process of molding.

Flow line A mark on a molded part made by the meeting of two flow fronts during molding. Also called a weld line.

Flow mark A wavy surface appearance on a molded part caused by the improper flow of the material into the mold.

Gate A small orifice that acts to restrict the plastic mix entering the mold cavity. When the cavity is filled, material in the gate solidifies and seals the cavity. The full mold under pressure will then have time to solidify completely while the injection plunger is being withdrawn. If the gate did not seal, the plunger would have to remain extended or the plunger withdrawal would discharge plastic mix from the cavity.

Gel A colloidal state consisting of interspersed solid and liquid, in which the solid particles are themselves interconnected in three dimensions.

Glass transition The change in an amorphous polymer or in amorphous regions of a partially crystalline polymer from a viscous condition to a hard and relatively brittle one. This transition generally occurs over a relatively narrow temperature region and is similar to the solidification of a liquid to a glassy state; it is not a phase transition. In the glass transition temperature region, properties such as hardness and brittleness as well as expansibility and specific heat undergo rapid changes.

Grain growth A phenomenon that occurs fairly close to but below the melting point of polycrystalline materials, in which the largest grains grow still larger while the smallest ones gradually diminish and disappear.

Granules Agglomerates intentionally formed by the addition of binder to promote flowability of the powder.

Guide pin A device that maintains the proper alignment of the force plug and the cavity as the mold closes.

Heat transfer Flow of heat by conduction, convection, or radiation.

Helix angle The angle of the flight at its periphery relative to a plane perpendicular to the screw axis.

Hold time control Determines the duration of the hold interval of the injection time.

Hot runner mold A mold in which the runners are insulated from the chilled cavities and remain hot so that the center of the runner never cools during a normal molding cycle.

Immiscible The inability of one phase to dissolve in a second phase to form a single solution.

Impregnation The partial filling of the open pore structure with another material.

Inclusion Embedded foreign material.

Inert Unreactive and stable.

Injection capacity The maximum volume of material that can be injected in one cycle — a measure of the geometry of the cylinder, plunger, and plunger stroke.

Injection molding The operation of forming parts by injecting a semi-fluid material into a cold mold of desired shape.

Injection pressure The amount of pressure placed on the plasticized material, which depends on the diameter of the screw or plunger, the diameter of the hydraulic injection cylinder, and the oil pressure.

Injection rate The maximum rate at which the injection cylinder can inject fully plasticized material on a single shot basis.

Injection speed profile Determines the velocity profile of the ram during the injection phase of the machine.

Insert An integral part of a ceramic molding consisting of other material that can be molded in position.

Isostatic A type of ceramic forming process in which the mold is flexible and pressure is applied hydrostatically.

Jetting Turbulent flow of mix from an undersize gate or thin section into a thick mold section, as opposed to laminar flow of material progressing radially from a gate to the extremities of the cavity.

Knit lines (also weld or flow lines) A flaw on a molded part made by the meeting of two flow fronts during the molding operation. It is caused by improper gate design, poor rheology of the mix, and/or improper molding conditions.

Knock-out stroke The maximum movement available to knock out or automatically lift the moldable part from the mold surface.

Land The horizontal bearing surface of a mold by means of which excess material (flash) escapes.

Length to diameter ratio The length to diameter ratio can be expressed in two ways: (a) L/D ratio based on the barrel length – the distance from the forward edge of the feed opening to the forward end of the barrel bore divided by the bore diameter and expressed as a ratio in which the diameter is reduced to 1 such as 20:1 or 24:1; and (b) L/D ratio based on the screw flight length – the distance from the rear edge of the feed opening to the forward end of the barrel bore divided by the bore diameter and expressed as a ratio in which the diameter is reduced to 1, such as 20:1 or 24:1.

Locating ring A ring that serves to align the nozzle of an injection cylinder with the entrance of the sprue bushing and the mold to the machine platen.

Loss on ignition The fraction or percentage weight loss of a material on heating in air from an initial defined state to a specified temperature such as 600°C, and holding there for a specified period such as 1 hour.

Manifold Refers to the distribution system that takes the single channel flow output of the injection cylinder and divides it to feed several injection nozzles.

Matrix In a polycrystalline and especially heterogeneous mass, the material between crystals or grains of the major component(s), which holds them together.

Maximum daylight The greatest distance the platens can be separated from each other.

Mechanical properties Those properties of a material that are associated with elastic or inelastic reaction when force is applied, or that involve the relation between stress and strain.

Mean diameter That diameter located at the centroid of the distribution of size.

Medium The organic solvents and resins used as a vehicle and temporary binder.

Melt fracture An instability in the melt flow through a cavity, starting at the entry to the cavity.

Melt index The amount, in grams, of a thermoplastic resin that can be forced through a 2.1 mm (0.0825 inch) orifice when subjected to a 2160 gram force for 10 minutes at 190°C.

Metastable material Unstable with respect to some transition, conversion, or reaction but stabilized kinetically either by rapid solidification or by some molecular characteristics.

Metering section of screw A relatively shallow portion of the screw at the discharge end with a constant depth and lead, and having a length of at least one or more turns of the flight.

Minimum mold thickness The difference between the maximum stroke and daylight.

Mold insert Part of a mold cavity that forms undercut or raised portions of a molded part.

Mold release agent A lubricant normally applied to the mold cavity to reduce part sticking and to facilitate demolding.

Molding pressure The pressure applied directly or indirectly on the mix to allow compaction.

Molecular weight distribution The ratio of the weight average molecular weight to the number average molecular weight, which gives an indication of the distribution.

Morphology The characteristic shape, form, or surface texture or contours of the crystals, grains, or particles of a material, usually on a microscopic scale.

Multiple cavity An operation that produces more than one part per molding cycle via several mold cavities per mold.

Newtonian fluid A fluid in which the shear stress is proportional to the shear rate. In this case the shearing force per unit area between two parallel planes of fluid in relative motion is proportional to the velocity gradient between the planes.

Nucleation The first step in the process of crystallization, in which the characteristic atomic arrangement is first established in a submicroscopic zone, cluster, or particle.

Olefins Unsaturated hydrocarbons with the general formula C_nH_{2n} (e.g., C_2H_4, ethylene).

Oxidation A type of chemical reaction characterized by the transfer of electrons from the material being oxidized to the oxidant, resulting in a positive increase in valence of the oxidized material.

Oxidizing agent A material that causes oxidation of an atom or ion and is itself reduced (gains electrons) in the process.

Oxidizing atmosphere A furnace atmosphere with an oversupply of oxygen that tends to oxidize the materials placed in it.

Packing rate A direct function of the injection ram velocity at the moment the cavity is filled. The packing rate should be controlled in an effort to control shear, which can cause lamination and stress.

Particle size The size or size range of a given or typical particle.

Particle size distribution A profile of the sizes of particles contained in a material. The quantities must be expressed on some basis, which may be total number, total surface, or total weight or volume of the particles.

Parting line A line appearing on the molded part, formed at the junction of the mold components at the primary mold opening.

Peak cavity pressure Defines part resolution, therefore should be kept high enough to ensure that all crevices in the part are filled properly. Control over peak cavity pressure provides control over common problems such as flashing, sink marks, and short shots.

Permeability The property of a porous material that permits a fluid to pass through it.

pH The negative logarithm of the hydrogen ion activity.

Phase Any homogeneous volume of a liquid, or specified crystalline solid, including its constant or fixed chemical identity.

Pinhole A defect characterized by a small conical hole that appears on the surface of molded parts after binder removal or sintering.

Pitch The distance from any point on the flight of a screw line to the corresponding point on an adjacent flight, measured parallel to the axis of the screw line or threading.

Plasticity The property of a material that allows it to be repeatedly deformed without rupture when acted upon by a force sufficient to cause deformation, and which allows it to retain its shape after the applied force has been removed.

Plasticizer A chemical that can be added to a molding mix to improve the compatibility of the components and the flowability of the mix.

Plasticizing capacity Capacity of the machine to plasticize material per hour. The plasticizing capacity per hour should be equal to the total weight of production per hour.

Platens Mounting plates of an injection molding machine to which the mold halves are attached.

Plunger The reciprocating metal part that forces the mix into the mold cavity.

Plunger machine This type of machine uses a plunger to force the material over a spreader or torpedo. The heat is supplied by resistance heaters. The material is heated by conduction and convection. The flow in the cylinder is basically laminar.

Polycrystalline Comprised of many crystals or crystallites intimately bonded together.

Polymers A long, chain-like structure in which the number of units comprising an individual chain varies from chain to chain.

Polymorph In crystals, one of two or more geometrical forms of the same material.

Pore A small opening, void, interstice, or channel within a consolidated solid mass; usually larger than atomic or molecular dimensions.

Pore size The characteristic or equivalent dimensions (or range of these dimension) of a pore or family of pores in a material.

Pore size distribution The volume fraction of all measurable pores in each of a set of equivalent diameter increments from the smallest to the largest present in a material.

Porosity The volume of pores contained in the solid within it.

Pressure control Determines the pressure at which boost cut-off occurs.

Primary recrystallization A process by which nucleation and growth of a new generation of strain-free grains occur in a matrix.

Pullback control Determines the distance of ram travel for melt decompression.

Pullback switch Selects the location of pullback either before or after plastication.

Ram Depending on machine type, either a screw or a plunger.

Reactivity The susceptibility of a material to chemical change or action.

Reciprocating screw A combination injection and plasticizing unit in which a material can be injected into a mold cavity by reciprocating the screw as an injection plunger.

Recrystallization A process, usually physical, by which one crystal species is grown at the expense of another or at the expense of others of the same substance but smaller in size.

Reducing atmosphere A furnace atmosphere that tends to remove oxygen from the materials placed in it.

Rheology Any or all of the mechanical properties of fluids, semifluids, or plastic mixes of particulate solids with liquids.

Runners The means of conveying a plastic mix from the large end of the sprue to a point just outside the mold cavity. The size and shape of the runners are important, since they must be large enough to convey the material rapidly to the mold cavity.

Sagger A fired ceramic container to hold a part during firing and to protect it from the flame.

Screw channel With the screw in the barrel, the space bounded by the surfaces of flights, the root of the screw, and the bore of the barrel. This is the space through which the mix is conveyed and pumped.

Screw flight The helical metal thread of the screw.

Screw machine This type of machine uses a rotating screw to plasticize the material. As the screw turns, the plasticized material is injected by bringing the screw forward, at which time it acts as a plunger.

Screw recovery rate Volume of material per second plasticized while the screw is running.

Shape factor A dimensionless ratio of characteristic lengths, surface areas, or volumes of the particles that is useful for characterizing when comparing particles that otherwise have similar physical properties.

Shot capacity The weight or volume of material that can be injected in one shot. Normally given in grams or cm^3 of polystyrene.

Shot size control Determines the plastication limit of the screw or plunger.

Shrinkage The fractional reduction in dimensions or volume of a part when subjected to drying, dewaxing, or sintering.

Sinter (Sintering) To densify, crystallize, bond together, and/or stabilize a particulate material or part by heating close to but below the melting point.

Slurry Any pourable suspension having a high content of insoluble particulate solids in a liquid medium.

Soaking To hold a furnace at one temperature for a certain duration of time.

Sol A colloidal suspension comprised of discrete or separate solid particles suspended in a liquid.

Spreader A constraint placed in the front end of the cylinder to force the material against the hot inner wall of the cylinder, which helps to evenly heat the material.

Sprue bushing A hardened steel inset in an injection mold that contains the tapered sprue hold and has a suitable seat for the nozzle of the injection cylinder. Sometimes called an adapter.

Sprues The path for the flow of the hot plasticized material from the nozzle to the runners that carry it to the cavity. The part of the mold that forms this path is called the sprue bushing. The top of the sprue bushing contains a depression that fits the nozzle.

Stoichiometric Having the precise weight relation of the elements in a chemical compound; or being in the same weight relation as the theoretical combining weight of the elements involved.

Sublimation The volatilization of a solid directly to the vapor state, without passing through the liquid state.

Surface area The area, per unit weight of a powder, of all external plus internal surfaces that are accessible to a penetrating gas or liquid.

Surface tension The property, due to molecular forces, by which the surface film of all liquids tends to bring the contained volume into a form having the least area.

Tap density The apparent density of a powder obtained when the receptable is tapped or vibrated during loading under specified conditions.

Theoretical density The density of a material calculated from the number of atoms per unit cell and measurement of the lattice parameters.

Thermal capacity The amount of heat that a ceramic product will absorb, usually expressed in joules per °C.

Thermal stress Internal stresses developed within a part by virtue of temperature gradients or differences and accompanying thermal expansion or contraction effects.

Thixotropy The property of suspensions of decreasing viscosity under increasing shearing force or speed.

Toggle action A mechanism that exerts pressure developed by the application of force on a knee joint. It is used as a method of closing a mold and also serves to apply pressure at the same time.

Toggle clamp A clamping unit with a toggle mechanism directly connected to the moving platen.

True density A term used when considering the density of a solid. It is defined as the ratio of the mass of the solid to its true volume.

Tumbling Finishing operation by which gates, flash, and fins are removed.

Vapor pressure The pressure exerted by the vapor of a solid or liquid when in equilibrium with the solid or liquid.

Vehicle The carrier of the ceramic powder in a suspension.

Vents Small channels suitably located in the molded part, which are formed at the junction of the mold components at the primary mold opening.

Viscoelastic A term that refers to plastic mixes that store and dissipate energy during mechanical deformation.

Viscosity A measure of the resistance of a fluid or semifluid material to motion in shear.

Void Any empty space in a solid material.

Volatile A relative term expressing the tendency to form vapor.

Warpage Dimensional distortion in a molded part.

Weld lines A mark on a molded part made by the meeting of two flow fronts during molding.

Wetting agent A chemical additive that reduces the surface tension of a fluid, inducing it to spread readily on a surface to which it is applied, thereby causing even wetting of the surface with the fluids.

Yield point The minimum stress at which continuous flow takes place. This yield point will be evident in tensile, compression, torsion, or shear tests, but will vary with the rate of stressing.

Index